GRAVITATIONAL WAVES

Sources and Detectors

EDOARDO AMALDI FOUNDATION SERIES

Vol. 1: Gravitational Wave Experiments
edited by E. Coccia, G. Pizzella & F. Ronga

Vol. 2: Gravitational Waves: Sources and Detectors
edited by I. Ciufolini & F. Fidecaro

Forthcoming

Vol. 3: Essays and Recollections on 20th Century Physics: A Selection of Historical Writings by Edoardo Amaldi
edited by G. Battimelli & G. Paoloni

Volume 2

PROCEEDINGS OF THE INTERNATIONAL CONFERENCE ON

GRAVITATIONAL WAVES

Sources and Detectors

Cascina (Pisa), Italy
19–23 March 1996

Editors

Ignazio Ciufolini
Istituto di Fisica dello Spazio Interplanetario
CNR, Frascati
Dipartimento Aerospaziale
Università "La Sapienza", Roma

Francesco Fidecaro
Università di Pisa
INFN Sezione di Pisa

Published by

World Scientific Publishing Co. Pte. Ltd.
P O Box 128, Farrer Road, Singapore 912805
USA office: Suite 1B, 1060 Main Street, River Edge, NJ 07661
UK office: 57 Shelton Street, Covent Garden, London WC2H 9HE

British Library Cataloguing-in-Publication Data
A catalogue record for this book is available from the British Library.

GRAVITATIONAL WAVES: SOURCES AND DETECTORS

ISBN 981-02-2854-6

This book is printed on acid-free paper.

Printed in Singapore by Uto-Print

Foreword

This volume contains the contributions to the "International Conference on Gravitational Waves: Sources and Detectors" that took place in Cascina near Pisa. A few miles from the conference place the walls of the central building of the Virgo interferometer are now being raised while CNRS and INFN physicists are working hard on the apparatus. It is a great pleasure to recall the extremely stimulating week which helped spreading to a wider community current knowledge and ideas on gravitational waves.

These Proceedings are part of the "Edoardo Amaldi Foundation" Series, as recognition of Prof. E. Amaldi's leading role in the development of fundamental physics in Italy and Europe.

Many institutions contributed to the success of the conference. The "Comune di Cascina", its Major Dr. Carlo Cacciamano, who has been supporting the Virgo project since its very beginning, and its staff provided excellent local support through Teatro Politeama and generously contributed in welcoming and entertaining the participants.

The University of Pisa and the Scuola Normale Superiore, the main local academic institutions generously sponsored the conference.

The Istituto Nazionale di Fisica Nucleare was the main promoter of the conference, together with the Consiglio Nazionale delle Ricerche.

Informatica Universitaria kindly made Macintoshs available to allow participants to connect to their home computer.

The Banca Popolare di Lajatico kindly offered its services on site.

Help came from the local high school Istituto Tecnico Commerciale A. Pesenti, enthusiastically organized by Prof. Ferdinando Davini with several students participating to the conference life.

Enterprises are made by people and it is a pleasure to thank the persons who contributed by their hard work to the conference organization and running.

Mrs. Angela Parini from Comune di Cascina helped us interfacing with local authorities.

The Conference Secretariat was attended by Carla Tazzioli and Claudia Tofani, ready to provide help to the participants.

Computers were set up and kept fully efficient by Silvia Arezzini, with support of the computing team of INFN-Pisa, Mauro Giannini, Tania Palla and Giuseppe Terreni.

The whole organization was firmly kept under control by our Scientific Secretary Mrs. Lucia Lilli.

Many other contributed that it is impossible to thank individually. To all an heartily thank you!

For the Organizing Committee
Francesco Fidecaro

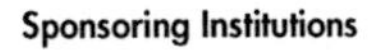

Sponsoring Institutions

Comune di Cascina
Istituto Nazionale di Fisica Nucleare
Consiglio Nazionale delle Ricerche
Università di Pisa
Scuola Normale Superiore, Pisa
Informatica Universitaria, Pisa
Banca Popolare di Lajatico

Under the Auspices of

Società Italiana di Fisica
Accademia del Lincei

International Advisory Committee

P. Bender (Univ. of Colorado, Boulder, USA)
B. Bertotti (Univ. di Pavia, Italy)
D. Blair (UWA, Nedlands, Australia)
V. Braginsky (Moscow State Univ., Russia)
A. Brillet (LAL, Orsay, France)
M. Cerdonio (Univ. di Padova, Italy)
T. Damour (IHES, Bures/Yvettes, France)
L. Grishchuk (Washington Univ., USA)
R. Matzner (Univ. of Texas, Austin, USA)
F. Pacini (Oss. Astr. di Arcetri, Italy)
G. Pizzella (Univ. di Roma II, Italy)
R. Price (Univ. of Utah, USA)
B. Schutz (Univ. of Wales, Cardiff, UK)
K. Thorne (Caltech, Pasadena, USA)

Local Organizing Committee

I. Ciufolini (CNR/Univ. di Roma I)
F. de Felice (Univ. di Padova)
A. Degasperis (Univ. di Roma I)
A. Di Giacomo, Chairman (Univ. di Pisa)
V. Ferrari (Univ. di Roma I)
F. Fidecaro (Univ. di Pisa)
A. Giazotto (INFN Pisa)
S. Solimeno (Univ. di Napoli)

Scientific Secretariat

L. Lilli (INFN Pisa)

Contents

Interferometric Detectors

Cosmology

Astrophysics

Data Analysis

Experimental Prototypes

Gravitational Waves: Theory

CALCULATING GRAVITATIONAL WAVEFORMS – BY ANY MEANS NECESSARY!

S. L. SHAPIRO
Center for Astrophysics & Relativity
326 Siena Drive, Ithaca, New York 14850

The merger of neutron star, black hole and neutron star-black hole binaries provide the most promising sources of gravitational radiation for detection by the LIGO/VIRGO laser interferometers now under construction. This fact has motivated several different theoretical studies of the inspiral and coalescence of compact binaries. Analytic analyses of the inspiral waveform have been performed in the Post-Newtonian (PN) approximation. Analytic and numerical treatments of the coalescence waveform from binary neutron stars have been performed using Newtonian hydrodynamics and the quadrupole radiation approximation. Numerical simulations of coalescing black hole and neutron star binaries are also underway in full general relativity. The flavor of each of these approaches will be described and their virtues and limitations summarized.

1 Introduction

Seeing the magnificent Leaning Tower for the first time reminded me how very apt it is for this beautiful and historic region to be once again at the forefront of research in experimental gravitation physics. Indeed, this conference has been called to celebrate the ongoing construction of new gravitational wave detectors, like VIRGO here in Pisa and LIGO in the United States. As a theorist, the best way that I know how to *celebrate* such a development is to *calculate*, which in this case means to calculate gravitational waveforms from promising astrophysical sources. The most promising sources for VIRGO and LIGO are coalescing binary star systems containing black holes (BH) and neutron stars (NS). At present, we do not possess a single, unified prescription for calculating gravitational waveforms over all the regimes and all the corresponding bands of detectible frequencies from such events. Instead, we must be crafty in breaking up the coalescence into several distinct epochs and corresponding frequency bands and employing appropriate theoretical tools to investigate each epoch separately. One of our immediate theoretical goals is to construct a smooth, self-consistent join between the different solutions for the different epochs. Ultimately, we may succeed in formulating a single computational approach that is capable by itself of tracking the entire binary coalescence and merger and determining the waveform over all frequency bands. But for now we must content ourselves with calculating waveforms by any means possible – by any means necessary!

It is useful to recall some of the vital statistics of the LIGO/VIRGO network now under construction (see Thorne [1] for an excellent review and references.) It consists of earth-based, kilometer-scale laser interferometers most sensitive to waves in the $10 - 10^3$ Hz band. The expected rms noise level has an amplitude $h_{rms} \lesssim 10^{-22}$. The most promising sources for such detectors are NS/NS, NS/BH and BH/BH coalescing binaries. The event rates are highly uncertain but astronomers estimate [2,3] that in the case of NS/NS binaries, which are observed in our own galaxy as radio pulsars, the rate may be roughly $(3/yr)(distance/200\ Mpc)^3$. For binaries containing black holes, the typical black hole mass range in the frequency range of interest is $2 - 300 M_\odot$. For typical NS/NS binaries, the total inspiral timescale across the detectible frequency band is approximately 15 mins. During this time the number of cycles of gravitational waves, $\mathcal{N}_{cyc}$, is approximately 16,000.

Although much of the current theoretical focus is directed toward LIGO and VIRGO, other detectors that will come on line in the future will also be important. For example, LISA is a proposed space-based, 5 million-kilometer interferometer that will be placed in heliocentric orbit. The relevant frequency band for LISA is $10^{-4} - 1$ Hz. The most promising sources in this band are short-period, galactic binaries of all types (main sequence binaries; white dwarf-white dwarf binaries, and binaries containing neutron stars and stellar-mass black holes) as well as supermassive BH/BH binaries. The typical black hole mass in a detectible BH/BH binary must be between $10^3 - 10^8 M_\odot$, where the upper mass limit is set by the lower bound on the observable frequency.

2 Attacking the Problem on Many Theoretical Fronts

Gravitational waveforms from coalescing compact binaries may be conveniently divided into two pieces [4]. The *inspiral* waveform is the low frequency component emitted early on, before tidal distortions of the stars become important. The *coalescence* waveform is the high frequency component emitted at the end, during the epoch of distortion, tidal disruption and/or merger. Existing theoretical machinery for handling the separate epochs differs considerably.

The inspiral waveform can be calculated using the Post-Newtonian (PN) approximation to general relativity. The PN formalism consists of a series expansion in the parameter $\epsilon \sim M/r \sim v^2$, where M is the mass of the binary, r is the separation and v is the orbital velocity. (Here and below $G = c = 1$.) This parameter is small whenever the gravitational field is weak and the velocity is slow. In this formalism, which is essentially analytic, the stars are treated as point masses. The aim of the PN analysis is to compute to $\mathcal{O}[(v/c)^{11}]$ in order that theoretical waveforms be sufficiently free of systematic errors to be reliable as templates against which the LIGO/VIRGO observational data can be compared [5]. For further discussion of the PN formalism and references, see Blanchet & Damour [6], Kidder, Will & Wiseman [7], and Apostolatos et al. [8].

The coalescence waveform is influenced by finite-size effects, like hydrodynamics in the case of neutron stars, and by tidal distortions. For binary neutron stars, many aspects of coalescence can be understood by solving the Newtonian equations of hydrodynamics while treating the gravitational radiation as a perturbation in the quadrupole approximation. Such an analysis is only valid when the two inequalities, $\epsilon \ll 1$ and $M/R \ll 1$ are both satisfied. Here R is the neutron star radius. Newtonian treatments of the coalescence waveform come in two forms: numerical simulations in 3+1 dimensions and analytic analyses based on triaxial ellipsoid models of the interacting stars. The ellipsoidal treatments can handle the influence of tidal distortion and internal fluid motions and spin, but not the final merger and coalescence. (For a detailed treatment and references, see Chandrasekhar [9]; Carter & Luminet [10]; Kochanek [11]; Lai, Rasio & Shapiro [12,13,14,15,16], hereafter LRS1-5 or collectively LRS; and Lai & Shapiro [17].) Numerical simulations are required to treat disruptions, ejection of mass and shock dissipation which usually accompany the merger. (For details, see Oohara & Nakamura [18]; Shibata, Nakamura & Oohara [19]; Rasio & Shapiro [20,21,22], hereafter RS1-3; Davies et al. [23]; Zughe, Centrella & McMillan [24], Ruffert, Janka & Schafer [25], and references therein.)

Fully relativistic calculations are required for quantitatively reliable coalescence waveforms. They are also required to determine those qualitative features of the final merger which can only result from relativistic gravitation (e.g. catastrophic collapse of merging neutron stars.) These calculations treat Einstein's equations numerically in 3+1 dimensions without approximation. In the case of neutron stars, the equations of relativistic hydrodynamics must be solved together with Einstein's field equations. (For earlier work, see articles in Smarr [26] and Evans, Finn & Hobill [27]; for recent progress see Matzner et al. [28] and Wilson & Mathews [29], and references therein.)

3 Phase Errors in the Inspiral Waveform

Measuring the binary's parameters by gravitational wave observations is accomplished by integrating the observed signal against theoretical templates [4]. For this purpose it is necessary that the signal and template remain in phase with each other within a fraction of a cycle ($\delta\mathcal{N}_{cyc} \lesssim 0.1$) as the signal sweeps through the detector's frequency band. To leading order we may treat the system as a point-mass, nearly-circular Newtonian binary spiraling slowly inward due to the emission of quadrupole gravitational radiation. In this limit the number of cycles spent sweeping through a logarithmic interval of frequency f is

$$\left(\frac{d\mathcal{N}_{cyc}}{d\ln f}\right)_0 = \frac{5}{96\pi}\frac{1}{M_c^{5/3}(\pi f)^{5/3}} \quad , \tag{1}$$

where the "chirp mass" M_c is given by $M_c \equiv \mu^{3/5}M^{2/5}$. Here μ is the reduced mass and M is the total mass of the binary. It is expected that LIGO/VIRGO measurements will be able to determine the chirp mass to within 0.04 per cent for a NS/NS binary and to within 0.3 per cent for a system containing at least one BH [1].

The PN formalism can be used to determine corrections to eqn 1 arising from PN contributions to the binary orbit. For example, suppose one of the stars has a spin **S** inclined at an angle i to the normal direction to the orbital plane. This spin induces a gravitomagnetic field which modifies the orbit of the companion. In addition, the wave emission rate, which determines the inspiral velocity, is augmented above the value due to the familiar time-changing quadrupole mass moment by an additional contribution from the time-changing quadrupole current moment. The result is easily shown to yield a "correction" to the Newtonian binary phase (eqn 1) given by

$$\frac{d\mathcal{N}_{cyc}}{d\ln f} = \left(\frac{d\mathcal{N}_{cyc}}{d\ln f}\right)_0 \left[1 + \frac{113}{12}\frac{S}{M^2}x^{3/2}cos\ i\right] \quad , \tag{2}$$

where $x \equiv (\pi M f)^{2/3} \approx M/r$ and where we have assumed that the mass of the spinning star is much greater than that of the companion. The frequency dependence of the correction term enables us in principle to distinguish this spin contribution from the Newtonian piece. In practice, it turns out that we may need to know independently the value of the spin in order to determine reliably the reduced mass μ (and thereby M, and the individual masses, since we already know M_c from the Newtonian part of eqn 2.) If we somehow know that the spin is small, we can determine μ to roughly 1 per cent for NS/NS and NS/BH binaries and 3 per cent for BH/BH binaries [1]. Not knowing the value of the spin worsens the accuracy of μ considerably, but this may be improved if wave modulations due to spin-induced Lens-Thirring precession of the orbit are incorporated [8]. This example illustrates how the PN formalism may be used to do classical stellar spectroscopy on binary systems containing compact stars.

Newtonian compressible ellipsoids can be used to analyze finite-size effects that lead to additional corrections to the phase of a NS/NS inspiral waveform [14]. Consider for definiteness two identical $1.4M_\odot$ neutron stars, each with radius $R/M = 5$ and supported by a stiff polytropic equation of state with adiabatic index $\Gamma = 3$. Track their orbit as they spiral inward from a separation $r_i = 70R$ to $r_f = 5R$, corresponding to a sweep over wave frequency from $f_i = 10$ Hz to $f_f = 522$ Hz (recall for Keplerian motion, $f \propto r^{-3/2}$.) To lowest Newtonian order, the total number of wave cycles emitted as the stars sweep through this frequency band is 16,098. If the two stars have zero spin, then the main hydrodynamic correction to the point-mass Newtonian result is due to the static Newtonian quadrupole interaction induced by the tidal field. The change in the number of cycles varies like $\delta\mathcal{N}_{cyc}^{(I)} \propto r^{-5/2} \propto f^{5/3}$ and therefore arises chiefly at large f (small r). Sweeping through the entire frequency band results in a small change $\delta\mathcal{N}_{cyc}^{(I)} \approx 0.3$; in the low frequency band from 10 Hz to 300 Hz, the change is only 0.1. Such a small change probably can be neglected in designing low-f wave templates.

Suppose instead that each NS has an intrinsic spin. In this case $\delta\mathcal{N}_{cyc}^{(S)} \propto r^{1/2} \propto f^{-1/3}$ and the change occurs chiefly at low f (large r). Now the quadrupole moments of the stars are induced by spin as well as by tidal fields. The change in the number of wave cycles as the orbit decays to r_f is $\delta\mathcal{N}_{cyc}^{(S)} \approx 9/P_{ms}^2$, where P_{ms} is the spin period in msec. Hence for rapidly spinning NS's with $P_{min} \lesssim 9$, the effect is potentially important and must be taken into account in theoretical templates.

Unlike many binaries consisting of ordinary stars, NS binaries are not expected to be corotating (synchronous) at close separation, because the viscosities required to achieve synchronous behavior are implausibly large [11,30,14]. Were this otherwise, the resulting corrections on the inspiral waveform phase evolution would be enormous and would dominate the low-f phase correction: $\delta\mathcal{N}_{cyc}^{(SS)} \approx 15$ in orbiting from $r = r_i$ to $r = r_f$.

4 Hydrodynamic Instabilities and Coalescence

In Newtonian gravitation, the motion of binary fluid stars in circular equilibrium is unstable when the orbital separation is sufficiently close. These global hydrodynamic instabilities can drive the binary system to rapid coalescence once the tidal interaction between the two stars becomes sufficiently strong. Newtonian hydrodynamic simulations have recently demonstrated the existence of these instabilities for binaries containing compressible fluid stars [20,21,22]. At about the same time, the original analytic work for binaries containing incompressible fluids [9] has been extended to compressible ellipsoid models by LRS [12,13,14,15,16]. The analytic work confirmed the existence of both secular and dynamical instabilites in the binary system; the numerical simulations followed the nonlinear growth of the dynamical instability and tracked the evolution of unstable binaries all the way to complete coalescence.

Consider a sequence of circular equilibrium binaries of fixed mass, parametrized by their separation. Construct a plot of the total energy of the binary system, or of the total angular momentum, as a function of the separation. The onset of instability along the sequence is located precisely at the turning point along each of these two equilibrium curves. The turning points on both curves occur at the same separation. The significance of these simultaneous turning points for coalescence is apparent when we calculate as a perturbation the rate of inward radial drift that gravitational radiation induces in the nearly circular orbit: $\dot{r} = -(dE_{GW}/dt)(dE_{eq}/dr)^{-1}$, where dE_{GW}/dt is the rate at which gravitational waves carry off energy and dE_{eq}/dr is the slope along the energy equilibrium curve. Clearly, as the binary approaches a turning point, $dE_{eq}/dr \rightarrow 0$ and the inward drift becomes huge, causing the two stars to plunge rapidly together in just a few orbital periods.

The physical origin of this instability is purely Newtonian and arises from the tidal field causing the effective potential between the two stars to rise more steeply than 1/r. This Newtonian, hydrodynamic instability is in addition to, and competes with, the familiar relativistic instability that causes, for example, the motion of test particles on circular orbits around black holes to be unstable when the separation is sufficiently small.

The hydrodynamic instability is particularly important for neutron stars with stiff equations of state, i.e. with adiabiatic index $\Gamma \gtrsim 2$, for which the onset of dynamical instability arises before the binary components come into contact. For identical neutron stars with $\Gamma = 2$ in synchronous orbit, it turns out [20] that when $r \lesssim 3R$, the circular orbit becomes unstable to radial perturbations and the two stars undergo rapid coalescence. The stars come into contact after about one orbital revolution. The emission of gravitational waves is strongest at this point. After two revolutions their cores have merged and the system resembles a single, elongated ellipsoid. Soon after, the configuration settles down to an axisymmetric rotating equilibriium object consisting of a cold, rigidly rotating core and a hot, differentially rotating halo. The evolution from a nonaxisymmetric to an axisymmetric object is fairly abrupt and results in a sharp drop in the wave amplitude. Identifying the frequency at which this drop occurs can be used to infer the radii of the coalescing stars. Knowing the masses of the stars from the inspiral waveform will then enable us to determine features of the nuclear equation of state that determine the NS mass-radius relation.

Further information can be extracted from the wave amplitudes near the end of the coalescence, since the properties of the waves depend sensitively on the stiffness of the equation of state. When $\Gamma \gtrsim 2.25$ the final merged configuration is not perfectly axisymmetric. Instead, it is triaxial and continues to radiate a persistent, periodic wave train [21]. The reason is that a polytropic fluid with $\Gamma > 2.25$, corresponding to polytropic index $n < 0.8$, can exist as a nonaxisymmetric, uniformly rotating ellipsoid in equilibrium (a compressible Jacobi ellipsoid.) Determining the presence or absence of this persistent radiation will provide a significant constraint on the stiffness of the nuclear equation of state.

5 Why Numerical Simulations In Full GR Are Necessary

Fully relativistic numerical simulations are clearly required to obtain *quantitatively* reliable coalescence waveforms. However, a numerical approach in full GR is also required for deciding between *qualitatively* different outcomes, even in the case of neutron stars. Consider, for example, the nearly head-on collision of two identical neutron stars moving close to free-fall velocity at contact. Assume that each star has a mass larger than $0.5M_{max}$, where M_{max} is the maximum mass of a cold neutron star. When the two stars collide, two recoil shocks propagate through each of the stars from the point of contact back along the collision axis. This shock serves to convert bulk fluid kinetic energy into thermal energy. The typical temperature is $kT \sim M/R$. What happens next? There are two possibilities. One possibility is that after the merged configuration undergoes one or two large amplitude oscillations on a dynamical timescale (msecs), the coalesced star, which now has a mass larger than M_{max}, collapses immediately to a black hole. Another possibility is that the thermal pressure generated by the recoil shocks is sufficient to hold up the merged star against collapse in a quasi-static, hot equilibrium state until neutrinos carry away the thermal energy on a neutrino diffusion timescale (10s of secs). The two outcomes are both plausible but very different. The implications for gravitational wave, neutrino and possibly gamma-ray bursts from NS/NS collisions are also very different for the two scenarios. Because the outcomes depend critically on the role of time-dependent, nonlinear gravitation, resolving this issue requires a numerical simulation in full GR.

Consider another puzzle: do sufficiently close binary neutron stars collapse to black holes *prior* to contact and merger? According to the preliminary and simplified numerical calculations in GR of Wilson and Matthews [29], the answer seems to be yes, at least if each neutron star has a mass very close to M_{max}. But according to the Newtonian ellipsoidal calculations of LRS [12] and Lai [31], the answer is no. The later calculations show that the effect of a tidal field is to stabilize a star against catastrophic collapse, not destabilize it, so that the maximum mass of a star in a binary orbit increases with decreasing separation. This may be another issue that requires an accurate, fully relativistic calculation with no assumed simplifications to find the correct answer.

6 Numerical Relativity: Present Status, Future Prospects

Calculations of coalescence waveforms from colliding black holes and neutron stars require the tools of numerical relativity – the art and science of solving Einstein's equations numerically on a spacetime lattice. Numerical relativity in 3+1 dimensions is in its infancy and is fraught with many technical complications. Always present, of course, are the usual difficulties associated with solving multidimensional, nonlinear, coupled PDE's. But these difficulties are not unique to relativity; they are also present in hydrodynamics, for example. But numerical relativity must also deal with special problems, like the appearance of singularities in a numerical simulation. Singularities are regions where physical quantities like the curvature (i.e., tidal field) or the matter density blow up to infinity. Singularities are always present inside black holes. Encountering such a singularity causes a numerical simulation to crash, even if the singularity is inside a black hole event horizon and causally disconnected from the outside world. Another special difficulty that confronts numerical relativity is the challenge of determining the asymptotic gravitational waveform which is generated during a strong-field interaction. The asymptotic waveform is just a small perturbation to the background metric and it must be determined in the wave zone far from the strong-field sources. Such a determination presents a problem of dynamic range: one wants to measure the waveform accurately far from the sources, but one must put most of the computational resources (i.e. grid) in the vicinity of those same sources, where most of the nonlinear dynamics occurs, Moreover, to determine the outgoing asymptotic emission, one must wait for the wave train to propagate out into the far zone, but by then, the simulation may be losing accuracy because of the growth of singularities in the strong-field, near zone.

Perhaps the most outstanding problem in numerical relativity is the coalescence of binary black holes. The late stages of the merger can only be solved by numerical means. To solve this relativistic two-body problem, the National Science Foundation is currently funding a "Grand Challenge Alliance" of numerical relativists and computer scientists at various institutions in the United States. Prior to 1995, no algorithm that could integrate two black holes in binary orbit long enough to get a gravitational wave out to 10 per cent accuracy existed, even in principle. That is because the multiple complications described above all conspired to make the integration of two black holes increasingly divergent at late times, well before the radiation content could be reliably determined. More recently, however, the Grand Challenge Alliance has reported several promising developments. In particular, new formulations of Einstein's field equations have been proposed[32,33]; that cast them is a flux-conservative, first order, hyberbolic form where the only nonzero characteristic speed is that of light. As a result of this new formulation, it may be possible to "cut-out" the interior regions of the black holes from the numerical grid and install boundary conditions at the hole horizons ("horizon boundary conditions".) Removing the black hole interiors is crucial since that is where the spacetime singularities reside, and they are the main sources of the computational inaccuracies. So now there is renewed hope that the binary black hole problem can be solved.

Acknowledgments

It is a pleasure to thank Dong Lai and Fred Rasio for several useful discussions. This work has been supported in part by NSF Grants AST 91-19475, AST 93-15375, and ASC/PHY 93-18152 (DARPA supplemented) and by NASA Grant NAGW-2364.

References

1. Thorne, K.S. 1995 in *Proceedings of Snowmass 94 Summer Study on Particle and Nuclear Astrophysics and Cosmology*, ed. E.W. Kolb & R. Peccei (World Scientific: Singapore), in press.
2. Phinney, E.S. *Astroph. Journ.* **380**, L17 (1991)
3. Narayan, R., Piran, T., & Shemi, A. *Astroph. Journ.* **379**, L17 (1991)
4. Cutler, C., Apostolatos, T.A., Bildsten, L., Finn, L.S., Flanagan, E.E., Kennefick, D., Markovic, D.M., Ori, A., Poisson, E., Sussman, G.J., and Thorne, K.S. *Phys. Rev. Lett.* **70**, 1984 (1993)
5. Cutler, C., & Flanagan, E.E. *Phys. Rev.* D **42**, 2658 (1994)
6. Blanchet, L. & Damour, T. *Phys. Rev.* D **46**, 4304 (1992)
7. Kidder, L.E., Will, C.M., & Wiseman, A.G. *Phys. Rev.* D **47**, 3281 (1993)
8. Apostolatos, T.A., Cutler, C., Sussman, G.J., and Thorne, K.S., *Phys. Rev.* D **49**, 6274 (1994)
9. Chandrasekhar, S. 1969, *Ellipsoidal Figures of Equilibrium* (New Haven: Yale University Press); revised Dover edition 1987
10. Carter, B., & Luminet, J.P. *Mon. Not. Royal Astr. Soc.* **23**, 212 (1985)
11. Kochanek, C.S. *Astroph. Journ.* **398**, 234 (1992)

12. Lai, D., Rasio, F.A., & Shapiro, S.L. *Astroph. Journ. Suppl.* **88**, 205 (1993) (LRS1)
13. Lai, D., Rasio, F.A., & Shapiro, S.L. *Astroph. Journ.* **406**, L63 (1993) (LRS2)
14. Lai, D., Rasio, F.A., & Shapiro, S.L. *Astroph. Journ.* **420**, 811 (1994) (LRS3)
15. Lai, D., Rasio, F.A., & Shapiro, S.L. *Phys. Rev.* D **423**, 344 (1994) (LRS4)
16. Lai, D., Rasio, F.A., & Shapiro, S.L. *Astroph. Journ.* **437**, 742 (1994) (LRS5)
17. Lai, D. & Shapiro, S.L. *Astroph. Journ.* **443**, 705 (1995)
18. Oohara, K. & Nakamura, T. *Progr. Theor. Phys.* **82**, 535 (1989)
19. Shibata, M., Nakamura, T., & Oohara, K. *Progr. Theor. Phys.* **88**, 1079 (1992)
20. Rasio, F.A & Shapiro, S.L. *Astroph. Journ.* **401**, 226 (1992) (RS1)
21. Rasio, F.A & Shapiro, S.L. *Astroph. Journ.* **432**, 242 (1994) (RS2)
22. Rasio, F.A & Shapiro, S.L. *Astroph. Journ.* **438**, 887 (1995) (RS3)
23. Davies, M.B., Benz, W., Piran, T., & Thielemann, F.K. *Astroph. Journ.* **431**, 742 (1994)
24. Zughe, X., Centrella, J.M., & McMillan, S.L.W. *Phys. Rev.* D **50**, 6247 (1994)
25. Ruffert, M., Janka, H.T., and Schafer, G. *Astrophys. Sp. Sci.* **231**, 423 (1995)
26. Smarr, L. ed, *Sources of Gravitational Radiation* (Cambridge: Cambridge University Press) 1979
27. Evans, C.R., Finn, L.S., & Hobill, D.W., eds, *Frontiers in Numerical Relativity*, (Cambridge: Cambridge University Press) 1989
28. Matzner, R.A., Seidel, H.E., Shapiro, S.L., Smarr, L., Suen, W.-M. Teukolsky, S.A. & Winicour, J. *Science* **270**, 941 (1995)
29. Wilson, J.R. and Mathews, G.J. *Phys. Rev. Lett.* **75**, 4161 (1995)
30. Bildsten, L. & Cutler, C. *Astroph. Journ.* **175**, 400 (1992)
31. Lai, D. 1996, *Phys.Rev.Letters*, submitted
32. Choquet-Bruhat, Y. & York, J.W. 1995, *C.R.Acad.Sci.Paris*, submitted
33. Bona, C., Masso, J., Seidel, E., & Stela, J. *Phys. Rev.* D **75**, 600 (1995)

Sources of Gravitational Waves

THE POPULATION OF COLLAPSED STARS: ASTROPHYSICAL EVIDENCE

FRANCO PACINI
Arcetri Astrophysics Observatory and Department of Astronomy and Space Sciences University of Florence (Italy)

1 Introduction

The phenomena which lead to the birth of collapsed stars and accompany their subsequent evolution are our best hope to detect high frequency gravitational waves. Because of this reason we shall outline here the related astrophysical evidence and estimate the proportion of collapsed objects among normal stars.

Stars have masses in the approximate range $10^{-2} \lesssim M \lesssim 10^2$ solar masses (solar mass $M_\odot = 2 \times 10^{33}$ g). Stars with masses of the order of (or less than) one solar mass live 10^{10} years or more. Massive objects burn their nuclear fuel very quickly and have much shorter lives, down to some 10^7 years or so.

Theory and observations have led astronomers to conclude that there are three different end-points of stellar evolution: white dwarfs, neutron stars, black holes. We shall not consider here the additional possibility of complete disruption of a massive star by thermonuclear explosion: this case — associated with some Supernovæ — is of little interest when discussing sources of gravitational waves.

The critical parameter which determines the final outcome in the life of a star is its mass. We note explicitly that one should not confuse the initial mass with that of the final end-point: severe mass loss often occurs during stellar evolution.

2 White dwarfs (WD)

In the case of an initial value $M < 8M_\odot$ a strong stellar wind in the late evolutionary stages produces a substantial mass reduction. At the end one is left with a degenerate core surrounded by a planetary nebula. In the core the pressure of the Fermi electrons can support the gravity as long as the mass is not larger than about $1.4M_\odot$ (Chandrasekhar's limit). The corresponding radius is about 10^9 cm, the density around $10^5 - 10^6$ g cm^{-3}.

White dwarfs were first discovered more than a century ago. It has been estimated that their total number in the galaxy is about 10^{10}, roughly 10% of the total stellar population. Of course, a large number of them — the old ones — have had time to cool and cannot be detected because of the very low luminosity.

The birth of a WD is a fairly gentle event. Because of this reason and the relatively weak gravitational field, single white dwarfs are not considered important as sources of gravitational waves. Close WD binaries may however be such a source, with detectable frequencies (determined by the orbital motion) and amplitudes. Because of the large number they would create a confusion limited background noise (see Hellings, this volume).

3 Neutron stars and black holes

An initial stellar mass larger than about $8M_\odot$ leads to a core in excess of $1.4M_\odot$. In this case the degenerate electron pressure cannot balance the gravity. In a matter of seconds, the density reaches values so high that the stellar material becomes composed mostly of neutrons through inverse β-decay.

Neutron stars would have nuclear densities (about 10^{14} g cm^{-3}), radius around 10 Km, maximum mass around $3M_\odot$.

The gravitational energy released during the collapse of the core — about 10^{53} erg — would cause the violent explosion of the outer stellar layers, an hypothesis first made by Baade and Zwicky in the '30s. Although a discussion of Supernovæ is outside the scope of our talk (see the article of Chiosi, this volume), we recall that today's astrophysics has confirmed this suggestion: type II Supernovæ are indeed connected with the collapse of the central core. Since the visible energy (luminosity, kinetic energy of the outer shell) is much less than the above mentioned 10^{53} erg, there is ample room for an energetic, invisible burst in the form of neutrinos and/or gravitational waves.

If the mass of the core is above $3M_\odot$, there can be no equilibrium configuration and the formation of a black hole seems unavoidable. In order to determine the number of black holes and neutron stars in the Galaxy we need to know the value of the initial stellar mass M_B above which the collapse of the core cannot be stopped by the neutron gas. Calculations of stellar structure and evolution indicate $M_B = 40 \pm 20M_\odot$. If we combine this result with the known distribution of stellar masses (Salpeter's mass function) $N(M) \propto M^{-2.5}$ we obtain a ratio between the number of neutron stars and black holes in the range 10–50.

The total number of neutron stars in the galaxy should be of order 10^9. Indeed Supernovæ occur — in galaxies like our own — roughly once every 30 years. If we extrapolate back in time, in our galaxy there have been about 5×10^8 Supernovæ. A somewhat higher frequency of events during the early evolution of the Galaxy is indicated by the cosmic abundance of heavy elements. This suggests that the number of past explosions should have been close to 10^9, roughly equal to the present number of neutron stars. If so, the corresponding number of black holes should be around 2×10^7–10^8.

The formation of neutron stars and black holes certainly leads to a burst of gravitational radiation. We note here that this burst is not necessarily unique. Indeed, during the collapse the rotational energy increases like R^{-2} (we assume conservation of angular momentum) while the gravitational binding increases like R^{-1}. The collapsing core may therefore become unstable against centrifugal forces and break into pieces. The collapse and/or coalescence of the fragments could, at least in principle, produce a sequence of bursts.

4 Pulsars and accreting X-ray sources

Neutron stars were discovered in 1968. Observations by Hewish, Bell and coworkers did show the existence in the sky of sources emitting very regular radio pulses, with periods around 1 second. At present, more than 600 pulsars have been discovered in our Galaxy. The pulses are a lighthouse effect associated with the rotation of a celestial body. Only neutron stars can have the extremely stable rotational periods in the observed range. The (small) observed secular increase of periods is easily understood as the gradual loss of rotational energy due to the presence of a strong magnetic field at the stellar surface, giving rise to an electromagnetic torque. Rotation periods in the range 1 ms up to seconds and magnetic fields around 10^{12} gauss had indeed been predicted by theorists because of conservation of angular momentum and magnetic flux in the collapsing stellar core.

Shortly after the discovery of the first pulsars, one of them was found to lie close to the center of the Crab Nebula. The period is 33 ms and its slowing down corresponds to an energy loss of 10^{38} erg s^{-1}, exactly what is required to energize the surrounding nebula.

Most pulsars have periods in the range 33 ms – 1 s and magnetic fields of order 10^{12} gauss. Their ages can be estimated from the ratio between period and period's derivative and are up to about 10^7 years. It is thought that this is the typical value for the lifetime of pulsars. Some interesting exceptions to the rule were found with the discovery of sources with periods in the millisecond range, much weaker magnetic fields (10^8 – 10^9 gauss), ages around 10^9 years or more. Unlike the normal pulsars which are almost always isolated, millisecond pulsars are often found in binary systems.

Various mechanisms have been proposed in order to explain how pulsars work. If one connects the pole to the equator of a magnetized rotating sphere through a nonrotating circuit, an electromotive force arises and the circuit is traversed by a current. In a laboratory experiment (say with a sphere of size ~ 10 cm, field 10^4 gauss, spinning frequency $\sim 10^3$ s^{-1}) the difference of potential between poles and equator is just a few volts. In the case of neutron stars (size $\sim 10^6$ cm, field $\sim 10^{12}$ gauss, spinning frequency ~ 1000 times a second) the difference of potential exceeds 10^{16} volts. Around the neutron star the resulting electric field is about 10^{10} volts cm^{-1} and the electric force on a charge largely exceeds the gravitational pull. The outer parts of the stellar surface cannot be in equilibrium and the particles are shot out along the magnetic field lines.

The charges drawn off the surface form a magnetosphere which can be divided in two regions. The first (corotating magnetosphere) contains the field lines which close before a critical distance $r_c = c/\Omega$. In this region the particles slide along the rigidly rotating field lines. The corotating magnetosphere cannot extend beyond the critical distance because otherwise the velocity Ωr would exceed the speed of light. The lines of force which pass beyond this distance define the so-called open magnetosphere: in this region there cannot be pure corotation and the plasma escapes freely under the influence of the electromagnetic field. The potential difference in the magnetosphere is available for an electrostatic acceleration up to very high energies.

An alternative possibility involves a neutron star rotating about an axis different from the magnetic axis. This system radiates low frequency waves at the basic rotation frequency Ω. The near-zone is similar to the one discussed for an aligned rotator, with the extra complication of time dependency. At $r \gg c/\Omega$ the electromagnetic field becomes a wave field. The energy loss is the flux of the Poynting's vector: $I\Omega\dot{\Omega} \approx B_c^2 c r_c^2$ where B_c is the field strength at r_c.

Low frequency electromagnetic waves with $f \equiv eB/mc\Omega \gg 1$ accelerate particles very efficiently. Since the gyrofrequency eB/mc is much larger than the wave frequency Ω, the particles move in a strong, nearly static, crossed electric and magnetic field. In a very short time ($\ll \Omega^{-1}$) they reach relativistic velocities along the direction of propagation and can ride the wave at constant phase. In the case of a plane wave, a particle acquires a Lorentz factor $\gamma = f^{2/3}$. In the Crab Nebula, at the beginning of the wave zone, $f \sim 10^{11}$ and the electrons could acquire an energy $\sim 10^{13}$ eV. A young pulsar could accelerate particles almost up to the highest energies found in the Crab Nebula and, perhaps, in the cosmic rays.

It would be easy to show that the relation $\dot{\Omega} = K\Omega^n$ entails a braking index n which depends on the geometry of the magnetic field outside the neutron star. For a dipole field $n = 3$; for a radial field $n = 1$. The actual braking index is known only in a few cases which yield $1.4 \lesssim n \lesssim 2.8$.

It is interesting to recall that a loss of rotational energy dominated by gravitational waves would correspond to $n = 5$. At least when the braking index has been measured, one can therefore exclude that gravitational waves are *at present* an important mechanism to slow down the rotation of neutron stars (at least in principle, the situation could have been different soon after the collapse).

After 10^7 years or so isolated pulsars become undetectable. The situation is different in binary systems, if the neutron star is accompanied by a less evolved, normal star. Matter falling from the normal star into the neutron star or into a black hole releases gravitational energy. If the rate of accretion is sufficiently high, the mechanism is self-regulatory since the heated gas radiates and exerts a pressure on the surrounding matter (Thompson scattering on the electrons). This leads to a critical luminosity (named after Eddington) which depends only upon the mass of the collapsed body $L \sim 10^{38} M/M_{\odot}$ erg s^{-1}. A good fraction of the stellar X-ray sources are observed to be close to this limit. The radiation should come out in the X-ray range: if we take an emitting surface with a radius of order 10^7 cm, a luminosity $\sim 10^{38}$ erg s^{-1} and we assume a black body spectrum, the temperature should be $T \sim 10^{7}\,^{\circ}$K. This corresponds to a peak of emission around 1 KeV. It can also be shown that accretion accelerates the rotation of the collapsed star.

The launch of the UHURU X-ray satellite in 1969 led to the discovery in the sky of many point-like X-ray sources and the subsequent optical studies have indicated that these are associated with binary systems.

Accretion can be effective around neutron stars as well as around black holes. A study of the dynamics of the system has, in some cases, indicated the presence of an obscure companion with $M > 3M_{\odot}$, the likely signature of the black hole. In some cases these compact sources show a periodic modulation of the X-ray emission, in other cases only irregular intensity fluctuations. The first case is attributed to the presence of a strongly magnetized rotating neutron star and to the periodic visibility of the region above the polar caps. It is interesting to note that these sources do show the secular spin-up expected as consequence of the accretion of gas which carries angular momentum.

We will not review here with the many beautiful observational results and theoretical understanding but we stress that both pulsars and X-ray sources are basically machines which release gravitational energy, unlike normal stars which exploit the nuclear binding energy. Pulsars do it in a highly non-thermal way through a combination of fast rotation and large scale electromagnetic fields. X-ray sources result instead from the presence of a very hot gas which is being accreted.

Accretion can only last as long as the companion is in the appropriate evolutionary stage. When matter is not supplied any longer, the X-rays disappear. What's left behind? The answer is simple: a rejuvenated neutron star, spinning very rapidly. It is the current view that the previously mentioned millisecond pulsars are born in this way and represent the final outcome of the evolutionary sequence. At this stage the magnetic fields have probably decayed to the observed values and the rate of release of rotational energy is small. This resurrection can therefore last for a very long time, comparable with the age of Galaxy.

5 Coalescence of collapsed stars. Gamma ray bursts. Gravitational waves

We have noted earlier that the known pulsars are unlikely to be strong sources of gravitational waves. Their present rotational energy loss is dominated by electromagnetic torques, as indicated by the value of the braking index. For millisecond pulsars the stability of the period implies that gravitational waves are not emitted at a significant rate. The situation could in principle have been a different in the very early phases, when a newly formed neutron star is likely to be endowed — for a short time — with large scale vibrations and possibly rotates close to the breaking limit $P \sim 1$ ms (frequency around 1000 Hz). If electromagnetic losses are not important in these early stages, there would not be subsequent observable effects. One cannot rule out this possibility with the available information about pulsars or the energetics of Supernovæ Remnants.

An important effect occurs in close binary systems where the orbital motion leads to the emission of gravitational waves. As well known, this is indeed the case of the source PSR 1913+16, one of the few binary systems associated with normal radio pulsars. Both components are neutron stars and the gradual changes of the orbital motion ($P \sim 8$ hours) indicates a progressive shrinking of the orbit, at the rate predicted by the theory of gravitational radiation. The corresponding frequency is of order 10^{-4} Hz but it is gradually increasing. It has been calculated that in about 10^8 years the two neutron stars will merge into a single object and one should expect a large final burst of gravitational waves at a frequency around 10^3 Hz (it is interesting to recall that during the last 1000 seconds or so the frequency emitted will sweep the band 10 Hz to 10^3 Hz!).

Estimates for the frequency of such mergings in our own galaxy are uncertain, perhaps about one every 10^5 years. This is based upon the statistics of binary pulsars and the expected lifetime against gravitational radiation.

Coalescence phenomena would probably manifest themselves in various ways. One possible manifestation are the well known gamma-ray bursts which were discovered more than 20 years ago. Our knowledge about their properties has greatly increased in recent years, when the Gamma-Ray Observatory (GRO) started to report an impressive number of events, in total more than 1200 (roughly, one per day). These are transient phenomena with very short lifetime and/or time structure (milliseconds up to — in some cases — several minutes). The typical energy range is between 10 KeV and 10 MeV. No associated emission has been detected at other wavelengths.

The short time-scale Δt suggests that they originate from very small sources although relativistic motions in the source can modify the quantitative requirement that their size should be less than $c\Delta t$. This had originally led to the suggestion by Ruderman and myself that gamma-ray bursts could be the result of sudden readjustments of the magnetosphere of old, dead neutron stars. This scenario could be compatible with the measured fluxes only if the sources were rather close to us (say, 10–100 parsec or so). We would then expect an isotropic distribution of bursts since the typical distance from us would be much less than the thickness of the galactic plane. This isotropy is confirmed by the GRO data but – at the same time – it is now clear that the observations contradict the expected distribution of events $N(> L) \propto L^{-3/2}$. Weak bursts are more rare than expected, something which can result from cosmological effects and suggests much larger distances.

The observed distribution of luminosities can actually be accommodated either in a model where bursts originate in the halo of the Galaxy (at a distance 100-300 Kpc) or at much larger cosmological distances (about 10^{28} cm, the Hubble radius). In the first case, each event would correspond to a total energy release about 10^{41} erg; in the second it must involve at least 10^{50}–10^{51} erg. In both cases, the very high intrinsic luminosity entails the initial presence of a compact "fireball", with a huge density of gamma rays, electrons and positrons.

If gamma ray bursts are an evidence for the merging of binary collapsed stars, there can be little doubt that these phenomena should also lead to a strong emission of gravitational waves. Such events may represent our best hope to observe high frequency waves with the presently planned interferometric instruments if (and when) they will be able to detect them at a distance of a few hundred Mpc. The expected rate of detections could then be a few per year.

THE QUASI-NORMAL MODES OF STARS AND BLACK HOLES

V. FERRARI
ICRA (International Center for Relativistic Astrophysics)
Dipartimento di Fisica "G.Marconi", Università di Roma, Rome, Italy

Non-radial oscillations of stars excited by external perturbations, are associated to the emission of gravitational waves. The characteristic eigenfrequencies of these oscillations, computed by using the relativistic theory of stellar perturbations, will be compared with those of black holes.

1 Introduction

The study of stellar oscillations started at the beginning of this century, when Shapley[1] (1914) and Eddington[2] (1918) suggested that the variability observed in some stars is due to periodic pulsations. The subsequent study of this phenomenon, carried out in the framework of the newtonian theory of gravity, has been a powerful tool in the investigation of stellar structure. In General Relativity, the interest in the theory of stellar pulsations is enhanced by the fact that a pulsating star emits gravitational waves with frequencies and damping times each belonging to characteristic "quasi-normal" modes. Since the fluid composing the star and the gravitational field are coupled, the emitted radiation carries information on the structure of the star, and also on the manner in which the gravitational field couples to matter. Conversely, for black holes the quasi-normal modes are purely gravitational, and the corresponding eigenfrequencies depend only on the parameters that identify the spacetime geometry: mass, charge and angular momentum. In sections 2, 3 and 4 of this lecture, I shall introduce the basic equations of the theory of stellar perturbations which has been developed in collaboration with S. Chandrasekhar [3,9], under the assumption of no rotation. In section 5 the characteristics of the spectrum of the quasi-normal modes of stars and black holes, and the information it gives on the nature and the structure of the source will be discussed.

2 The perturbed spacetime

As a consequence of a perturbation, all metric functions change by an infinitesimal amount with respect to their unperturbed values, and, if we are dealing with a star, each element of fluid suffers an infinitesimal displacement from its equilibrium position, identified by the lagrangian displacement

$\vec{\xi}$. Consequently, the thermodynamical variables ϵ and p, respectively the energy-density and the pressure, also change by an infinitesimal amount. Our analysis will presently be restricted to the study of adiabatic, axisymmetric perturbations of stars composed by a perfect fluid, and we shall assume that all perturbed quantities have a time dependence $e^{i\sigma t}$. The perturbed quantities are determined by solving Einstein's equations coupled to the hydrodynamical equations for a star, while for a black hole only Einstein's equations for the metric perturbations need to be considered. In order to separate the variables, all tensors can be expanded in tensorial spherical harmonics, and the azimuthal number m can be set to zero (axisymmetic perturbations). These harmonics belong to two different classes depending on the way they transform under the parity transformation $\theta \to \pi - \theta$ and $\varphi \to \pi + \varphi$. In particular those that transform like $(-1)^{(\ell+1)}$ are said to be *axial*, and those that transform like $(-1)^{(\ell)}$ are said to be *polar*. Consequently, the perturbed equations split into two distinct sets the *axial* and the *polar*, each belonging to different parities. If we choose the following line-element, appropriate to describe an axially symmetric, time-dependent spacetimes, [a]

$$ds^2 = e^{2\nu}(dt)^2 - e^{2\psi}(d\varphi - q_2 dx^2 - q_3 dx^3 - \omega dt)^2 - e^{2\mu_2}(dx^2)^2 - e^{2\mu_3}(dx^3)^2, \quad (1)$$

we find that the *axial* equations involve the perturbations of the off-diagonal components of the metric, i.e. $\{\delta\omega, \delta q_2$ and $\delta q_3\}$, and that the *polar* equations involve the diagonal part of the metric $\{\delta\nu, \delta\mu_2, \delta\psi, \delta\mu_3\}$, coupled to the thermodynamical variables $\{\delta\epsilon, \delta p, \vec{\xi}\}$ in the case of stars.

3 A Schroedinger equation for the axial perturbations

The equations for the axial perturbations can be considerably simplified by introducing, after separating the variables, a new function $Z_\ell(r)$, constructed from the radial part of the axial metric components, and which satisfies the following Schroedinger-like equation

$$\frac{d^2 Z_\ell^{ax}}{dr_*^2} + [\sigma^2 - V_\ell(r)] Z_\ell^{ax} = 0, \quad (2)$$

where $r_* = \int_0^r e^{-\nu+\mu_2} dr$. For a black hole [10]

$$V_{\ell BH}(r) = \frac{e^{2\nu}}{r^3}[l(l+1)r - 6Mr], \qquad \text{and} \qquad e^{2\nu} = 1 - \frac{2M}{r}, \quad (3)$$

[a] It may be noted that with our choice of the gauge the number of free functions is seven. The extra degree of freedom which we allow will be eliminated by imposing boundary conditions suitable to the problem on hand.

and for a star [4]

$$V_{\ell Star}(r) = \frac{e^{2\nu}}{r^3}[l(l+1)r + r^3(\epsilon - p) - 6m(r)], \qquad \nu_{,r} = -\frac{p_{,r}}{\epsilon + p}. \tag{4}$$

Outside the star ϵ and p are zero and eq. (4) reduces to eq. (3), also known as the Regge-Wheeler potential.

Thus the axial perturbations of black holes and stars are fully described by a Schroedinger-like equation with a potential barrier that depends, respectively, on the black hole mass, and on how the energy-density and the pressure are distributed inside the star in its equilibrium configuration. It should be stressed that the axial perturbations of stars are not coupled to any fluid pulsation: *they are pure gravitational perturbations, and do not have a newtonian counterpart.*

4 The polar perturbations

The expansion in tensorial spherical harmonics (with $m = 0$) shows that the polar metric functions and the thermodynamical variables have the following angular dependence

$$\begin{aligned} \delta\nu &= N_\ell(r)P_l(\cos\theta)e^{i\sigma t} & \delta\mu_2 &= L_\ell(r)P_l(\cos\theta)e^{i\sigma t} \\ \delta\mu_3 &= [T_\ell(r)P_l + V_\ell(r)P_{l,\theta,\theta}]e^{i\sigma t} & \delta\psi &= [T_\ell(r)P_l + V_\ell(r)P_{l,\theta}\cot\theta]e^{i\sigma t}, \\ \delta p &= \Pi_\ell(r)P_l(\cos\theta)e^{i\sigma t} & 2(\epsilon+p)e^{\nu+\mu_2}\xi_r(r,\theta)e^{i\sigma t} &= U_\ell(r)P_l e^{i\sigma t} \\ \delta\epsilon &= E_\ell(r)P_l(\cos\theta)e^{i\sigma t} & 2(\epsilon+p)e^{\nu+\mu_3}\xi_\theta(r,\theta)e^{i\sigma t} &= W_\ell(r)P_{l,\theta}e^{i\sigma t}, \end{aligned} \tag{5}$$

where $P_l(\cos\theta)$ are the Legendre polynomials. After separating the variables the relevant Einstein's equations become

$$\begin{cases} (T_\ell - V_\ell + N_\ell)_{,r} - \left(\frac{1}{r} - \nu_{,r}\right)N_\ell - \left(\frac{1}{r} + \nu_{,r}\right)L_\ell = 0, \\ V_{\ell,r,r} + \left(\frac{2}{r} + \nu_{,r} - \mu_{2,r}\right)V_{\ell,r} + \frac{e^{2\mu_2}}{r^2}(N_\ell + L_\ell) + \sigma^2 e^{2\mu_2 - 2\nu}V_\ell = 0, \end{cases} \tag{6}$$

$$\begin{cases} -(T_\ell - V_\ell + L_\ell) = W_\ell & (= 0 \text{ for B.H.}), \\ \left[\frac{d}{dr} + \left(\frac{1}{r} - \nu_{,r}\right)\right](2T_\ell - kV_\ell) - \frac{2}{r}L_\ell = -U_\ell & (= 0 \text{ for B.H.}), \\ \frac{1}{2}e^{-2\mu_2}\left[\frac{2}{r}N_{\ell,r} + \left(\frac{1}{r} + \nu_{,r}\right)(2T_\ell - kV_\ell)_{,r} - \frac{2}{r}\left(\frac{1}{r} + 2\nu_{,r}\right)L_\ell\right] + \\ \frac{1}{2}\left[-\frac{1}{r^2}(2nT_\ell + kN_\ell) + \sigma^2 e^{-2\nu}(2T_\ell - kV_\ell)\right] = \Pi_\ell & (= 0 \text{ for B.H.}), \end{cases} \tag{7}$$

where $k = l(l+1)$, and $2n = (l-1)(l+2)$. After some manipulation, the hydrodynamical equations and the conservation of barion number give the following expression for the hydrodynamical quantities

$$\Pi_\ell = -\frac{1}{2}\sigma^2 e^{-2\nu}W_\ell - (\epsilon+p)N_\ell, \qquad E_\ell = Q\Pi_\ell + \frac{e^{-2\mu_2}}{2(\epsilon+p)}(\epsilon_{,r} - Qp_{,r})U_\ell, \tag{8}$$

$$U_\ell = \frac{[(\sigma^2 e^{-2\nu} W_\ell)_{,r} + (Q+1)\nu_{,r}(\sigma^2 e^{-2\nu} W_\ell) + 2(\epsilon_{,r} - Qp_{,r})N_\ell](\epsilon + p)}{[\sigma^2 e^{-2\nu}(\epsilon + p) + e^{-2\mu_2}\nu_{,r}(\epsilon_{,r} - Qp_{,r})]}, \tag{9}$$

where

$$Q = \frac{(\epsilon + p)}{\gamma p}, \qquad \gamma = \frac{(\epsilon + p)}{p}\left(\frac{\partial p}{\partial \epsilon}\right)_{entropy=const} \tag{10}$$

and γ is the adiabatic exponent (defined in ref. [3], equation (106)).

For a black hole, a suitable reduction of eqs. (6) and (7), with W_ℓ, U_ℓ, Π_ℓ set equal zero, shows that the new function

$$Z_\ell^{pol}(r) = \frac{r}{nr + 3M}\left(3MV_\ell(r) - rL_\ell(r)\right), \tag{11}$$

satisfies the following wave equation

$$\frac{d^2 Z_\ell^{pol}(r)}{dr_*^2} + [\sigma^2 - V_{BH}]Z_\ell^{pol}(r) = 0, \tag{12}$$

where

$$V_{BH}(r) = \frac{2(r-2M)}{r^4(nr+3M)^2}[n^2(n+1)r^3 + 3Mn^2r^2 + 9M^2nr + 9M^3]. \tag{13}$$

Thus, as for the axial perturbations, the equations for the polar perturbations of a Schwarzschild black hole reduce to a single Schroedinger-like equation, but with a different potential barrier. Equation (12) with the potential (13) is known as the Zerilli equation [11], and it will also governe the metric perturbations in the exterior of a non-rotating star. The functions Z_ℓ^{ax} and Z_ℓ^{pol} contain all information on the gravitational waves emerging at infinity. In fact, it has been shown that the imaginary and the real part of the Weyl scalar Ψ_0, which represents the outgoing part of the radiative field (cfr. [12] eqs. 345 and 353), can be expressed in terms of Z_ℓ^{ax} and Z_ℓ^{pol}, respectively.

It is now interesting to see how eqs. (6)-(7) and the hydrodynamical equations (8,9) can be reduced if the perturbed object is a star. One may try to operate on these equations in a way similar to that used to find equation (12), hoping to find again a Schroedinger-like equation, possibly with some source in terms of the fluid variables. Unfortunately this is not possible, since the Schroedinger equation for black holes arises by virtue of the equilibrium equations, that are very different in the case of a star. In addition, this fact was to be expected, as already in newtonian theory the equations for the polar perturbations are described by a fourth order linear differential system. However a remarkable simplification is still possible. The first of eqs. (7) and eqs. (8,9) show that the fluid variables $[W_\ell, U_\ell, E_\ell, \Pi_\ell]$ can be expressed as a combination of the metric perturbations $[T_\ell, V_\ell, L_\ell, N_\ell]$ and their first derivatives. Therefore, after their direct substitution on the right hand side of the last three eqs. (7) we obtain a set of new equations which involves only the perturbations of the metric functions $[T_\ell, V_\ell, L_\ell, N_\ell]$. The final set is

$$\left\{\begin{array}{l} X_{\ell,r,r} + \left(\frac{2}{r} + \nu_{,r} - \mu_{2,r}\right) X_{\ell,r} + \frac{n}{r^2} e^{2\mu_2}(N_\ell + L_\ell) + \sigma^2 e^{2(\mu_2-\nu)} X_\ell = 0, \\ (r^2 G)_{\ell,r} = n\nu_{,r}(N_\ell - L_\ell) + \frac{n}{r}(e^{2\mu_2} - 1)(N_\ell + L_\ell) + \\ r(\nu_{,r} - \mu_{2,r})X_{\ell,r} + \sigma^2 e^{2(\mu_2-\nu)} r X_\ell, \\ -\nu_{,r} N_{\ell,r} = -G_\ell + \nu_{,r}[X_{\ell,r} + \nu_{,r}(N_\ell - L_\ell)] + \\ \frac{1}{r^2}(e^{2\mu_2} - 1)(N_\ell - rX_{\ell,r} - r^2 G_\ell) - \\ e^{2\mu_2}(\epsilon + p)N_\ell + \\ \frac{1}{2}\sigma^2 e^{2(\mu_2-\nu)} \left\{ N_\ell + L_\ell + \frac{r^2}{n} G_\ell + \frac{1}{n}[rX_{\ell,r} + (2n+1)X_\ell] \right\}, \\ L_{\ell,r}(1 - D) + L_\ell \left[\left(\frac{2}{r} - \nu_{,r}\right) - \left(\frac{1}{r} + \nu_{,r}\right) D \right] + \\ X_{\ell,r} + X_\ell \left(\frac{1}{r} - \nu_{,r}\right) + DN_{\ell,r} + N_\ell \left(D\nu_{,r} - \frac{D}{r} - F \right) + \\ \left(\frac{1}{r} + E\nu_{,r}\right) \left[N_\ell - L_\ell + \frac{r^2}{n} G_\ell + \frac{1}{n}\left(rX_{\ell,r} + X_\ell\right) \right] = 0, \end{array}\right. \tag{14}$$

where

$$\left\{\begin{array}{ll} A = \frac{1}{2}\sigma^2 e^{-2\nu}, & B = \frac{e^{-2\mu_2}\nu_{,r}}{2(\epsilon+p)}(\epsilon_{,r} - Qp_{,r}), \\ D = 1 - \frac{A}{2(A+B)} = 1 - \frac{\sigma^2 e^{-2\nu}(\epsilon+p)}{\sigma^2 e^{-2\nu}(\epsilon+p) + e^{-2\mu_2}\nu_{,r}(\epsilon_{,r} - Qp_{,r})}, & \\ E = D(Q-1) - Q, & \\ F = \frac{\epsilon_{,r} - Qp_{,r}}{2(A+B)} = \frac{2[\epsilon_{,r} - Qp_{,r}](\epsilon+p)}{2\sigma^2 e^{-2\nu}(\epsilon+p) + e^{-2\mu_2}\nu_{,r}(\epsilon_{,r} - Qp_{,r})}, & \end{array}\right. \tag{15}$$

and V_ℓ and T_ℓ have been replaced by X_ℓ and G_ℓ defined as

$$\left\{\begin{array}{l} X_\ell = nV_\ell \\ G_\ell = \nu_{,r}\left[\frac{n+1}{n} X_\ell - T_\ell\right]_{,r} + \frac{1}{r^2}(e^{2\mu_2} - 1)[n(N_\ell + T_\ell) + N_\ell] \\ + \frac{\nu_{,r}}{r}(N_\ell + L_\ell) - e^{2\mu_2}(\epsilon + p)N_\ell + \frac{1}{2}\sigma^2 e^{2(\mu_2-\nu)}[L_\ell - T_\ell + \frac{2n+1}{n} X_\ell]. \end{array}\right. \tag{16}$$

Equations (14) describe the perturbations of the gravitational field in the interior of the star, with no reference to the motion of the fluid. Once these equations have been solved, the fluid variables can be obtained in terms of the metric functions from the first of eqs. (7) and eqs. (8,9). This fact is remarkable: it shows that all the information on the dynamical evolution of a physical system is encoded in the gravitational field, a result which expresses the physical content of Einstein's theory of gravity. Moreover, it should be stressed that the decoupling of the equations governing the metric perturbations from the equations governing the hydrodynamical variables is possible in general, and *requires no assumptions on the equation of state of the fluid.*

Thus, if we are interested exclusively in the study of the emitted gravitational radiation, we can solve the system (14) and disregard the fluid behaviour.

Equations (14) have to be integrated for each value of the frequency from $r = 0$, up to the boundary of the star. There the spacetime becomes a vacuum sperically symmetric spacetime, and the perturbed metric functions match continuously with the metric functions that describe the polar perturbations of a Schwarzschild black hole, i.e. eqs (11,12,13). Thus the boundary conditions appropriate to the problem are

$$
\begin{array}{ll}
i) & \text{all functions are regular at} \quad r = 0, \\
ii) & \delta p = 0 \qquad \text{at the boundary of the star} \\
iii) & \text{all functions and their first derivatives are continuous at the} \\
 & \text{boundary of the star.}
\end{array}
\tag{17}
$$

5 The characteristic frequencies of the quasi-normal modes

The concept of quasi-normal modes plays a central role in the theory of perturbations of stars and black holes. In newtonian theory the oscillations of a perturbed star can be decomposed into normal modes, i.e. solutions of the perturbed equations that satisfy the boundary conditions (17) i), ii), and that correspond to a discrete set of real eigenfrequencies. Their relativistic generalization are the quasi-normal modes, and in this case the characteristic frequencies are complex, since the imaginary part is the inverse of the damping time associated to the emission of gravitational waves. Although the completeness of the quasi-normal modes has never been proved, numerical simulations show that an initial perturbation will, during the very last stages, decay as a superposition of these pure modes, and that a large fraction of the radiation will be emitted at the corresponding frequencies. The boundary conditions that identify the quasi-normal modes of a star are that, in addition to (17), at radial infinity only pure outgoing waves must prevail. The role of the equations in the interior of the star is that of providing the initial conditions for the integration of the Zerilli or the Regge-Wheeler equation in the exterior. Since a polar perturbation excites the fluid motion, the amount of energy which leaks out of the star in the form of gravitational waves depends on the exchange of energy between the fluid and the gravitational field. Conversely, an axial perturbation does not excite any fluid motion, and the boundary conditions depend only on the shape of the potential of the wave-equation, i.e. on how the energy-density and the pressure are distributed in the equilibrium configuration. Thus, the eigenfrequencies of the axial quasi-normal modes carry information essentially

Table 1: The complex characteristic frequencies of the quasi-normal modes of a Schwarzschild black hole.

	$M\sigma_0 + iM\sigma_i$		$M\sigma_0 + iM\sigma_i$
$\ell = 2$	0.3737+i0.0890	$\ell = 3$	0.5994+i0.0927
	0.3467+i0.2739		0.5826+i0.2813
	0.3011+i0.4783		0.5517+i0.4791
	0.2515+i0.7051		0.5120+i0.6903

on the structure of the star, and the polar, in addition, elucidate the manner in which the fluid and the gravitational field couple at supernuclear regimes.

For a black hole, the quasi-normal modes are defined to be solutions of the wave-equations that satisfy the boundary conditions of a *pure outgoing wave at infinity* and of a *pure ingoing wave at the horizon* (no radiation can emerge from the horizon). The corresponding frequencies are characteristic of many different processes involving the dynamical perturbations of black holes, and are the same both for the polar and for the axial perturbations, i.e. *the two potential barriers (3) and (13) are isospectral.* In 1975 Chandrasekhar and Detweiler [13] computed the first few eigenfrequencies of a Schwarzschild black hole, and subsequently Leaver [14] determined the next values with very high accuracy. He showed that, for a given ℓ, $M\sigma_0$ decreases with the order of the mode, and approaches a non-zero constant value, while $M\sigma_i$ increases, i.e. the damping time decreases. In Table 1 we show the first four values, respectively for $\ell = 2$ and $\ell = 3$. For example, remembering that $1M_\odot = 1.48 \cdot 10^5 cm$ and assuming that the black hole mass is $M = nM_\odot$, the conversion to physical unities gives the following values of the frequency and damping time

$$\begin{aligned} \nu_0 &= \frac{c}{2\pi n \cdot M_\odot (M\sigma_0)} = \frac{32.26}{n} (M\sigma_0)\, kHz, \\ \tau &= \frac{nM_\odot}{(M\sigma_i)c} = \frac{n \cdot 0.4937 \cdot 10^{-5}}{(M\sigma_i)}\, s. \end{aligned} \tag{18}$$

In order to compare the frequencies at which black holes and stars emit gravitational waves, we shall first consider, as an example, the polar perturbations of three models of star with a polytropic equation of state

$$p = K\rho^{1+\frac{1}{m}}, \qquad m = 1, \qquad K = 100 \quad km, \tag{19}$$

identified by different values of the central density. The corresponding mass, radius and surface gravity are given in Table 2. The polar quasi-normal modes

Table 2: Parameters of the three models of polytropic stars used to compute the polar eigenfrequencies

ρ in gr/cm^3	$\frac{M}{M_\odot}$	R in km	$\frac{2M}{R}$
$3 \cdot 10^{15}$	1.266	8.861	0.422
$6 \cdot 10^{15}$	1.35	7.413	0.538
10^{16}	1.3	6.465	0.594

Table 3: The characteristic frequencies and damping times of the $\ell = 2$ s and **w** polar modes of polytropic stars, compared with the first three eigenfrequencies of a Schwarzschild black hole with the same mass

	s-modes		**w**-modes		black hole	
$\frac{2M}{R}$	ν_0 (kHz)	τ (s)	ν_0 (kHz)	τ (s)	ν_0 (kHz)	τ (s)
0.422	3.0366	0.076	13.1556	$2.42 \cdot 10^{-5}$	9.5226	$7.02 \cdot 10^{-5}$
	6.7384	5.642	22.3438	$1.83 \cdot 10^{-5}$	8.8346	$2.28 \cdot 10^{-5}$
	10.1980	0.077	31.2207	$1.26 \cdot 10^{-5}$	7.6726	$1.31 \cdot 10^{-5}$
0.538	3.9166	0.060	12.4960	$3.65 \cdot 10^{-5}$	8.9300	$7.49 \cdot 10^{-5}$
	7.9610	0.623	19.4390	$2.30 \cdot 10^{-5}$	8.2848	$2.43 \cdot 10^{-5}$
	11.8669	0.035	26.3559	$1.94 \cdot 10^{-5}$	7.1952	$1.39 \cdot 10^{-5}$
0.594	4.5310	0.061	10.8420	$6.20 \cdot 10^{-5}$	9.2735	$7.21 \cdot 10^{-5}$
	8.7109	0.151	16.9960	$3.27 \cdot 10^{-5}$	8.6035	$2.34 \cdot 10^{-5}$
	12.7429	0.035	22.5540	$2.59 \cdot 10^{-5}$	7.4719	$1.34 \cdot 10^{-5}$

of a star belong essentially to two different classes
i) slowly-damped modes, or s-modes,
ii) highly-damped modes, or **w**-modes,
and the values of the first three eigenfrequencies of the $\ell = 2$ s- [15] and **w**-modes [16] are shown in Table 3, compared with the polar eigenfrequencies of a Schwarzschild black hole having the same mass.

The damping time τ indicates how fast the energy is dissipated in the form of gravitational waves, and since the τ's associated to the **w**-modes are of the same order of magnitude both for stars and black holes, (note also that they both decrease with the order of mode), it is natural to interpret the **w**-modes as being essentially modes of the gravitational field. However, since the boundary conditions to be imposed at the surface of the star and at the black hole horizon are different, the real part of the eigenfrequency, ν_0, will, in general, be different: higher for a star than for a black hole with the same mass, and increasing with the order of mode rather than decreasing. The s-polar modes have a different physical origin. They are essentially fluid pulsations whose energy is dissipated in the form of gravitational radiation at a rate which depends on how strong is the coupling between the fluid and the gravitational field. Thus, the values of the damping times are considerably longer that those of the **w**-modes. The frequency of the fundamental mode is smaller than that of a black hole with the same mass, and increases with the compactness of the star, because the time scale of these processes is related to the speed of acoustic waves in the fluid.

Let us now consider the axial perturbations. Since they do not excite any motion in the fluid, one may expect that slowly damped axial quasi-normal modes should not exist. However, this is not the case for the following reason. The slowly damped quasi-normal modes associated to the Schroedinger-like equation (2) with the potential barrier (4), are the equivalent of the quasi-stationary states that one encounters in quantum mechanics in the study of the emission of α-particles by a radioactive nucleus, also described by a Schroedinger equation. In that case σ^2 is replaced by the energy E and the potential barrier is suitable for the problem on hand. The boundary conditions for the two problems are the same: regularity of the wave function at the center, and pure outgoing waves emerging at infinity. In a quasi-stationary state E is allowed to be complex: $\Re E$ is the energy of the α-particle, and $\Im E$ is the inverse of the mean life-time (Γ) of the particle (the inverse of the damping time in our context). It is known from atomic physics that a quasi-stationary state will exist if the potential barrier has a minimum followed by a maximum, and if the potential well is sufficiently deep. For a star, the potential barrier should be considered in two regions: the interior $r < r_1$, where it depends on ϵ and p, and the exterior $r > r_1$, where it reduces to the barrier of a Schwarzschild black hole which has a maximum at $r = 3M$. If the radius of the star is smaller than $3M$ and if the star is very compact, the potential well in the interior may be deep enough to allow the existence of one or more quasi-normal s-mode. This conjecture can easily be proved, and in Table 4 we show the eigenfrequencies of the first four s- [6] and **w**-modes [17] computed for the very simple models of homogenous stars with decreasing values of the ratio R/M, i.e. increasing compactness. It emerges that if $R/M > 2.4$ the depth of the potential well in the interior is not sufficient to allow the existence of an s-mode, and only the **w**-modes survive. However, if $R/M < 2.4$, the s-modes appear, and their number is finite and increases with the compactness of the star, as well as the damping times.

The spectrum of the quasi-normal modes, whose main properties we have described, gives important information on the nature of the perturbed source:

Table 4: The characteristic frequencies and damping times of the first four $\ell = 2$, s and w axial modes of homogenoeus stars, with $M = 1.35 M_\odot$, and different values of R/M. The data are compared with the eigenfrequencies of a black hole with the same mass.

	s-modes		w-modes		black hole	
R/M	ν_0 (kHz)	τ (s)	ν_0 (kHz)	τ (s)	ν_0 (kHz)	τ (s)
2.4	8.6293	$1.52 \cdot 10^{-3}$	11.1738	$1.70 \cdot 10^{-4}$	8.9300	$7.49 \cdot 10^{-5}$
	–	–	14.2757	$8.03 \cdot 10^{-5}$	8.2848	$2.43 \cdot 10^{-5}$
	–	–	18.2232	$5.70 \cdot 10^{-5}$	7.1952	$1.39 \cdot 10^{-5}$
	–	–	22.6669	$4.88 \cdot 10^{-5}$	6.0099	$0.95 \cdot 10^{-5}$
2.3	5.6153	0.54	11.1084	$3.02 \cdot 10^{-4}$		
	7.5566	$1.16 \cdot 10^{-2}$	13.0403	$1.73 \cdot 10^{-4}$		
	9.3319	$1.02 \cdot 10^{-3}$	15.1512	$1.28 \cdot 10^{-4}$		
	–	–	17.4412	$1.06 \cdot 10^{-4}$		
2.28	4.4333	10.8	10.4128	$5.45 \cdot 10^{-4}$		
	6.0168	$2.50 \cdot 10^{-1}$	11.9074	$2.91 \cdot 10^{-4}$		
	7.5462	$1.44 \cdot 10^{-2}$	13.4813	$2.07 \cdot 10^{-4}$		
	8.9891	$1.83 \cdot 10^{-3}$	15.1428	$1.67 \cdot 10^{-4}$		
2.26	2.6041	$5.38 \cdot 10^{3}$	10.7852	$7.60 \cdot 10^{-4}$		
	3.5427	$1.69 \cdot 10^{2}$	11.6922	$5.34 \cdot 10^{-4}$		
	4.4802	$1.22 \cdot 10^{1}$	12.6138	$4.22 \cdot 10^{-4}$		
	5.4127	$1.37 \cdot 10^{-1}$	13.5512	$3.56 \cdot 10^{-4}$		

1) If the axial and the polar spectra coincide, the source is a black hole. *This is a very strong signature.* In a suitably choosen TT-gauge the axial and the polar part of the metric tensor are respectively

$$h^{ax}_{\mu\nu} = \begin{pmatrix} (t) & (r) & (\varphi) & (\vartheta) \\ 0 & 0 & 0 & 0 \\ 0 & 0 & 0 & 0 \\ 0 & 0 & h^{ax}_{\vartheta\vartheta} & h^{ax}_{\vartheta\varphi} \\ 0 & 0 & h^{ax}_{\varphi\vartheta} & h^{ax}_{\varphi\varphi} \end{pmatrix}, \quad h^{pol}_{\mu\nu} = \begin{pmatrix} (t) & (r) & (\varphi) & (\vartheta) \\ 0 & 0 & 0 & 0 \\ 0 & h^{pol}_{rr} & h^{pol}_{r\vartheta} & 0 \\ 0 & h^{pol}_{\vartheta r} & h^{pol}_{\vartheta\vartheta} & 0 \\ 0 & 0 & 0 & 0 \end{pmatrix}, \qquad (20)$$

thus, the detection of these two components of the emitted radiation will provide a direct evidence of the existence of black holes.

2) If the source is a star, the presence of the s-modes in the axial spectrum indicates that the star has a very compact core, while their number is directly related to the value of the ratio R/M. The question whether stars with a core compact enough to allow the existence of axial s-modes can exist in nature is

open, and it will probably receive an answer when axial gravitational waves will be observed.

Can the quasi-normal modes be excited? In the case of black holes we know they can in a variety of situations, for example when a gravitational wave-packet is scattered on the potential barrier, or when a mass $m_0 << M$ is captured by the black hole. In this case, the integration of the Zerilli and the Regge-Wheeler equations with the source term given by the stress-energy tensor of the infalling mass allows to compute the waveform and the energy emitted in these processes. It has been shown (see ref. [18] for an extensive bibliography on the subject) that the burst of gravitational waves ends in a ringing tail emitted when the particle coaleshes into the black hole ($2 < \frac{r}{M} < 4.5$). This part of the signal can be fitted with a linear superposition of quasi-normal modes. For a particle falling radially the total radiated energy is $\Delta E \sim 0.01 \left(\frac{m_0^2}{M}\right)$, which can be increased by up to a factor of 50 if the particle has an initial angular momentum.

In the case of a star it has been shown (see K. Kokkotas' paper in this volume) that both the **s**- and the **w**-axial modes can be excited if a gravitational wave-packet is scattered by the potential barrier, but much remains to be done in more realistic situations like the capture of infalling masses. For a star, these kind of calculations are complicated by the fact that we do not know how the mass m_0 interacts with the fluid composing the star after it crosses the surface. A preliminary integration of the axial[19] and the polar equations[15] with a source due to an infalling mass, and performed by truncating the integration when m_0 reaches the surface of the star, shows that indeed both the **s**- and the **w**-axial modes are excited, and that a considerable fraction of the emitted energy goes into the **w**-modes. Further work on this subject is in progress.

I would like to conclude this lecture by stressing an interesting aspect of the theory of perturbations: although it is based on the simplifying assumption that the perturbations of the physical quantities are small with respect to their unperturbed values, nevertheless, the results that one obtains by using this assumption are, to some extent, general. For example, in 1985 Stark and Piran[20,21] computed the energy spectrum emitted when an axisymmetric distribution of rotating polytropic fluid collapses to form a black hole, and they showed that it is very similar to that one obtains by integrating the Zerilli or the Regge-Wheeler equations when a mass falls in. In particular, the relevant contribution to the emitted energy is given at those frequencies at which the newborn black hole oscillate, namely at the frequencies of the quasi-normal modes.

References

1. H. ShapleyAp. J. **40**, 448 (1914)
2. A.S. EddingtonM.N.R.A.S. **79**, 2 (1918)
3. S.Chandrasekhar, V. Ferrari *Proc. R. Soc. Lond.* **A428**, 325 (1990)
4. S.Chandrasekhar, V. Ferrari *Proc. R. Soc. Lond.* **A432**, 247 (1991)
5. S.Chandrasekhar, V. Ferrari *Proc. R. Soc. Lond.* **A433**, 423 (1991)
6. S.Chandrasekhar, V. Ferrari *Proc. R. Soc. Lond.* **A434**, 449 (1991)
7. S.Chandrasekhar, V. Ferrari, R. Winston *Proc. R. Soc. Lond.* **A434**, 635 (1991)
8. S.Chandrasekhar, V. Ferrari *Proc. R. Soc. Lond.* **A437**, 133 (1992)
9. V. Ferrari *Phil. Trans. R. Soc. Lond.* **A340**, 423 (1992)
10. T.Regge, J.A. Wheeler *Phys. Rev.* **108**, 1063 (1957)
11. F.J. Zerilli*Phys. Rev.* **D2**, 2141 (1970)
12. S.Chandrasekhar*The mathematical theory of black holes*Oxford: Claredon Press (1983)
13. S.Chandrasekhar, S.L.Detweiler*Proc. R. Soc. Lond.* **A344**, 441 (1975)
14. E.W. Leaver*Proc. R. Soc. Lond.* **A402**, 285 (1985)
15. V. Ferrari, F. Perrotta *in preparation*
16. K.D. Kokkotas, B.F. Schutz *Proc. Mon. Not. R. Astron.Soc.* **255**, 119 (1992)
17. K.D. Kokkotas*Mon. Not. R. Astron. Soc.* **268**, 1015 (1994)
18. V.Ferrari *Proceedings of the 7th Marcel Grossmann Meeting* ed. by Ruffini R. & Kaiser M., World Scientific Publishing Co Pte Ltd, 1995
19. A. Borrelli, V. Ferrari *in preparation*
20. R.F. Stark, T. Piran*Phys. Rev. Lett.* **55 n. 8**, 891 (1985)
21. R.F. Stark, T. Piran *Proceedings of the 4th Marcel Grossmann Meeting* ed. by R. Ruffini Elsevier Science Publishers B.V. 327 1986

GRAVITATIONAL WAVE EMISSION DURING THORNE-ŻYTKOW OBJECT FORMATION AND GRAVITATIONAL WAVE 'PULSARS'

K.A. POSTNOV, S.N. NAZIN

Sternberg Astronomical Institute, Universitetskij pr., 13, Moscow, Russia

Gravitational radiation during a Thorne-Żytkow object (TZO) formation is computed for hydrodynamical treatment of neutron star spiral-in into a realistic red giant envelope given by Terman et al. [1]. It is shown that if strong hypercritical regime of accretion is set in onto a neutron star with a strong magnetic field $\gtrsim 10^{13}$ G, a possible neutrino emission anisotropy would cause gravitational radiation emission modulated with the neutron star spin frequency harmonics (a 'gravitational wave pulsar'). Such 'pulsars' may also arise during supernova explosions and binary neutron star merging. The computed gravitational wave signal could be detected in future gravitational wave experiments (LISA, LIGO).

1 Gravitational radiation from spiralling-in neutron star

Thorne-Żytkow -objects (neutron star (NS) cores inside red giant (RG) envelope) were first considered by Thorne & Żytkow [2]. TZO may result from the common envelope evolution in massive binary systems [3]. A crude estimation shows that the formation rate of TZO in the Galaxy can be $\geq 10^{-4}\text{yr}^{-1}$ and about 100-200 TZO may simultaneously exist at any given time in the Galaxy [4].

We calculated gravitational wave emission in the leading quadrupole order during NS spiralling-in to the envelope of a red giant with account of the results of 3D hydrodynamical treatment of the spiral-in process into real models of red giant [1]. We obtained that the characteristic frequencies of the waveforms range from 10^{-5} to 5×10^{-3} Hz with the dimensionless amplitude h_+ of order 3×10^{-23} at a distance of 10 kpc. The computed gravitational wave signal falls within the limits of the LISA detector sensitivity.

2 Gravitational wave pulsar: gravitational wave emission due to neutrino asymmetry in a strong magnetic field

Neutron stars with very strong magnetic fields ($> 10^{12} - 10^{13}$ G) are known to exist in nature. During the spiral-in process a strong Bondi-Hoyle accretion of matter on to the NS may occur under some conditions [5]. This results in the temperature increase close to the NS surface and the possibility of electron-positron pair creation at $T > 0.5$ MeV, which annihilate to produce neutrino emission. If such a neutrino cooling is set in, the accretion rates may reach $\simeq 1$ $M_\odot$ yr^{-1}, with the neutrino luminosity[5,6,7,8] $L_\nu \approx 10^{45}$ erg s^{-1}.

Neutrino emission in a strong magnetic field may be anisotropic due to quantum-electrodynamic effects if the electrons are strongly quantized (i.e. populate the ground Landau level). This is the case under physical conditions near the NS surface during the hypercritical accretion of 1 $M_\odot$ yr^{-1} (density $\rho \sim 10^3 - 10^4$ g cm^{-3}, temperature $T \simeq 0.5$ MeV). Therefore, one may expect a quadrupole-like anisotropy of the neutrino emission[9] in the direction Ω in the form

$$f(\Omega) = \frac{15}{4\pi(8k^2 + 20k + 15)}(1 + k\sin^2\chi)^2 , \tag{1}$$

where k is the anisotropy coefficient, χ is the angle between the Ω and the magnetic field direction.

As was first shown by Epstein[10], the neutrino emission anisotropy may cause gravitational wave emission. Recently, the anisotropic neutrino recoil was applied by Burrows & Hayes[11] to explain high space velocities of young pulsars (up to 500 km/s)[12]. They also showed that a characteristic pulse of gravitational radiation with 'memory' appears as a result of the anisotropic collapse.

Therefore, if a rotating NS with strong magnetic field spirals-in to the red supergiant envelope and the conditions for neutrino loss dominated accretion are satisfied, one may expect a 'gravitational wave pulsar' to appear at a frequency close to the NS spin frequency ω. The h_+ polarization of the gravitational wave emitted during this process from a NS lying at a distance r is

$$\frac{\partial h_+(t,\mathbf{x})}{\partial t} = \frac{k}{32k^2 + 80k + 60}\frac{G}{c^4}\frac{L_\nu(t - r/c)}{r}\sum_{i=0}^{4}\Phi_i^+ \cos i\omega(t - r/c) , \tag{2}$$

Functions Φ_i^+ are dependent on both the angle between the NS spin axis and the direction to observer (ξ) and the angle between the NS spin axis and the dipole magnetic field axis (β).

The amplitude of the modulation is of order

$$rh \simeq k\frac{G}{c^4}\frac{L_\nu}{\omega} \approx 10^{-5}\text{cm}\left(\frac{k}{0.1}\right)\left(\frac{L_\nu}{10^{45}\text{erg/s}}\right)\left(\frac{P}{1s}\right) , \tag{3}$$

which for a typical galactic distance of 10 kpc falls below the sensitivity limit of the LIGO detector.

However, the 'gravitational wave pulsar' may appear during the formation of a NS with strong magnetic field in a supernova explosion. Neutrino luminosities during the collapse may reach $L_\nu \sim 10^{54}$ erg/s, so taking the spin period of NS at birth $P = 1$ ms and assuming $k = 0.1$ we get from equation (3) $rh \sim 10$ cm, which falls within the limits of the first-order LIGO detector sensitivity. For an advanced LIGO sensitivity of $\sim 10^{-22}$, such "pulsars" may be detected from any part in the Galaxy.

Another implication may be merging binary NS (or NS and black hole) when one of the component has a strong magnetic field. Due to a huge tidal heating of close uncorotated NS [13], temperatures of order 10^{10} K may be achieved and an effective neutrino cooling via URCA-processes at a rate $\sim 10^{52} - 10^{53}$ erg/s may occur. As NS never reach corotation during the spiral-in [14,13], the waveforms from the inspiralling NS binaries will be modulated at the spin period of the magnetized component P with amplitude given by equation (3), which for $P = 1$ s is $rh \sim 5 \times 10^2$ cm and thus may be detected from 1 Mpc distances.

The neutrino emission anistropy may also appear when the optical depth of the NS with respect to neutrino absorption exceeds unity [15]. Then if the neutrino is generated in two regions near the NS polar caps, the resulting emission will also be modulated with the NS spin period. The neutrino emission anisotropy in this case is dipole-like, but this does not change our main conclusions.

References

1. J.L. Terman *et al.*, *ApJ* **445**, 367 (1995).
2. K.S. Thorne and A.N. Żytkow, *ApJ* **212**, 832 (1977).
3. B. Paczyński, in *Structure and Evolution of close binary systems*, IAU Symp. 73, Eds. Egglton P.P. *et al.*, (Dordrecht, 1976).
4. P. Podsiadlowski *et al.*, *MNRAS* **274**, 485 (1995).
5. R.A. Chevalier, *ApJ* **411**, L33 (1993).
6. S.A. Colgate, *ApJ* **163**, 221 (1971).
7. J.C. Houck and R.A. Chevalier, *ApJ* **376**, 234 (1991).
8. Ya.B. Zel'dovich *et al.*, *Soviet Astron.* **16**, 209 (1972).
9. D.G. Yakovlev, private communication, 1996.
10. R. Epstein, *ApJ* **223**, 1037 (1978).
11. A. Burrows and J. Hayes, SISSA preprint # astro-ph/9511106, (1995).
12. A.G. Lyne and D.R. Lorimer, *Nature* **369**, 127 (1994).
13. P. Mészáros and M.J. Rees, *ApJ* **397**, 570 (1992).
14. L. Bildsten and C. Cutler, *ApJ* **400**, 145 (1992).
15. V.M. Lipunov, *Astrophysics of Neutron Stars*, (Springer:Berlin, 1992).

GRAVITATIONAL RADIATION FROM NONSPHERICAL EVOLUTION OF PRE-SN

A.F. ZAKHAROV
Institute of Theoretical and Experimental Physics,
B. Cheremushkinskaya, 25, 117259, Moscow

We consider the gravitational radiation during nonspherical evolution of pre-SN. The scenario of Supernovae evolution was considered by Imshennik recently. Unlike the gravitational radiation analysis, which was considered by Imshennik & Popov in frameworks of Peters & Mathews formalism, the gravitational radiation is analysed in framework of $(PN)^{5/2}$ - approximation by Damour - Deruelle & Lincoln - Will. It is shown, that the eccentricity is more than 0.1 at the moment of filling by a low mass component of a Roche's lobe, thus the conclusion by Imshennik and Popov is incorrect (that final eccentricity is less than 0.1). If SN lies in Large Magellanic Cloud ($R = 50Kpc$), then we have the following estimation for amplitude of gravitational waves $h \approx 8 \times 10^{-20}$. The frequency of emitted gravitational waves is about $1kHz$. Thus the source of the gravitational radiation may be observed using VIRGO detector.

The model of nonspherical symmetrical evolution of pre-SN is considered in the paper [1] . We remind main stages of the scenario.

Stage I. This stage consists a formation of a rotating protoneutron star as a result of the gravitational collapse.

Stage II. The formed protoneutron star can be unstable or becomes unstable if there are small perturbations, therefore a close binary system of neutron stars is formed, we define the system parameters from the laws of conservation of mass and angular momentum.

Stage III. There is mutual decreasing of a distance between components of a binary system (since there is a gravitational radiation), until filling of a low mass component of a Roche's lobe.

Stage IV. There are mass losses of a low mass component and the unstable neutron star with the mass about $0.1M_\odot$ is formed.

Stage V. The unstable neutron star with mass $0.1M_\odot$ explodes and emits the energy about 10^{51} ergs, according to calculations by Blinnikov et all [3].

Thus SN (or pre - SN) is the powerful source of photon and neutrino radiation as well source of gravitational radiation. That is the attractive feature of the scenario.

Therefore we consider the gravitational radiation of binary system of neutron stars, and the parameters of a system are equal to values, which were considered by Imshennik [1], namely, mass of a protoneutron star is equal $m = M_t = 2M_\odot$, and the momentum moment is equal $J_0 = 8.81 * 10^{49}$ ergs / s. Unlike the paper by Imshennik & Popov [2], where the gravitational radiation was considered in framework of Peters & Mathews formalism [5, 4], the gravitational radiation is analysed in the framework of approach of Damour - Deruelle [6] using expressions of Lincoln & Will [7].

We use the equations for description of motion of coalescing binary system of neutron stars, and the equations include (post $)^{5/2}$ - Newtonian correction terms, since only the $(\mathrm{post})^{5/2}$ - Newtonian correction term represents the dominant - radiation - reaction effects [6, 7].

According to the approach of Lincoln & Will we give definition of quasicircular orbits of a binary system [7]. It is easy to see, that the solution of equations of motion with vanishing eccentricity is impossible, since at $e = 0$ we have from equations of motion of the binary system that $\frac{de}{d\phi} \neq 0$ (where ϕ is polar angle, e - eccentricity). At first we define of quasicircular orbit in the framework of $(\mathrm{PN})^2$ - approach as follows $\frac{d\omega}{d\phi} = 1$, i.e. the particle and osculating ellipse are turned with the same velocity (ω - the angle between line of node and the pericentric line). We have $f = \phi - \omega = const$ in the case, using the choice of initial value ϕ it is possible to consider the following value $f = \pi$. Thus, we obtain in the framework of $(\mathrm{PN})^2$ - approach, that $\frac{dp}{d\phi} = \frac{de}{d\phi} = 0$ ($p = a(1-e^2)$, a - semimajor axis). It is easy to see, that we have in the $(\mathrm{PN})^2$ - approach

$$e \approx (3-\eta)(m/p) - (15 + \frac{17}{4}\eta + 2\eta^2)(m/p)^2, \tag{1}$$

where $\eta = (1-m_2)m_2/m^2$, m_2 is a mass of a low mass component. It is easy to see also that the approximation for e is the solution also in $(\mathrm{PN})^{5/2}$ - approach

$$f = \pi + \frac{64}{5}\frac{\eta}{3-\eta}u^{3/2}. \tag{2}$$

We have the following equation in the $(\mathrm{PN})^{5/2}$ - approach for following coordinate $u \equiv \frac{m}{p}$ which depends from the angle ϕ

$$\frac{du}{d\phi} \approx 16u^{7/2}[4/5 + (1-\eta)u]. \tag{3}$$

It is easy to see that the inequality $\frac{dr}{du} < 0$ is valid during an evolution of a binary system. Therefore a distance between components continuously decreases. Clearly, that inequality is invalid at any point of an elliptical orbit (if we assume an elliptical motion of a binary system). Really, as far as

$$\hat{r} \equiv \frac{r}{m} \approx [1 + (3-\eta)u - (6 + \frac{41}{4}\eta + 2\eta^2)u^2]/u,$$

$$\frac{d\hat{r}}{du} = \frac{1}{u^2} - (6 + \frac{41}{4}\eta + 2\eta^2) < 0,$$

$\frac{dr}{du} < 0$, thus a binary system moves on a orbit as a spiral so, that distance between components continuously decreases.

We considered evolution of binary system which emits gravitational waves. We use the approach of quasicircular orbits. Really, Lincoln & Will shown, that the solution of equations of motion of a binary system is general enough, since general solution evolves into quasicircular orbits [7]. The rapid decreasing of eccentricity to small values was shown in the paper [2].

We consider the binary system evolution until filling by a low mass component of a Roche's lobe, according to Imshennik's model [1] . The critical value of orbit radius is connected with radius of low mass component - R_2. Namely, we determine the value from the approximation [8]

$$\frac{r_{cr}}{R_2} = \frac{1}{0.52(m_2/m)^{0.44}}. \tag{4}$$

In this case radius of a Roche's lobe is determined as sphere radius , and the sphere has a Roche's lobe volume.

We choose values of mass of a protoneutron star and a momentum moment, which are equal to corresponding values from the papers [1,2], namely $M = M_t = 2M_\odot$, and the momentum moment is equal $J_0 = 8.81 * 10^{49}$ ergs/s. We recall, that limiting value ($\delta = m_2/m = 0.205$ ($\eta = 0.163$) [1]) was obtained from a condition we that the time of decreasing distance between of components of a binary system (which is connected with an emission of gravitational waves) is equal about 1 hour, therefore if $\eta = 0.2$, then the value of this constant is about the limiting value.

Radius of a low mass star with mass $m_2 = 1.1 * 10^{33} g$ (similarly to paper [2]) is equal about 13.32 km. We assume, that there is an coalescence of a binary system according to approach of quasicircular orbits.

It is known that the amplitudes waves forms at Earth will have the amplitude [7,9]

$$h_{obs} \approx 7 \times 10^{-23}(4\eta)\frac{m}{2.8M_\odot}\frac{100Mpc}{R}, \tag{5}$$

therefore if SN lies in Large Magellanic Cloud ($R = 50Kpc$), then $h \approx 8\times 10^{-20}$. The frequency of emitted gravitational waves is about $1kHz$ [9]. Thus the source of the gravitational radiation may be observed using VIRGO detector at least, in principle. We obtain that the eccentricity is about 0.12 when the distance is minimal [9]. Thus, conclusion of Imshennik & Popov [2], that the eccentricity is smaller than 0.1 at the final moment, is incorrect. There is the natural consequence of non vanishing post - Newtonian parameter value (which is about 7%). The final eccentricity is about 0.11 in the case of the limiting value of the mass ratio $\delta = m_2/m = 0.205$. Nevertheless, we have decreasing of an eccentricity for binary systems with small post - Newtonian parameters, for example such as binary pulsar PSR 1913 + 16, where a post-Newtonian parameter about 10^{-6}.

We remark, that the considered system is the sample, in which post-Newtonian parameter is not too small, especially at the moment of minimal distance between components and certainly it is necessary to take into account post - Newtonian terms. If we consider vanishing value of post - Newtonian parameter in the problem there is a possibility to obtain the incorrect conclusions, similarly to results of the paper [2] , namely the conclusion about monotonous reduction of eccentricity during the evolution and the conclusion that the final value of eccentricity is less than 0.1.

More detailed discussion of the problem was presented at the paper [9].

Acknowledgments

I would like to express my gratitude to Organizing Committee of the International Conference on Gravitational Waves for the invitation to present the paper and for the warm hospitality at Pisa.

The work was supported in part by ESO grant B-07-036.

References

1. Imshennik V.S. *Pis'ma v Astron. Zhurn.* **18**, 489 (1992).
2. Imshennik V.S. and Popov D.V. *Pis'ma v Astron. Zhurn.* **20**, 620 (1994).
3. Blinnikov S.I. *et al.*, *Astron. Zhurn.* **67**, 1181 (1990).
4. Peters P.C. and Mathews J. *Phys. Rev.* **131**, 435 (1963).
5. Peters P.C. *Phys. Rev.* **136**, B1224 (1964).
6. Damour T. and Deruelle N. *Phys. Lett.* A **87**, 81 (1981).
7. Lincoln C. and Will C. *Phys. Rev.* D **42**, 1123 (1990).
8. Masevich A.G. and Tutukov A.V. *Stellar Evolution: Theory and Observations*, (Nauka, Moscow, 1988).
9. Zakharov A.F. *Astron. Zhurn.* **73**, 632 (1996).

THE EVOLUTION OF RADIATING DEBRIS TRAPPED BY A BLACK HOLE

L.GERGELY and Z. PERJÉS

KFKI Research Institute for Particle and Nuclear Physics, Budapest 114, P.O.Box 49, H-1525 Hungary

We consider the gravitational radiation backreaction on a test particle orbiting a spinning black hole. We want to see if the initial, adiabatic development of the orbit converges to any special set of Carter orbits in the Lense-Thirring approximation. Thus we work out the energy and angular momentum losses for the situation when the plane of the orbit rotates about the symmetry axis of the black hole with a constant angular velocity Ω. The losses will slowly evolve the constants of the motion. The terms independent of Ω reproduce the results of Peters and Mathews. This is a complete description of the orbit, and provides a prediction of the values of the constants of the motion of the Carter orbits at the next stage of the evolution.

1 Introduction

The determination of the signal template of the radiating debris trapped by a black hole is of great importance for the forthcoming interferometric gravitational observatories. One wants to describe the evolution of the orbit of a dust particle in the neighborhood of a black hole under the influence of gravitational radiation backreaction. In a highly idealized picture, the orbit of the particle is a Carter geodesic [1] characterized by four constants of the motion. In the spirit of perturbation theory, the radiation backreaction effects may be taken into account by evolving these constants of the motion, thus picturing the trajectory of the particle by a sequence of the unperturbed orbits.

As described in Shapiro's lecture, the history of the orbit may be divided in separate epoches. It has been pointed out by Thorne that the description of the generic orbit in the strong field region of a Kerr black hole is difficult due to the special nature of the separation constant. There is a considerable ease in the computation of special orbits such as equatorial or quasicircular. Several papers have obtained the signals from such special orbits (Shibata [2], Ryan [3], Kidder, Will and Wiseman [4,5], Blanchet, Damour and Iyer [6]). In order that any of these special orbits can be credited as significant contributor to the gravitational signal, one has to prove that the Carter constants evolve to such values.

In our work, we have taken up the evolution of the orbit in the far field region of the black hole. The trajectory of the idealized, nonradiating particle is described in the Lense-Thirring picture [7,8] as an ellipse the plane of which rotates about a momentary axis $\mathbf{\Omega}$. The Lense-Thirring approximation is linear in the rotation terms, thus the (additive) Schwarzschild effects may be neglected in this treatment.

We are interested in the cumulative effects of the secular motion, neglecting minute deviations from the exact orbit. Thus we work out the quadrupole formula and angular momentum losses for a plane of the orbit rotating about the symmetry axis of the black hole with an constant angular velocity Ω. We assume that Ω is small, thus we keep terms at most linear in Ω.

2 The Losses

We consider the orbit of a small particle of mass m_2 about a body with mass m_1 and angular momentum vector $\mathbf{J}$. In the Lense-Thirring approximation [7], the Lagrangian of the system has the form [8]

$$\mathcal{L} = \mu\dot{r}^2/2 - gm_1m_2/r + \delta\mathcal{L}, \qquad \delta\mathcal{L} = 2g\mu/(c^2r^3)\mathbf{J}\cdot(\dot{\mathbf{r}}\times\mathbf{r}) \tag{1}$$

where $\delta\mathcal{L}$ is the perturbation term due to rotation, $r = |\mathbf{r}|$, $\mu = m_1m_2/(m_1 + m_2)$ is the reduced mass, and a dot denotes d/dt. Hence the orbital momentum $\mathbf{L} = \mathbf{r}\times\mathbf{p}$ and Lenz-Runge vector $\mathbf{A} = \mathbf{p}\times\mathbf{L}/\mu - gm_1m_2\mathbf{r}/r$, satisfy the equations of motion, respectively

$$\dot{\mathbf{L}} = \frac{2g}{c^2r^3}\mathbf{J}\times\mathbf{L}\,, \qquad \dot{\mathbf{A}} = \frac{2g}{c^2r^3}\mathbf{J}\times\mathbf{A} + \frac{6g}{c^2r^5\mu}(\mathbf{J}\cdot\mathbf{L})\,\mathbf{r}\times\mathbf{L}\,. \tag{2}$$

Here $\mathbf{p} = \partial\mathcal{L}/\partial\dot{\mathbf{r}}$ is the momentum of the orbiting mass.

We average over one period $T = 2\pi a^{3/2}/[g(m_1+m_2)]^{1/2}$ of the motion on the ellipse of major axis a and eccentricity e. We thus obtain $\langle\dot{\mathbf{L}}\rangle = \mathbf{\Omega}\times\mathbf{L}$, $\langle\dot{\mathbf{A}}\rangle = \mathbf{\Omega}\times\mathbf{A}$ where $\mathbf{\Omega} = \Omega(\mathbf{J}/J - 3\mathbf{L}/L\cos\theta)$, $\Omega = 2gJ/[c^2a^3(1-e^2)^{3/2}]$ and $\cos\theta = (\mathbf{J}\cdot\mathbf{L})/JL$.

The equation of the elliptic orbit is $\dot{\psi} = [g(m_1+m_2)a(1-e^2)]^{1/2}/r^2$ where ψ is the angle subtended by the direction of the particle with the x_1 axis corotating with the Lenz-Runge vector.

For a particle with coordinates r_i and mass tensor $Q_{ij} = \mu r_ir_j$, the changes of the energy and angular momentum in the quadrupole approximation are [10],

$$\frac{dE}{dt} = \frac{g}{5c^5}\left(\frac{d^3Q_{ij}}{dt^3}\frac{d^3Q_{ij}}{dt^3} - \frac{1}{3}\frac{d^3Q_{ii}}{dt^3}\frac{d^3Q_{jj}}{dt^3}\right), \qquad \frac{dL_i}{dt} = -\frac{2g}{5c^2}\epsilon_{ijk}\frac{d^2Q_{jr}}{dt^2}\frac{d^3Q_{kr}}{dt^3} \tag{3}$$

The contribution of the velocity quadrupole terms to the losses is negligible [5]. Evaluating the right-hand sides, dropping higher-order terms in Ω, and averaging over the period T, the losses take the form

$$\begin{aligned}
\left\langle \frac{dL_z}{dt} \right\rangle &= -\frac{4g^{7/2}m_1^2m_2^2(m_1+m_2)^{1/2}}{5c^5a^{7/2}(1-e^2)^2}(8+7e^2)\cos\theta - 4\frac{g^3m_1^2m_2^2}{a^2c^5e^2}\Omega \\
&\times\left(\frac{e^2+2}{(1-e^2)^{1/2}} - 2 - 2[1-(1-e^2)^{1/2}]^2\sin^2\theta\cos^2(3\Omega t_0\cos\theta) \right. \\
&- \left. \cos^2\theta\frac{21e^2+2-2(2e^2+1)(1-e^2)^{1/2}}{(1-e^2)^{1/2}} \right) \qquad (4) \\
\left\langle \frac{dL}{dt} \right\rangle &= -4\frac{g^{7/2}m_1^2m_2^2(m_1+m_2)^{1/2}}{5c^5a^{7/2}(1-e^2)^2}(8+7e^2) \\
&+16 \frac{g^3m_1^2m_2^2}{c^5a^2(1-e^2)^{1/2}}[5-(1-e^2)^{1/2}]\Omega\cos\theta \qquad (5) \\
\left\langle \frac{dE}{dt} \right\rangle &= -\frac{32}{5}\frac{g^4m_1^2m_2^2(m_1+m_2)}{c^5a^5(1-e^2)^{7/2}}(1+\frac{73}{24}e^2+\frac{37}{96}e^4) \\
&-24 \frac{g^{7/2}m_1^2m_2^2(m_1+m_2)^{1/2}}{5c^5a^{7/2}(1-e^2)^2}(8+7e^2)\Omega\cos\theta\,. \qquad (6)
\end{aligned}$$

Among the losses, the only quantity depending on the time variable t_0 of the periastron shift is $\langle\frac{dL_z}{dt}\rangle$. In the limiting case of circular orbits, $e\to 0$, the dependence of $\langle\frac{dL_z}{dt}\rangle$ on t_0 vanishes. Despite the appearance of factors e^2 in some denominators, the limiting values of the losses are regular. The terms independent of Ω agree with the results of Peters and Mathews [9,10].

References

1. B.Carter, *Phys. Rev.* **174**, 1559 (1968).
2. M.Shibata, *Phys. Rev.* D **50**, 6297 (1994).
3. F.Ryan, *Phys. Rev.* D **52**, R3159 (1995).
4. L.Kidder, *Phys. Rev.* D **52**, 821 (1995).
5. L.Kidder, C.Will and A.Wiseman, *Phys. Rev.* D **47**, 4183 (1993).
6. L.Blanchet, T. Damour and B. Iyer, *Phys. Rev.* D **51**, 5360 (1995).
7. H. Thirring and J. Lense, *Phys. Zeitschr.* **19**, 156 (1918), English translation: *Gen.Rel.Gravitation*, **16**, 711 (1984).
8. Landau and Lifsitz, The Classical Theory of Fields (Pergamon, Oxford, 1975).
9. P.C. Peters and J.Mathews, *Phys. Rev.* **131**, 435 (1963).
10. P.C. Peters, *Phys. Rev.* **136**, B1124 (1964).

GRAVITATIONAL WAVEFORMS FOR MERGERS OF NEUTRON STARS WITH BLACK HOLES

W. KLUŹNIAK, W.H. LEE
University of Wisconsin-Madison, Physics Dept.,
1150 University Ave., Madison, WI, 53706, USA

We present preliminary Newtonian results for the merger of a black hole and a synchronized neutron star. The computations of the dynamical instability of the polytropic star and of the ensuing mass transfer were carried out using a smooth particle hydro code. The gravitational waveforms computed in the quadrupole approximation are presented and compared with those for the merger of a double neutron star system.

1 Introduction

Coalescing neutron binaries are of course prime sources for future gravitational wave detectors—the discovery of the Hulse–Taylor type neutron star binaries, and the observation of their orbital decay leave no doubt as to the reality of such events. On theoretical grounds, it is thought that close binaries composed of a black hole (BH) and a neutron star (NS) also exist. It has been estimated [6,4] that the merger rate of these BH-NS binaries is comparable to the rate of double NS mergers, i.e. it is $\geq 10^{-5}$/yr per galaxy. Thus, coalescing BH-NS binaries are of great interest to observers of gravity waves. BH-NS mergers are also thought [7] to be the leading candidate sources of gamma-ray bursts. In contrast, recent calculations [5,9] indicate that coalescing NS-NS binaries are not a likely source of the observed gamma-ray bursts.

If gamma-ray bursts do indeed originate in NS-BH mergers, one can hope that, in the foreseeable future, a gravitational wave impulse will be detected simultaneously with a burst of neutrinos [1] directly preceding a gamma-ray burst. The vision of such a threefold coincidence motivates the present work.

2 Numerical Method

The numerical simulations presented here have been performed using a smooth particle hydrodynamics (SPH) code. This technique is fully Lagrangian and eliminates the need for a grid to carry out calculations. It is ideal for the study of complicated fluid flows in three dimensions. For the present work we have developed our own SPH code, which will be described elsewhere. [3] A tree algorithm was implemented to optimize the computation of long-range interactions.

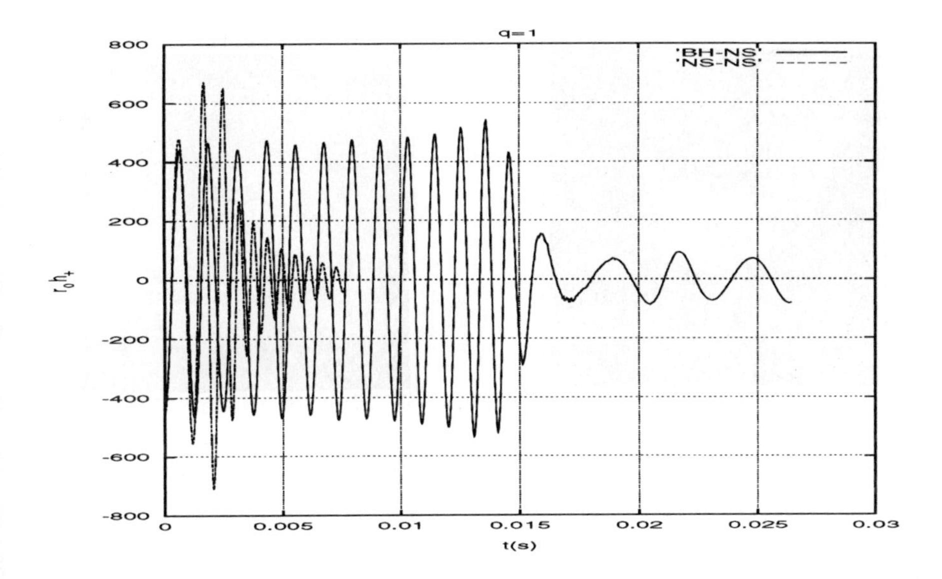

Figure 1: Waveforms for q=1, NS-NS (dotted line), and BH-NS (solid line).

The neutron star is modeled as a polytrope of index Γ=3 of mass of $1.4M_\odot$ and unperturbed radius 13.4 km. We use 2176 particles of unequal masses. The black hole is represented [2] by a point mass with an absorbing spherical boundary at the Schwarzschild radius. Momentum is strictly conserved. The dynamical calculation is fully Newtonian. We compute the gravitational radiation waveforms in the quadrupole approximation.

3 Binary Mergers

To determine the hydrodynamical instability limit [8] we allow the binaries to evolve once a complete sequence of *synchronized* equilibrium configurations for a range in binary separations is constructed [8] for each value of the mass ratio $q = M_{NS}/M_{BH}$. The initial evolution is reminiscent of the double NS case discussed by Rasio & Shapiro—a star of initial distance to the BH slightly less than the critical separation will be rapidly deformed and will transfer mass to the BH on the orbital timescale.

The NS-NS merger [8] differs from the BH-NS case. [2] This is seen by comparing the final density profiles (Fig. 2). Incidentally, the results we obtained

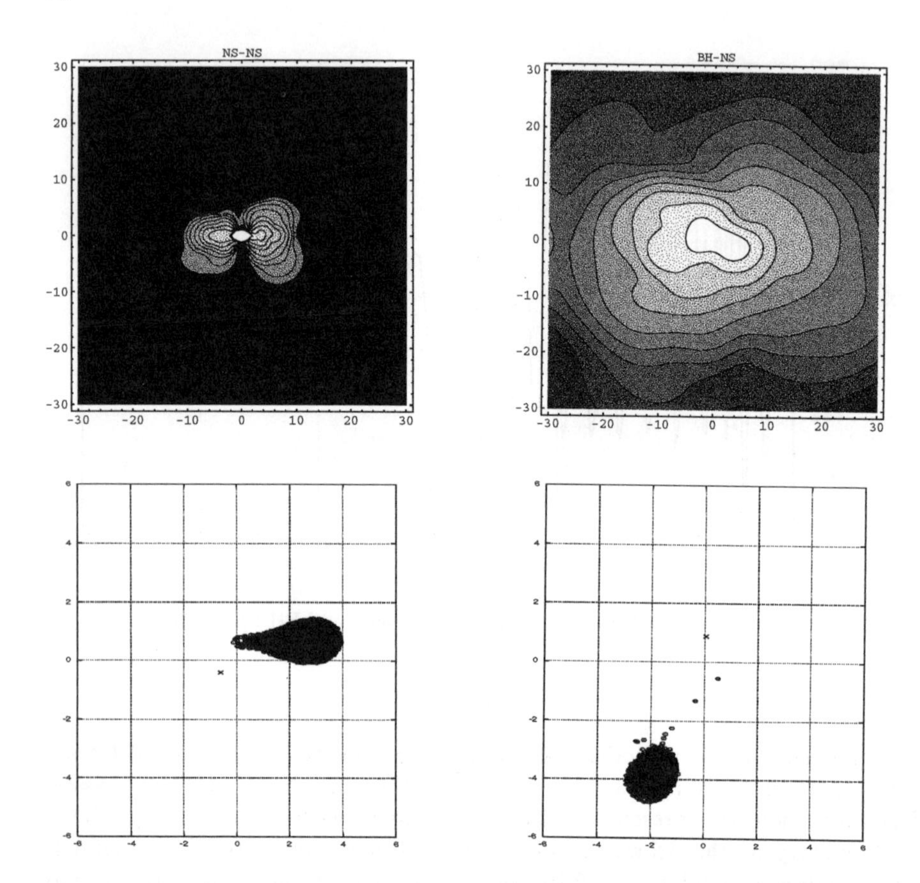

Figure 2: Top: Final density profiles in the plane perpendicular to the initial orbital plane. Bottom: Particle position snapshots on the orbital plane for $q = 0.31$ at t=0.5P and t=1.25P. P is the initial orbital period and distance is in units of the unperturbed stellar radius.

to date do not support gamma-ray burst models [7,10], which require that an accretion torus will be formed upon disruption of the NS, and that no baryons will be present along the axis of the system. Note that the gravitational waveforms of the BH-NS merger are clearly different from those of the NS-NS event (Fig. 1).

The results vary with q. Fig. 3 shows the waveform for $q = 0.3$. As illustrated in Fig. 2 (bottom) with two snapshots of the SPH particle positions projected onto the orbital plane, the dynamical instability in this case leads to a brief accretion event rather than to a complete merger.

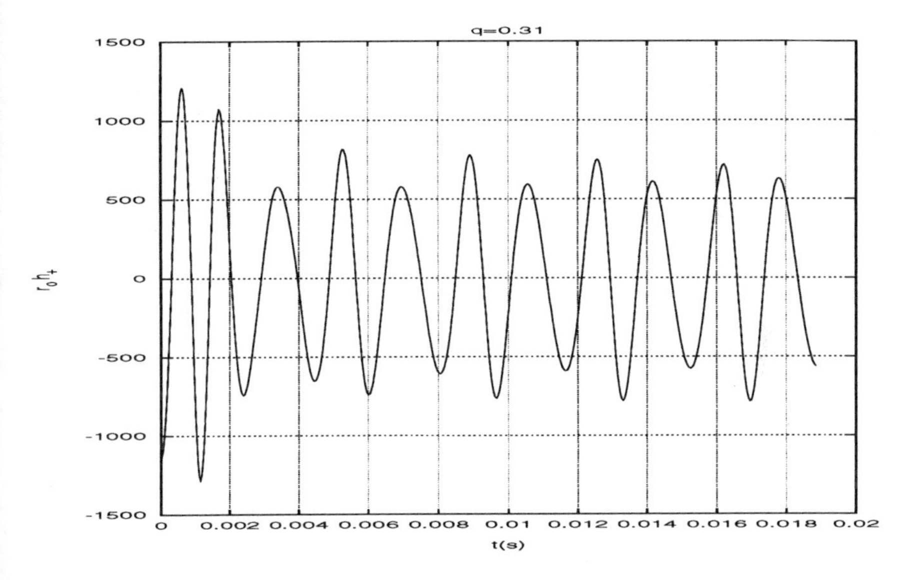

Figure 3: Waveform for q=0.31.

Acknowledgments

We gratefully acknowledge support of UNAM, NASA, KBN and the Sloan Foundation.

References

1. W. Kluźniak in Neutrino Telescopes, M. Baldo-Ceolin ed., Venezia (1994), p. 421.
2. W.H. Lee and W. Kluźniak, *Acta Astronomica* **45**, 705 (1995).
3. W.H. Lee, 1996, in preparation
4. V.M. Lipunov *et al.*, *Astrophys. J.* **454**, 593 (1995).
5. G. Mathews in Gamma-ray Bursts, C. Kouveliotou ed., AIP, in press.
6. R. Narayan, T. Piran and A. Shemi, *Astrophys. J.* **379**, L17 (1991).
7. B. Paczyński, *Acta Astronomica* **41**, 157 (1991).
8. F.A. Rasio and S.L. Shapiro, *Astrophys. J.* **432**, 242 (1994).
9. M. Ruffert in Gamma-ray Bursts, C. Kouveliotou ed., AIP, in press.
10. H.J. Witt *et al.*, *Astrophys. J.* **422**, 219 (1994).

Numerical Relativity

GRAVITATIONAL WAVES FROM ISOLATED NEUTRON STARS

E. GOURGOULHON, S. BONAZZOLA
Département d'Astrophysique Relativiste et de Cosmologie,
UPR 176 C.N.R.S.,
Observatoire de Paris,
F-92195 Meudon Cedex, France

Continuous wave gravitational radiation from isolated rotating neutron stars is discussed. The general waveform and orders of magnitude for the amplitude are presented for various known pulsars. The specific case of gravitational radiation resulting from the distortion induced by the stellar magnetic field is presented. Finally some preliminary results about the signal from the whole population of neutron stars in the Galaxy are discussed.

1 Introduction

As compact objects with significant internal velocities, neutron stars constitute a priori valuable gravitational wave sources. Besides catastrophic events such as their coalescence [1] or their gravitational collapse [2], these stars can be interesting sources of continuous wave gravitational radiation, provided they deviate from axisymmetry. Whereas the accompanying lecture [3] focuses on mechanisms, such as spontaneous symmetry breaking, which imply accretion from a companion, the present lecture is devoted to radiation from isolated rotating neutron stars.

2 General considerations

2.1 A general formula for gravitational emission by a rotating star

When dealing with neutron stars, the classical *weak-field* quadrupole formula for gravitational wave generation is not valid. For highly relativistic bodies, Ipser [4] has shown that the leading term in the gravitational radiation field h_{ij} is given by a formula which is structurally identical to the quadrupole formula for weak-field sources [5], the Newtonian quadrupole being simply replaced by Thorne's quadrupole moment [6] $\mathcal{I}_{ij}$. The non-axisymmetric deformation of neutron stars being very tiny, the total Thorne's quadrupole can be linearly decomposed into the sum of two pieces: $\mathcal{I}_{ij} = \mathcal{I}_{ij}^{\rm rot} + \mathcal{I}_{ij}^{\rm dist}$, where $\mathcal{I}_{ij}^{\rm rot}$ is the quadrupole moment due to rotation and $\mathcal{I}_{ij}^{\rm dist}$ is the quadrupole moment due to the process that distorts the star, for example an internal magnetic field [7], anisotropic stresses from the nuclear interactions or irregularities in the solid crust[8]. Let us make the assumption that the distorting process has a privileged direction, i.e. that two of the three eigenvalues of $\mathcal{I}_{ij}^{\rm dist}$ are equal. Let then α be the angle between the rotation axis and the principal axis of $\mathcal{I}_{ij}^{\rm dist}$ which corresponds to the non-degenerate eigenvalue. The two modes h_+ and $h_\times$ of the gravitational radiation field in a transverse traceless gauge derived from the Thorne-Ipser quadrupole formula are[7]

$$h_+ = h_0 \sin\alpha \left[\frac{1}{2}\cos\alpha \sin i \cos i \cos\Omega t - \sin\alpha \frac{1+\cos^2 i}{2} \cos 2\Omega t\right] \quad (1)$$

$$h_\times = h_0 \sin\alpha \left[\frac{1}{2}\cos\alpha \sin i \sin\Omega t - \sin\alpha \cos i \sin 2\Omega t\right] , \quad (2)$$

where i is the inclination angle of the "line of sight" with respect to the rotation axis and

$$h_0 = \frac{16\pi^2 G}{c^4} \frac{I\,\epsilon}{P^2\, r} , \quad (3)$$

where r is the distance of the star, $P = 2\pi/\Omega$ is the rotation period of the star, I its moment of inertia with respect of the rotation axis and $\epsilon := -3/2\mathcal{I}_{\hat{z}\hat{z}}^{\rm dist}/I$ is the *ellipticity* resulting from the distortion process.

It may be noticed that Eqs. (1)-(3) are structurally equivalent to Eq. (1) of Zimmermann & Szedenits[9], although this latter work is based on a different physical hypothesis (Newtonian precessing *solid* star).

From formulæ (1)-(2), it appears clearly that there is no gravitational emission if the distortion axis is aligned with the rotation axis ($\alpha = 0$ or π). If both axes are perpendicular ($\alpha = \pi/2$), the gravitational emission is monochromatic at twice the rotation frequency. In the general case ($0 < |\alpha| < \pi/2$), it contains two frequencies: Ω and 2Ω. For small values of α the emission at Ω is dominant. Numerically, Eq. (3) results in the amplitude

$$h_0 = 4.21 \times 10^{-24} \left[\frac{\rm ms}{P}\right]^2 \left[\frac{\rm kpc}{r}\right] \left[\frac{I}{10^{38}\ {\rm kg\, m^2}}\right] \left[\frac{\epsilon}{10^{-6}}\right] . \quad (4)$$

Note that $I = 10^{38}\ {\rm kg\, m^2}$ is a representative value for the moment of inertia of a $1.4\, M_\odot$ neutron star. In the following, I is systematically set to this value.

The values of h_0 resulting from Eq. (4) are given in Table 1 for two young rapidly rotating pulsars (Crab and Vela), the nearby pulsar Geminga[10] and two millisecond pulsars: the second[a] fastest one, PSR 1957+20, and the nearby millisecond pulsar PSR J0437-4715[11]. At first glance, millisecond pulsars

[a] the "historical" millisecond pulsar PSR 1937+21, which is the fastest one, is not considered for it is more than twice farther away.

Table 1: Gravitational wave amplitude h_0 on Earth as a function of the ellipticity ϵ for five pulsars.

name	rotation period P [ms]	distance r [kpc]	GW amplitude h_0
Crab	33	2	$1.9 \times 10^{-27}(\epsilon/10^{-6})$
Vela	89	0.5	$1.1 \times 10^{-27}(\epsilon/10^{-6})$
Geminga	237	0.16	$4.7 \times 10^{-28}(\epsilon/10^{-6})$
PSR B1957+20	1.61	1.5	$1.1 \times 10^{-24}(\epsilon/10^{-6})$
PSR J0437-4715	5.76	0.14	$9.1 \times 10^{-25}(\epsilon/10^{-6})$

seem to be much more favorable candidates than the Crab or Vela. However, in Table 1, ϵ is in units of 10^{-6} and the very low value of the period derivative $\dot{P}$ of millisecond pulsars implies that their ellipticity is at most 2×10^{-9}, as we shall see in § 2.3.

2.2 Detectability by VIRGO

Let us give a crude estimate of the minimum amplitude h_0 detectable by the VIRGO interferometer. Whereas the expected amplitude is very weak, as compared with other astrophysical processes such coalescences or gravitational collapses, one can take advantage of the permanent character of the signal to increase the signal-to-noise ratio by increasing the observing time. Indeed for an integration time T, the signal-to-noise ratio reads

$$\frac{S}{N} = \frac{h_0}{\tilde{h}(f)}\sqrt{T} \, , \tag{5}$$

where $\tilde{h}(f)$ is VIRGO sensitivity (square root of the noise spectral density) at the frequency f. The miminum values of h_0 leading to $S/N = 1$ when $T = 3$ yr are given in Table 2. In view of these values, we shall take as a basis for our discussion that *in order to be detectable by VIRGO, a rotating neutron star must produce* $h_0 > 10^{-26}$.

2.3 Upper bounds on gravitational radiation from pulsars

An absolute upper bound on h_0 can be derived by assuming that the observed slowing down of the pulsar (the so-called $\dot{P}$) is entirely due to the energy carried away by gravitational radiation. Let us stress that this is not a realistic assumption since most of the $\dot{P}$ is thought to result instead from losses via

Table 2: Minimal amplitude h_0 detectable ($S/N = 1$) by VIRGO within 3 years of integration. The values of $\tilde{h}(f)$ have been taken from Giazotto et al.[12]

frequency f [Hz]	sensitivity $\tilde{h}$ [$\mathrm{Hz}^{-1/2}$]	detectable amplitude $\min h_0$
10	10^{-21}	10^{-25}
30	10^{-22}	10^{-26}
100	3×10^{-23}	3×10^{-27}
1000	3×10^{-23}	3×10^{-27}

Table 3: Absolute upper bounds on the ellipticity and the GW amplitude derived from the measured spin-down rate of pulsars.

name	GW frequencies		max. ellipticity	max. GW amplitude
	f [Hz]	$2f$ [Hz]	$\epsilon_{\max}$	$h_{0,\max}$
Vela	11	22	1.8×10^{-3}	1.9×10^{-24}
Crab	30	60	7.5×10^{-4}	1.4×10^{-24}
Geminga	4.2	8.4	2.3×10^{-3}	1.1×10^{-24}
PSR B1509-68	6.6	13.2	1.4×10^{-2}	5.8×10^{-25}
PSR B1706-44	10	20	1.9×10^{-3}	4.2×10^{-25}
PSR B1957+20	621	1242	1.6×10^{-9}	1.7×10^{-27}
PSR J0437-4715	174	348	2.9×10^{-8}	2.6×10^{-26}

electromagnetic radiation and/or magnetospheric acceleration of charged particles — at least for Crab-like pulsars. However, this provides an upper bound on the ellipticity ϵ and the gravitational wave amplitude h_0. The resulting values are given in Table 3. The five first entries in this Table correspond to the five highest values of $h_{0,\max}$ among the 706 pulsars of the catalog by Taylor et al.[13,14].

New et al.[15] have recently pointed out that if the mean ellipticity of pulsars is taken to be of the order of the $\epsilon_{\max}$ of millisecond pulsars, i.e. $\epsilon \sim 10^{-9}$ (cf. Table 3), then the Crab pulsar reveals to be a much worse candidate than PSR B1957+20, as it can be seen by setting $\epsilon = 10^{-9}$ in Table 1: $h_0^{\mathrm{Crab}} \simeq 2 \times 10^{-30}$ versus $h_0^{1957+20} \simeq 10^{-27}$.

Table 4: Minimum values of the ellipticity required to produce $h_0 = 10^{-26}$ on Earth.

name	min ϵ	
Crab	5.3×10^{-6}	$= 7 \times 10^{-3}\,\epsilon_{\rm max}$
Vela	9.1×10^{-6}	$= 5 \times 10^{-3}\,\epsilon_{\rm max}$
Geminga	2.1×10^{-5}	$= 9 \times 10^{-3}\,\epsilon_{\rm max}$
PSR B1957+20	9.1×10^{-9}	$> \epsilon_{\rm max}$
PSR J0437-4715	1.1×10^{-8}	$= 0.4\,\epsilon_{\rm max}$

2.4 Ellipticity required for a detectable amplitude

Given the threshold $h_0 = 10^{-26}$ for detectability by the VIRGO interferometer (§ 2.2), one can consider the corresponding value of ϵ resulting from Eq. (4) and compare it with the maximum value given by the pulsar slowing down (Table 3). The results are presented in Table 4. From that it can be concluded that an ellipticity as small as 10^{-8} leads to a detectable amplitude for the nearby millisecond pulsar PSR J0437-4715, whereas the Crab or Vela pulsar should have an ellipticity of the order 10^{-5}, which is about one percent of the maximum allowable ellipticity as given by the spin-down rate (§ 2.3).

3 The specific case of magnetic field induced distortion

In this section, we consider the specific example when the distortion results from the neutron star's magnetic field. In this case the ellipticity is expressible as $\epsilon = \beta\mathcal{M}^2/\mathcal{M}_0^2$, where $\mathcal{M}$ is the magnetic dipole moment, $\mathcal{M}_0 = 2.6 \times 10^{32}$ A m^2 and β a dimensionless coefficient which measures the efficiency of the magnetic structure in distorting the star. For an incompressible fluid Newtonian body endowed with a uniform magnetic field [16], $\beta = 1/5$. For a given pulsar, $\mathcal{M}$ can be inferred from the value of $P\dot{P}$.

We have developed a numerical code to compute the deformation of magnetized neutron stars within general relativity [7,17]. The solutions obtained are fully relativistic and self-consistent, all the effects of the electromagnetic field on the star's equilibrium (Lorentz force, spacetime curvature generated by the electromagnetic stress-energy) being taken into account. The magnetic field is axisymmetric and poloidal.

The reference (non-magnetized) configuration is taken to be a $1.4\,M_\odot$ static neutron star built with the equation of state UV_{14}+TNI of Wiringa, Fiks & Fabrocini [18]. Various magnetic field configurations have been considered; the most representative of them are presented hereafter.

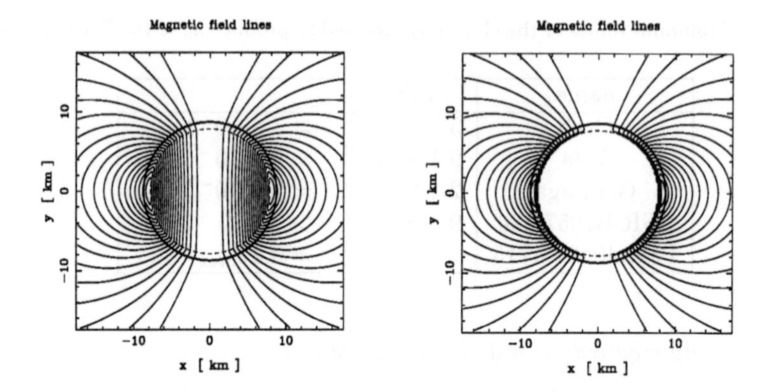

Figure 1: Magnetic field lines generated by a current distribution localized in the crust of the star (left) or exterior to a type I superconducting core (right). The thick line denotes the star's surface and the dashed line the internal limit of the electric current distribution (left) and the external limit of the superconducting region (right). The distortion factor corresponding to these configurations is $\beta = 8.84$ (left) and $\beta = 157$ (right).

Let us first consider the case of a perfectly conducting interior (normal matter, non-superconducting). The simplest magnetic configuration compatible [17] with the MHD equilibrium of the star results in electric currents in the whole star with a maximum value at half the stellar radius in the equatorial plane. The computed distortion factor is $\beta = 1.01$, which is above the 1/5 value of the uniform magnetic field/incompressible fluid Newtonian model [7] but still very low. Another situation corresponds to electric currents localized in the neutron star crust only. Figure 1 presents one such configuration: the electric current is limited to the zone $r > r_* = 0.9\, r_{\rm eq}$. The resulting distortion factor is $\beta = 8.84$.

In the case of a superconducting interior, of type I (which means that all magnetic field has been expulsed from the superconducting region), the distortion factor somewhat increases. In the configuration depicted in Fig. 1, the neutron star interior is superconducting up to $r_* = 0.9\, r_{\rm eq}$. For $r > r_*$, the matter is assumed to be a perfect conductor carrying some electric current. The resulting distortion factor is $\beta = 157$. For $r_* = 0.95\, r_{\rm eq}$, β is even higher: $\beta = 517$.

The above values of β, of the order $10^2 - 10^3$, though much higher than in the simple normal case, are still too low to lead to an amplitude detectable by the first generation of interferometric detectors in the case of the Crab or Vela pulsar, which would require [7] $\beta > 10^4$. It is clear that the more disordered the magnetic field the higher β, the extreme situation being reached by a stochastic magnetic field: the total magnetic dipole moment $\mathcal{M}$ almost vanishes, in agreement with the observed small value of $\dot{P}$, whereas the mean value of B^2 throughout the star is huge. Note that, according to Thompson & Duncan [19], turbulent dynamo amplification driven by convection in the newly-born neutron star may generate small scale magnetic fields as strong as 3×10^{11} T with low values of $B_{\rm dipole}$ outside the star and hence a large β. In order to mimic such a stochastic magnetic field, we have considered the case of counter-rotating electric currents. The resulting distortion factor can be as high as $\beta = 5.7 \times 10^3$.

If the neutron star interior forms a type II superconductor, the magnetic field inside the star is organized in an array of quantized magnetic flux tubes, each tube containing a magnetic field $B_c \sim 10^{11}$ T. As discussed by Ruderman [20], the crustal stresses induced by the pinning of the magnetic flux tubes is of the order $B_c B/2\mu_0$, where B is the mean value of the magnetic field in the crust ($B \sim 10^8$ T for typical pulsars). This means that the crust is submitted to stresses $\sim 10^3$ higher than in the uniformly distributed magnetic field (compare $B_c B/2\mu_0$ with $B^2/2\mu_0$). The magnetic distortion factor β should increase in the same proportion. We have not done any numerical computation to confirm this but plan to study type II superconducting interiors in a future work.

4 Gravitational radiation background from the whole population of neutron stars in the Galaxy

4.1 The squared signal

Let N be the total number of neutron stars in our Galaxy. The response of an interferometric detector to the gravitational wave field $(h^i_+, h^i_\times)$ [cf. Eqs. (1)-(2)] emitted by the $i^{\rm th}$ neutron star is

$$h_i(t) = F^i_+(t)\, h^i_+(t) + F^i_\times(t)\, h^i_\times(t) \ , \tag{6}$$

where $F^i_+(t)$ and $F^i_\times(t)$ are beam-pattern factors which depend on the direction of the star with respect to the detector arms. They vary with time because of the Earth rotation and revolution around the Sun. This results in an amplitude modulation [21,7] as well as a Doppler shift [22,23] of the signal. The time-average of the total (i.e. the sum on all the galactic neutron stars) is zero but not the time-average of the *squared* total signal:

$$\langle h^2 \rangle = \frac{1}{\tau} \int_{t_0}^{t_0+\tau} \left[\sum_{i=1}^{N} h_i(t) \right]^2 dt \ . \tag{7}$$

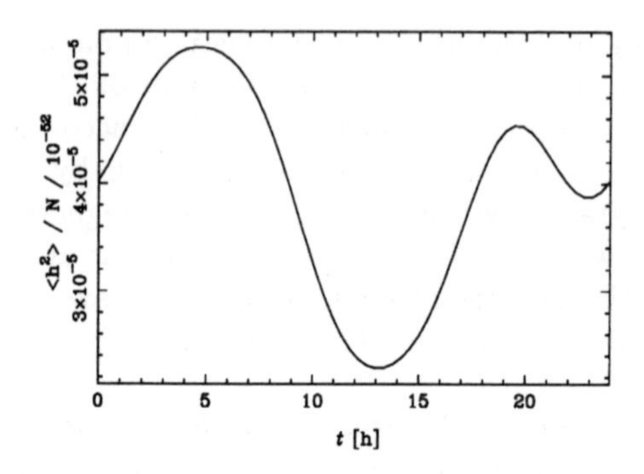

Figure 2: Total squared signal (divided by N) from a distribution of N neutron stars concentrated in the galactic disk with a height scale of 0.5 kpc. The average rotation period is assumed to be $\bar{P} = 5$ ms and the mean ellipticity $\bar{\epsilon} = 10^{-8}$.

The key-point is that if the galactic neutron star distribution is not isotropic, $\langle h^2 \rangle$ will exhibit temporal variation, with a period of one sidereal day. The precise shape of the signal depends on the neutron star distribution. An example [24] corresponding to a disk distribution is presented in Fig. 2.

It can be shown [24] that the signal-to-noise ratio for detecting this signal is

$$\frac{S}{N} = \frac{\langle h^2 \rangle}{2\tilde{h}^2} \left(\frac{T}{2\Delta\nu} \right)^{1/2} , \tag{8}$$

where $\Delta\nu$ is the frequency bandwidth and T is the observation time. It appears that such a signal might be detected if the number of neutron stars falling in the frequency range of the VIRGO detector is larger than $\sim 10^6 - 10^7$.

4.2 Number of rapidly rotating neutron stars in the Galaxy

From the star formation rate and the outcome of supernova explosions, the total number of neutron stars (NS) in the Galaxy has been estimated [25–26] to be about 10^9. The number of *observed* NS is much lower than this number: ~ 700 NS are observed as radio pulsars [13–14], ~ 150 as X-ray binaries, among which ~ 30 are X-ray pulsars [27–28], and a few as isolated NS, through their X-ray emission [29].

For our purpose, the relevant number is given by the fraction of these $\sim 10^9$ NS which rotates sufficiently rapidly to emit gravitational waves in the frequency bandwidth of VIRGO-like detectors. Putting the low frequency threshold of VIRGO to $f_{\text{min}} = 5$ Hz (optimistic value !), and using the fact that the highest gravitational frequency of a star which rotates at the frequency f is $2f$ (cf. § 2.1), this means the rotation period of a detectable NS must be lower than $P_{\text{max}} = 0.4$ s. The NS that satisfy to this criterion can be divided in three classes: (C1) young pulsars which are still rapidly rotating (e.g. Crab or Vela pulsars); (C2) millisecond pulsars, which are thought to have been spun up by accretion when being a member member of a close binary system (during this phase the system may appear as a low-mass X-ray binary); (C3) NS with $P < 0.4$ s which do not exhibit the pulsar phenomenon.

The number of millisecond pulsars in the Galaxy is estimated [30] to be of the order $N_2 \sim 10^5$. The number of *observed* millisecond pulsars ($P < 10$ s) is about 50 and is continuously increasing.

The number of young (non-recycled) rapidly rotating NS is more difficult to evaluate. An estimate can be obtained from the fact that the observed non-millisecond pulsars with $P < 0.4$ s represent 28 % of the number of catalogued pulsars and that the total number of active pulsars in the Galaxy is around 5×10^5. The number of rapidly rotating non-recycled pulsars obtained in this way is $N_1 \sim 1.4 \times 10^5$.

Adding N_1 and N_2 gives a number of $\sim 2 \times 10^5$ NS belonging to the populations (C1) and (C2) defined above. This number can be considered as a lower bound for the total number of NS with $P < 0.4$ s. The final figure depends on the number of members of the population (C3). This latter number is (almost by definition !) unknown. All that can be said is that it is lower than the total number of NS in the Galaxy ($\sim 10^9$).

Acknowledgments

This work has benefited from discussions with F. Bondu, A. Giazotto, P. Haensel, P. Hello and S.R. Valluri.

References

1. Blanchet L., this volume.
2. Stark R.F., Piran T., *Phys. Rev. Lett.* **55**, 891 (1985).
3. Bonazzola S., Gourgoulhon E., this volume.
4. Ipser J.R., *Astrophys. J.* **166**, 175 (1971).
5. Misner C.W., Thorne K.S., Wheeler J.A., *Gravitation* (Freeman, New York, 1973).
6. Thorne K.S., *Rev. Mod. Phys.* **52**, 299 (1980).

7. Bonazzola S., Gourgoulhon E., *Astron. Astrophys.*, in press (preprint: *astro-ph/9602107*).
8. Haensel P., in *Astrophysical Sources of Gravitational Radiation (Les Houches 1995)*, eds. J.-A. Marck, J.-P. Lasota, to be published.
9. Zimmermann M., Szedenits E., *Phys. Rev.* D **20**, 351 (1979).
10. Caraveo P.A., Bignami G.F., Mignani R., Taff L.G., *Astrophys. J.* **461**, L91 (1996)
11. Johnston S. et al., *Nature* **361**, 613 (1993).
12. Giazotto A. et al., in *First Edoardo Amaldi Conference on Gravitational Wave Experiments*, eds. E. Coccia, G. Pizzella, F. Ronga (World Scientific, Singapore, 1995).
13. Taylor J.H., Manchester R.N., Lyne A.G., *Astrophys. J. Suppl.* **88**, 529 (1993).
14. Taylor J.H., Manchester R.N., Lyne A.G., Camilo F., unpublished work (1995).
15. New K.C.B., Chanmugam G., Johnson W.W., Tohline J.E., *Astrophys. J.* **450**, 757 (1995).
16. Gal'tsov D.V., Tsvetkov V.P., Tsirulev A.N., *Zh. Eksp. Teor. Fiz* **86**, 809 (1984); English translation in *Sov. Phys. JETP* **59**, 472 (1984).
17. Bocquet M., Bonazzola S., Gourgoulhon E., Novak J., *Astron. Astrophys.* **301**, 757 (1995).
18. Wiringa R.B., Fiks V., Fabrocini A., *Phys. Rev.* C **38**, 1010 (1988).
19. Thompson C., Duncan R.C., *Astrophys. J.* **408**, 194 (1993).
20. Ruderman M., *Astrophys. J.* **382**, 576 (1991).
21. Jotania K., Dhurandhar S.V., *Bull. Astr. Soc. India* **22**, 303 (1994)
22. Jotania K., Valluri S.R., Dhurandhar S.V., *Astron. Astrophys.* **306**, 317 (1996)
23. Grave X. et al., this volume.
24. Bonazzola S., Gourgoulhon E., Giazotto A., in preparation.
25. Timmes F.X., Woosley S.E., Weaver T.A., *Astrophys. J.* **457**, 834 (1996).
26. Pacini F., this volume.
27. White N.E., Nagase F., Parmar A.N., in *X-ray binaries*, eds. Lewin W.H.G., van Paradijs J., van den Heuvel E.P.J. (Cambridge University Press, Cambridge, 1995).
28. van Paradijs J., in *The lives of the neutron stars*, eds. Alpar M.A., Kiziloglu Ü., van Paradijs J. (Kluwer Academic Publishers, Dordrecht, 1995).
29. Walter F.M., Wolk S.J., Neuhäuser R., *Nature* **379**, 233 (1996).
30. Bhattacharya D., in *X-ray binaries*, eds. Lewin W.H.G., van Paradijs J., van den Heuvel E.P.J. (Cambridge University Press, Cambridge, 1995).

BLACK HOLE COLLISIONS: A COMPUTATIONAL GRAND CHALLENGE

PABLO LAGUNA
Department of Astronomy & Astrophysics and
Center for Gravitational Geometry & Physics
Penn State University,
University Park, PA, 16802, USA

The coalescence of binary black holes epitomizes the ultimate test of Einstein's theory of gravity and has fundamental implications beyond the enclosure of general relativity. In astrophysics, black hole collisions may become the most promising source of gravitational radiation observable by detectors currently under construction. Within computational physics, a numerical solution to this problem constitutes a Grand Challenge demanding frontier advances in computer hardware and numerical algorithms. A consortium of research groups in the United States, funded by the National Science Foundation, has been established with the specific goal of simulating the collision of black holes and the gravitational radiation produced by such events. This paper reviews the current status of this effort and the major obstacles faced by this Alliance.

1 Introduction

The remainder of this century and the first decade in the next millennium could easily become and be remembered as the golden era in classical gravitation. In the observational arena, a new window in astronomy will be open when gravity wave detectors currently under construction become fully operational. In the field of computational gravitation or numerical relativity, the successful evolution of the coalescence of binary black holes will hopefully not only uncover new phenomena arising from the nonlinear nature of Einstein's equations, but it will also have a fundamental impact on the understanding of the gravitational radiation produced by such events. This period, of course, would be crowned if a satisfactory quantization of gravity is found!

Advances or discoveries in science are, in some instances, the result of certain degree of luck or the product of truly exceptional minds. In most cases, however, progress in science depends on the level of maturity of the ingredients that lead to such discoveries. This is certainly the case for finding astrophysically relevant solutions to Einstein's field equations, such as the coalescence of black holes. Although posed approximately sixty years ago, the relativistic Kepler problem has only been tractable in the last two decades thanks to the increase in speed and memory of computers, as well as to the sophistication of numerical techniques. These factors have provided a natural platform for the emergence of computational gravitation or numerical relativity.

The National Science Foundation in the United States recognized three years ago the problem of the coalescence of black holes as a "Computational Grand Challenge." An *Alliance* of research groups was formed[1] with the common goal of developing, within five years, a numerical code capable of performing simulations of inspiraling black hole binaries and constructing the emitted gravitational radiation. The institutions members of this Alliance are:

- Center for Astrophysics & Relativity (Stuart Shapiro)
- Cornell University (Saul Teukolsky)
- National Center for Supercomputing Applications (Faisal Saied, Paul Saylor, Edward Seidel, Larry Smarr)
- Northwestern University (Lee Samuel Finn)
- Penn State University (Pablo Laguna)
- Syracuse University (Jeffrey Fox)
- University of North Carolina, Chapel Hill (Charles R. Evans, James York)
- University of Pittsburgh (Jeffrey Winicour)
- University of Texas, Austin (James Browne, Matt Choptuik, Richard Matzner[a])

There are in addition two associate members: University of South Africa (Nigel Bishop) and Washington University, St. Louis (Wai-Mo Suen).

The purpose of this article is to report on the current progress of the Alliance's effort and state the questions that remain to be answered. The timing seems to be appropriate for a report of this nature since the Alliance has reached the middle point of its proposed lifetime. Of course, the hope within its members is that this coherent effort will continue after the original goal of colliding black holes has been reached.

Given the nature of these proceedings, it is impractical to review every single aspect that the Alliance has touched upon. Instead, I will concentrate on general issues regarding the following three areas:

- Numerical evolution of Einstein equations.

[a]Lead Principal Investigator

- The vicinity of black holes: Black hole collisions "without" black holes.
- Outer boundary conditions: Construction of gravitational wave forms.

The last two points involve the design and application of boundary conditions; after all, the "physics" in any problem is imprinted in its boundary conditions. This is particularly true in the collision of black holes, to the point that boundary conditions have an impact on the choice of formulations of Einstein equations adequate for the dynamics of spacetimes containing singularities.

2 Geometrodynamics

The most commonly used approach to evolving spacetimes numerically is geometrodynamics, namely the history of the geometry of spacelike hypersurfaces. Under this 3+1 or ADM[2] approach, six of the Einstein's equations are recast as evolution equations for the intrinsic, spatial metric g_{ij} of the hypersurfaces and their extrinsic curvature or second fundamental form K_{ij}. In the absence of matter fields, these equations read

$$\begin{aligned} \hat{\partial}_o g_{ij} &= -2\alpha K_{ij} \\ \hat{\partial}_o K_{ij} &= -\nabla_i \nabla_j \alpha + \alpha(R_{ij} - 2K_{ik}K^k_j + K_{ij}K), \end{aligned} \tag{1}$$

where R_{ij} is the 3-dimensional Ricci tensor and $\hat{\partial}_o \equiv \partial_t - \pounds_\beta$ with $\pounds_\beta$ the Lie derivative in the spacelike hypersurface. The freedom of choosing coordinates is encapsulated in the lapse function α (time) and the shift vector β^i (space).

The remaining four Einstein equations constitute constraints on the initial data (g_{ij}, K_{ij}):

$$\begin{aligned} R + K^2 - K_{ij}K^{ij} &= 0 \\ \nabla_j K^j_i - \nabla_i K &= 0. \end{aligned} \tag{2}$$

York's conformal approach[3] provides an elegant procedure for solving these constraint equations. Although initial data have been obtained for systems of orbiting and spinning black holes,[4,5] there is currently a considerable effort to make a connection with Post-Newtonian evolution of black holes.[6] That is, the goal is to construct initial data that will help to bridge the Post-Newtonian and full non-linear evolutions.

Even though the most advanced example of black hole collisions produced so far by the Alliance, 3-dimensional head-on collisions,[7] was performed using an ADM code, it is becoming apparent that the standard ADM formulation of Einstein's equations is not completely suitable to evolve black hole spacetimes. The root of the problem can be traced back to the known fact that, for arbitrary gauge conditions, the standard 3+1 formulation does not lead to hyperbolic evolution equation; that is, equations for which the mathematical characteristics agree with the physical speed of propagation of the field variables.

Partially motivated by the necessity of imposing boundary conditions in the vicinity of the black holes, boundary condition that require knowledge of the physical characteristic of the system (see next section), hyperbolic formulations of Einstein's equations have been proposed. The idea of rewriting Einstein's equations in a hyperbolic form is not new[8,9] and there is, by no means, a unique prescription. However, there are some hyperbolic formulations that seem to exhibit, by construction, certain numerical advantages.

The Alliance has singled out two of those hyperbolic formulations from which codes are going to be built upon. One of these procedures was developed by Bona and Masso.[10,11] The original motivation was to rewrite Einstein's equations in a first-order, flux conservative, hyperbolic form, so powerful machinery from computational fluid dynamics could be used. By introducing a special choice of variables, using the constraints and imposing a class of gauge choices, they were able to write Einstein's equations in the form

$$\partial_t\phi + \partial_i F^i(\phi) \quad = \quad S(\phi) \tag{3}$$

where $\phi = \{g_{ij}, K_{ij}, \partial_i g_{jk}\}$.

Even that one does not expect to encounter geometry "shocks," having a system of equations like (3), that admits the implementation of numerical algorithms to resolve sharp features, could become advantageous. Furthermore, since by construction the system (3) can be diagonalized, it is possible to obtain the eigenfields propagating along characteristic surfaces. It is the use of those eigenfields what, in principle, should facilitate the task of imposing boundary conditions based on the causal structure of the spacetime.

The other hyperbolic formulation that the Alliance is pursuing has been developed by Choquet-Bruhat and York.[12,13] Under this approach, Einstein's equations take the form

$$\begin{aligned}\hat{\partial}_o g_{ij} &= -2\alpha K_{ij} \\ \Box K_{ij} &= Q_{ij}(\alpha, \beta, \mathbf{g}, \mathbf{K}, \partial_i \mathbf{K}),\end{aligned} \tag{4}$$

where $\Box \equiv -\alpha^{-1}\hat{\partial}_o\alpha^{-1}\hat{\partial}_o + \nabla^k\nabla_k$. This system of equations possesses a complete spatial coordinate freedom, i.e., allows an arbitrary choice of the shift vector; the equations involve fields with direct connection with gauge invariant radiation extraction, and they can also be brought into a first-order, flux-conservative, symmetric hyperbolic form. As with the Bona/Masso system, this system also requires a class of "harmonic"-like lapse function conditions.

A point that has triggered a debate within the Alliance is whether a numerically successful hyperbolic reduction of Einstein's equations should have the speed of light as the only non-vanishing characteristic speed; that is indeed the case for the Choquet-Bruhat/York formulation. On the other hand, the Bona/Masso approach yields characteristic speeds different from the speed of light for some choices of shift vector. Although one would be tempted to prefer, from physical grounds, formulations with c as the only characteristic speed, in the Bona/Masso method the characteristic speeds that differ from c are *gauge speeds.* ¿From the point of view of implementing outgoing boundary conditions, what seems to be important is the knowledge of the characteristic speeds of the eigenfuctions, and not so much their value. This issue will be resolved once a complete implementation of both approaches is completed.

3 Excising Black Holes

The history of numerically evolving spacetimes containing black holes can be summarized as the continuous battle of handling the singularities these objects represent. Since numerical computations cannot be performed at spacetime singularities, the general approach to deal with black hole singularities has been to avoid them or remove them from the computational domain.

Until recently, avoiding singularities was the most popular approach. The general idea behind this method was to take advantage of the freedom of choosing spacelike hypersurfaces, represented by the lapse function, and numerically construct a foliation that avoids the singularities in the black hole interiors.[3] The hope was to slow down or even halt the evolution in strong curvature regions and proceed, at the same time, with the evolution in the remaining parts of the system. This procedure, although perfectly consistent from the mathematical point of view, exhibits serious numerical drawbacks. Constructing foliations that wrap around singularities has the disadvantage that the resulting "stretching" of the hypersurfaces translates into the appearance of sharp features in the metric variables. The computational burden to resolve this gauge-induced growth of the field variables becomes rapidly prohibitive.

A different approach to black hole dynamics has recently received considerable attention.[14,15] The idea here is to get rid off the singularities altogether by removing a region inside the black hole horizon containing the singularity.[16,17] For an observer outside the horizon, according to cosmic censorship, this situation should be indistinguishable from that in which the singularity is present.

A numerical implementation of this idea for excising black holes from the computational domain requires: (A) finding the horizon of the black hole and (B) imposing outgoing, into the hole, boundary conditions. Since event horizons require knowledge of the complete future development, they cannot be used to identify the region needed to be removed. Fortunately, there is an alternative solution: the use of apparent horizons instead. Apparent horizons do not require knowledge of the spacetime development. They are defined from information contained within a single spacelike hypersurface. Furthermore, their existence implies the presence of an enclosing event horizon, assuming, as we do, that cosmic censorship holds.

Once the apparent horizon has been found and a region inside or near the horizon containing the singularity has been removed, the next task is to impose outgoing boundary conditions. Owing to the one-way membrane nature of the horizon and trapped surfaces within it, the task of imposing outgoing boundary conditions is, in principle, facilitated because the light-cone are tilted inwards in such a way that only information inside of the computational domain is needed to update the information on the boundaries, that is, a boundary condition without a boundary condition. A stable finite difference scheme needs to be designed that takes into account the tilting of the light-cones. An extremely important feature of apparent horizon boundary conditions is that these conditions do not anchor black holes to the computational grid. One can, in principle, carry out an evolution in which the black holes float in the computational grid. The implementation of apparent horizon boundary conditions constitutes the major obstacle that the Alliance is currently facing.

4 Radiation Extraction and Outer Boundary Conditions

The numerical modeling of radiative systems involves the challenge of imposing boundary conditions that correctly mimic the energy loss by radiation and the proper $1/r$ asymptotic falloff of the radiation fields. This issue has tremendous importance since, after all, one of the main objectives of the Alliance is to determine the gravitational radiation at some large distance, many wavelengths away from the source. Radiative boundary conditions are not trivial to construct, even in "simple" situations such as the dynamics of scalar fields in fixed background geometries.[18] The problem is exacerbated when dealing with the full set of Einstein equations.

The Alliance has undertaken two closely related approaches that differ only in the way that the wave forms are computed. The goal is to construct an outer boundary module, which provides both, boundary conditions and extraction of gravitational waveforms. The general idea behind this module is to attach, at the outer boundary of the computational domain where the collisions take place, a computationally inexpensive, exterior evolution scheme that propagates the gravitational fields up to the location where waveforms are constructed. Hence, this approach not only requires developing an exterior evolution code, but also a matching infrastructure at a worldtube where the interior and exterior evolutions are joined. This matching worldtube serves the double purpose of: (A) *extracting* data from the interior evolution and feeding it to the exterior evolution, and (B) *injecting* exterior evolution data into the interior evolution, thus providing boundary conditions for the interior numerical code.

As mentioned above, two approaches are being considered in constructing exterior evolutions. One approach[20,21] is based on the idea that, beyond a certain distance from the sources, the spacetime geometry can be approximated by non-spherical perturbations of the Schwarzschild spacetime. An important issue in this approach is to construct, for all the multipoles of both parities, "wave equations" involving variables that are gauge-invariant variables to first order in the perturbative parameter. Once these gauge invariant quantities are constructed, the wave equations can be used to evolve initial data and extract radiation wave forms. The initial data and boundary conditions are obtained from the numerically evolved spatial metric and extrinsic curvature. Special care should be taken when setting up the initial data or boundary conditions. It is required that the data is given on surfaces of approximately constant Schwarzschild time. This procedure has been successfully tested for Misner data representing black holes at a moment of time symmetry.[21]

The second approach to the exterior evolution involves a formulation that views Einstein's equations as a characteristic initial value problem, namely the foliation of spacetime is given by a family of null cones emanating from a central timelike geodesic.[22] Characteristic formulations become inappropriate in the vicinity of strong fields because of the presence of caustics; on the other hand, a null approach, by construction, naturally adapts to radiative systems. This is the reason for combining a Cauchy evolution, in the region where strong fields develop, with a characteristic exterior evolution to handle the radiation emitted. In the matching of Cauchy with characteristic evolutions, the null cones do not emerge from a a central timelike geodesic; instead, the null foliation is anchored to a timelike worldtube whose interior is evolved by a Cauchy method. A full implementation of a mixed, Cauchy+characteristic, evolution is still under development. At present time, a 3-dimensional characteristic code is being developed and has been calibrated in the perturbative regime off Schwarzschild.[23]

5 Conclusions

The frontiers in numerical relativity are expanding to the point that explorations of complex, strongly gravitating systems are becoming a reality. One of the main driving engines pushing these limits has been the effort in solving the two black hole collision problem.

Significant progress has been made towards the ultimate goal of numerically evolving inspiraling black holes throughout their coalescence. Novel approaches for handling singularities and imposing radiative boundary conditions have been proposed; data structures for computational domains containing black holes have been designed, as well as powerful numerical algorithms; even the mathematical structure of Einstein's equations has been reconsidered. Over all, the previous understanding of the numerical approach to gravity has undergone a significant revolution.

In spite of successes such as head-on collisions of black holes,[7] the simulation of inspiraling coalescences remains to be tackled. The general consensus is, however, that if the current pace of effort and progress continues, the Kepler problem in general relativity will be a solved problem before the next millennium.

Acknowledgments

Work supported in part by NSF grants PHY/ASC-9318152 (ARPA supplemented), PHY-9357219 (NYI) and PHY-9309834. Special thanks to R.Matzner for comments and helpful discussions during the preparation of the manuscript.

References

1. http://jean-luc.ncsa.uiuc.edu/GC/GC.html
2. R. Arnowitt, S. Deser and C.W. Misner in *Gravitation*, edited by L. Witten (Wiley, New York, 1962).
3. J.W. York, in *Sources of Gravitational Radiation*, edited by L. Smarr (Cambridge University Press, Cambridge, England, 1979).
4. G.B. Cook, *Phys. Rev.* D **44**, 2983 (1991)
5. G.B. Cook, M.W. Choptuik, M.R. Dubal, S. Klasky, R.A. Matzner and S. Oliveira, *Phys. Rev.* D **47**, 1471 (1993)
6. G.B. Cook, *Phys. Rev.* D **50**, 5025 (1994)
7. P. Anninos, D. Hobill, E. Seidel, L. Smarr, and W.-M Suen, NCSA preprint (1996).
8. A.E. Fischer and J.E. Marsden, *Commun. Math. Phys.* **28**, 1 (1972)

9. Y. Choquet-Bruhat and T. Ruggeri, *Commun. Math. Phys.* **89**, 269 (1983)
10. C. Bona and J. Massó *Phys. Rev. Lett.* **68**, 1097 (1992)
11. C. Bona, J. Massó , E. Seidel and J. Stela, *Phys. Rev. Lett.* **51**, 1639 (1995)
12. Y. Choquet-Bruhat and J.W. York, *C. R. Acad. Sci. Paris*, **321**, 1089 (1995).
13. A. Abrahams, A. Anderson, Y. Choquet-Bruhat, and J.W. York, *Phys. Rev. Lett.* **75**, 3377 (1995)
14. E. Seidel and W.-M. Suen, *Phys. Rev. Lett.* **69**, 1845 (1992)
15. M. Alcubierre and B. Schutz, *J. Comp. Phys.* **112**, 44 (1994)
16. W. Unruh, *Class. Quantum Grav.* **14**, 1119 (1987) as cited in by Thornburg.
17. Thornburg, Ph.D. thesis, University of British Columbia, 1993.
18. P. Papadopoulos and P. Laguna, submitted *Phys. Rev.* D (1996).
19. A. Abrahams and C. Evans, *Phys. Rev.* D **46**, 4117 (1992)
20. A. Abrahams and C. Evans, *Phys. Rev.* D **46**, 4117 (1992)
21. A. Abrahams and R. Price, *Phys. Rev.* D in press (1996)
22. R. Isaacson, J. Welling and J. Winicour, *J. Math. Phys.* **24**, 7 (1983)
23. N.T. Bishop, R. Gomez, L. Lehner and J. Winicour, submitted *Phys. Rev.* D (1996).

Interferometric Detectors

STATE OF THE ART OF THE VIRGO EXPERIMENT

B. Caron, A. Dominjon, C. Drezen, R. Flaminio, X. Grave, F. Marion, L. Massonet, C. Mehemel, R. Morand, B. Mours, V. Sannibale, M. Yvert
LAPP–Annecy, Annecy, France

D. Babusci, S. Bellucci, S. Candusso, G. Giordano, G. Matone
INFN–Frascati, Frascati, Italy

J.–M. Mackowski, L. Pinard
IPN–Lyon, Lyon, France

C. Boccara, P. Gleizes, V. Loriette, J.–P. Roger
ESPCI–Paris, Paris, France

P. G. Pelfer, R. Stanga
University of Firenze and INFN–Firenze, Firenze, Italy

F. Barone, E. Calloni, L. Di Fiore, M. Flagiello, F. Garufi, A. Grado, M. Longo, M. Lops, L. Milano, S. Marano, S. Solimeno
University of Napoli and INFN–Napoli, Napoli, Italy

M. Barsuglia, B. Bhawal, F. Bondu, A. Brillet, V. Brisson, F. Cavalier, M. Davier, H. Heitmann, P. Hello, P. Heusse, J.–M. Innocent, L. Latrach, F. Le Diberder, C. Nary Man, P. Marin, M. Pham–Tu, M. Taubman, E. Tournier, J.–Y. Vinet
LAL–Orsay, Paris–Sud, France

G. Cagnoli, L. Gammaitoni, J. Kovalik, F. Marchesoni, M. Punturo
INFN–Perugia, Perugia, Italy

M. Beccaria, M. Bernardini, S. Braccini, C. Bradaschia, C. Casciano, G. Cella, A. Ciampa, E. Cuoco, G. Curci, R. De Salvo, R. Del Fabbro, A. Di Virgilio, D. Enard, I. Ferrante, F. Fidecaro, A. Gaddi, A. Gennai, A. Giazotto, G. Gorini, P. La Penna, G. Losurdo, S. Mancini, H. B. Pan, A. Pasqualetti, D. Passuello, R. Poggiani, P. Popolizio, F. Raffaelli, A. Vicere', Z. Zhang
University of Pisa and INFN–Pisa, Pisa, Italy

F. Bronzini, V. Ferrari, E. Majorana, P. Puppo, P. Rapagnani, F. Ricci
University of Roma and INFN–Roma, Roma, Italy

L. Holloway
University of Urbana, Urbana, USA and INFN–Pisa, Pisa,Italy

The status report of the VIRGO experiment is presented. The experiment has been approved in September 1993 and is now in the construction stage. Its aim is the detection of gravitational waves over a broad frequency range (from 10 Hz to 10 kHz) using a Michelson interferometer equipped with Fabry–Perot cavities 3 km long. The experiment will be installed in Cascina near Pisa and is planned to be operative during year 2000. The planned sensitivity is $\tilde{h} = 10^{-21}/\sqrt{Hz}$ at 10 Hz and $\tilde{h} = 3 \times 10^{-23}/\sqrt{Hz}$ at 500 Hz.

presented by A. Giazotto

1 Introduction

The gravitational waves (GW) are predicted by the theory of General Relativity [1]. There is actually an indirect evidence of their existence given by the observation of the binary pulsar system PSR 1913+16 [2].

The direct observation of the GW, beside being a relevant test of General Relativity, will start a new picture of the Universe. In fact, GW carry complementary information with respect to electromagnetic waves since the GW are not absorbed by matter and are emitted from massive sources in strong gravity conditions.

The aim of the VIRGO experiment is the direct detection of the GW and, in joint operation with other similar detectors, to perform gravitational waves astronomical observations. VIRGO is designed for a broadband detection, from 10 Hz to 10 kHz. The capability of low frequency operation should give the best possibilities for detecting gravitational radiation from monochromatic sources and coalescing binaries.

VIRGO consists of both a facility and a detector (Fig. 1):

- the facility consists of a pair of 3 km long tunnels, a large vacuum system and vacuum towers for containing the suspensions of the optical components;
- the detector includes the laser system, the interferometer, the seismic isolation system and the electronics.

VIRGO is a large international collaboration with physicists and engineers from nine different laboratories in France and in Italy. The share of responsibilities is as follows:

- **Annecy**: detection bench, vacuum towers, data acquisition, simulation;
- **Firenze**: accelerometers and inertial damping;

- **Frascati:** interferometer alignment;
- **Lyon:** mirror coatings;
- **Napoli:** digital filters, environmental monitoring, data storage and archiving;
- **Orsay:** laser, laser stabilization, injection bench, 3 km vacuum tube, baffles, global control;
- **Paris:** mirror metrology;
- **Perugia:** suspension thermal noise measurement, suspension materials;
- **Pisa:** infrastructure, pumping system, 3 km vacuum tube, seismic isolation system, vacuum compatibility of materials;
- **Roma:** mirror suspension, alignment.

2 Gravitational waves sources

In the weak field approximation the GW perturb the metric of the space–time as follows [3]:

$$g_{\mu\nu} = \eta_{\mu\nu} + h_{\mu\nu} \tag{1}$$

where $h_{\mu\nu}$ is the perturbation to the flat space–time metric $\eta_{\mu\nu}$ and $|h_{\mu\nu}| \ll 1$. The GWs have two states of polarizations h_+, $h_\times$ with quadrupolar patterns.

The perturbation h causes a change ΔL of the distance L between two free masses, such that:

$$\Delta L = \frac{hL}{2} \tag{2}$$

in the hypothesis that the wavelength λ of the GW is much larger than the mass separation L. At the moment terrestrial GW sources are not conceivable and only astrophysical sources in strong gravity conditions can be considered. Mainly five source types are considered promising:

- **supernovae:** the collapse of type II supernovae into neutron stars or black holes is predicted to produce short radiation bursts. The theoretical predictions suffer from the uncertainty in the degree of asymmetry and in the velocity of the collapse[3]. For a supernova exploding in the VIRGO cluster (10 Mpc) a value of $h \sim 10^{-21}$ is expected, with a rate of a few per year.

- **coalescing binary systems:** since nearly one half of the galactic stars come in binary systems, a number of systems with neutron stars or black holes as components can be expected. The energy loss by gravitational radiation can cause the coalescence of the system within a time much shorter than the age of the universe. The process produces the emission of GW with progressively increasing frequency up to a maximum just before coalescence. A few events per year are expected within a 100 Mpc radius[4].

- **pulsars:** asymmetries in the mass distribution of pulsars are believed to produce GW emission at twice the rotation frequency with h ranging from 10^{-28} to 10^{-24} [3]. The monochromaticity of GW allows long integration times of the signals. On the other hand, this is a symptom of the low power emitted as GW.

- **relic radiation:** it is important to mention this kind of radiation because it can give information on the very instant of the Big Bang. GWs, due to their low interaction cross section with matter, can carry the imprinting of this event. The detection of relic radiation can only be made in coincidence with other detectors.

- **neutron stars:** the predicted number of rotating neutron stars is very large ($10^8 \div 10^{10}$); it has been proposed to detect the collective GWs emitted by the whole ensemble using a non–linear analysis[5].

3 VIRGO: the detector

The main features of the laser interferometer VIRGO will be described here, referring the reader to [6] for a detailed discussion. The conceptual scheme of the detector is shown in Fig. 2.

A GW propagating orthogonally to the plane of the Michelson interferometer will alternatively stretch one arm and shrink the other one: thus there will be an oscillating change of the relative phase of the light in the two arms.

With the use of Fabry–Perot cavities in the arms, the optical path length is increased from 3 km to 120 km. A feedback system controls the positions of the mirror in such a way that the interferometer is kept in destructive interference condition, that is operating on the dark fringe. This both maximizes sensitivity and allows power recycling. The GW signal is extracted from the current circulating in the feedback loop.

Various sources of noise affect the sensitivity of VIRGO. The sources and the proposed approaches to reduce the noise will be discussed now.

3.1 Shot noise

The shot noise is the limiting factor to the sensitivity in the frequency region above a few hundreds Hz. It is due to the statistical fluctuations in the number of detected photons. It is inversely proportional both to the square root of the efficiency of the photodetector η and of the beam power P:

$$\tilde{h}_{sn} = \frac{\lambda}{8FL}(1 + (\frac{2\omega FL}{\pi c})^2)^{\frac{1}{2}}\sqrt{\frac{h\nu}{\eta P}} \tag{3}$$

where λ, ν are the wavelength and frequency of the laser, L the interferometer arm length, h the Planck constant, F the cavity finesse. In VIRGO the use of a 25 W Nd:Yag ring at 1.06 μm laser is planned. The recycling technique is used to increase the laser power, by adding a further mirror between the laser and the beam splitter. The necessity of high gain factors puts stringent requests on the mirror losses. For this purpose a R&D program is actually in progress about the mirror coatings [7]. On a 5 cm diameter mirror scattering and absorption have been reduced down to a few ppm. High efficiency InGaAs photodetectors are actually tested in Annecy [8].

3.2 Laser frequency and power fluctuations

Since small asymmetries between the two arms of the interferometer are unavoidable, the fluctuations in the laser frequency must be kept as small as possible. To this end, a program of R&D is actually in progress in Orsay [9].

The laser is locked by the injection–locking technique to an ultra–stable laser with 1 W power. The spectral density of the laser frequency fluctuations is shown in Fig. 3. Shot noise limit is exactly reached at 20 kHz, while excess noise from 10 to 200 Hz is mainly seismic.

The power fluctuations are reduced by an active control system to $\frac{\Delta \tilde{P}}{P} \leq 10^{-7} \frac{1}{\sqrt{Hz}}$ (to be compared with the required value of $10^{-8} \frac{1}{\sqrt{Hz}}$). The overall scheme of the laser and input optics of VIRGO is shown in Fig. 4.

3.3 Scattered light

A portion of the light scattered off the mirror can rejoin the main beam after multiple reflections on the walls of the beam pipe with a phase change due to the different optical path. The vibration of the pipe due to seismic noise makes the phase difference dependent on time. The use of suitable absorbing baffles in the beam pipe (Fig. 5) should allow to reduce this contribution to the level [10]:

$$\tilde{h}_{sl} = 10 \times 10^{-24} (\frac{10}{f})^2 Hz^{-\frac{1}{2}} \tag{4}$$

which is well below the VIRGO requirements. At the moment both glass and metal baffles are under consideration.

3.4 Vacuum requirements

The VIRGO interferometer will be operated under vacuum. The statistical fluctuation of the residual gas density can induce a fluctuation of the index of refraction thus of the interferometer phase difference. A vacuum level of $10^{-8} \div 10^{-9}$ mbar is planned. Two parallel R&D programs about the vacuum tube have been completed in Pisa (Straight Reinforced Tube) and in Orsay (Corrugated Tube) [11]. The first one has been selected for the final realization (Fig. 6). The Orsay group is in charge of the quality control at the factory and the Pisa group of the final assembling.

The presence of optical elements inside the vacuum environment demands a low contamination from hydrocarbon, this means an hydrocarbon partial pressure $< 10^{-13}$ mbar. A R&D program is in progress in Pisa about this topic.

3.5 Seismic noise

The ground vibrations are several orders of magnitude larger than the displacements due to the expected GW signal. This effect is strongly reduced by suspending each optical component to a seven stage cascade of resonators, the *superattenuator* (SA, Fig. 7).

A new SA equipped with metal blade springs will replace the originally proposed gas spring SA since it is more stable with temperature and easier to be controlled. The SA is developed in Pisa. Measurements performed with one stage and two stages have shown that each stage allows nearly 40 dB attenuation both in vertical and horizontal direction above 10 Hz (see Fig. 8) [12].

The resonances of the SA at very low frequency (< 3 Hz) produce quite large oscillations of the mirror which must be damped to keep the interferometer working on the dark fringe condition. A six degrees of freedom sensor system measures the oscillations with respect to ground and damps the SA movements down to an acceptable level. Work is actually in progress in Pisa on developing a pre–isolator stage for the SA. The method used is that of an inverted pendulum where low values of resonant frequencies (~ 50 mHz) are readily achievable. Prototypes have been successfully operated with suspended masses of 650 Kg at a frequency of 30 mHz.

3.6 Newtonian noise

Seismic waves passing close to the interferometer produce density fluctuations in the earth, which produce fluctuating gravitational forces on the test masses faking a stochastic background. This effect is well described by a model[13] with a trasnfer function $T(f) = \frac{\tilde{x}(f)}{\tilde{X}(f)}$ from the horizontal seismic displacement $\tilde{X}(f)$ to the GW induced displacement $\tilde{x}(f)$. The transfer function has a behaviour $T(f) \propto \frac{1}{\omega^2}$ above all the normal modes of the SA.

3.7 Thermal noise

The VIRGO sensitivity at low frequencies is limited by the thermal noise acting on the superattenuator and mirror system. The thermal noise is related to the dissipation in the system through the Fluctuation–Dissipation theorem, which states that stochastic forces arise with spectral density:

$$\tilde{F}^2(\omega) = 4k_BTR(\omega) \tag{5}$$

where $R(\omega)$ is the real part of the mechanical impedance $Z(\omega) = F(\omega)/v(\omega)$. The force causes Brownian motion of the system. Due to operation in vacuum, the dissipation in VIRGO is coming from the internal friction in the suspension materials. The dominant contribution in such a multistage system is given by the dissipation in the last stage. The last stage of the suspension named *marionetta* provides to the steering and alignment of the optical components; it is equipped with metal blades which suspend the mirror wires. The marionetta development in Pisa has allowed to show that with this configuration the vertical thermal noise contribution of the last stage of the superattenuator is reduced [14]. A R&D program to study the suitable geometries and materials for the mirror suspension is in progress [15]. The Fabry–Perot cavities mirrors are kept as massive as possible (20 to 40 Kg) to reduce the thermal noise.

3.8 Virgo sensitivity

The overall VIRGO sensitivity is shown in Fig. 9.

It is evident that the use of the SA decreases the seismic noise down to negligible levels. Various mechanical resonances contribute to the thermal noise in the frequency region of interest: the high frequency tail of the pendulum mode, the low frequency tail of the mirror internal modes and the narrow resonances of the violin modes. The contribution of the vertical modes is well below the pendulum one at low frequencies.

We point out that the lowest detectable frequency in VIRGO is 4 Hz. This is the point where the contribution of the seismic noise crosses the contibution from Newtonian noise. It is evident that there is no room for further improvements of the seismic attenuation due to the Newtonian noise contribution.

Summarizing, dominant sources of noise are: the seismic noise below 4 Hz, the thermal noise of the suspension system and of the mirror up to a few hundreds Hz and the shot noise above. The sensitivity in the high frequency region is of the order of $3 \times 10^{-23}/\sqrt{Hz}$.

3.9 Electronics and controls

The optical components of the interferometer undergo position and angle fluctuations due to seismic and thermal noise which can be damped by active locking methods. A global control system keeps the interferometer locked to the working point: the Fabry–Perot cavities and the recycling cavity are kept in resonance and the light output from the beam–splitter mantained at the dark fringe. The active controls will be performed with digital servo loops.

3.10 Data management

A dedicated simulation program has been developed in Annecy to describe the behaviour of the VIRGO apparatus taking into account the the estimated noise sources and the GW produced by the various sources [16]. The expected acquisition rate is of the order of 1 MB/s [17]. The concept of trigger on interesting events is currently under study. We plan a data analysis in coincidence with other similar detectors such as the LIGO project and with neutrino and gamma rays detectors.

3.11 Buildings

The site acquisition is actually in progress. The buildings construction is foreseen in one year time. At middle 1997 the installation of towers, pumping system, superattenuators, lasers, optics and electronics should start.

3.12 The 1998 interferometer

The construction of the VIRGO interferometer in Cascina will require about 5 years. We will first build the central part of the system within the middle of 1998. During the next two years, while the construction of the 3 km arms is going on, we will have the opportunity to install a recycled Michelson interferometer in the final position in the central building, and to test all the critical subsystems of VIRGO and their final functions: laser, detection optics, seismic isolation system, controls, alignment and data management. In this way, we expect to start data acquisition with full sensitivity by the end of 2000, a few months only after the construction is completed.

4 Conclusions

We have presented the status of the VIRGO experiment. Most of the critical R&D has been very successful and we expect to start running the instrument at the final sensitivity in year 2001.

1. C. W. Misner, K, S. Thorne and J. A. Wheeler, *Gravitation*, W. H. Freeman and Company, San Francisco (1973)
2. J. H. Taylor and J. M. Weisberg, *Astrop. J.* **345** (1989) 434
3. *Three hundreds years of gravitation*, Ch. 9: Gravitational radiation by K. S. Thorne, ed. by S. W. Hawking and W. Israel, Cambridge University Press (1987)
4. B. F. Schutz, *Nature* **323** (1986) 310
5. S. Bonazzola, A. Giazotto and E. Gorgoulhon, to be published
6. *VIRGO: proposal for the construction of a large interferometric detector of gravitational waves* (1989); *VIRGO: Final Conceptual Design* (1992); *VIRGO: Final Design* (1995); A. Giazotto, *Phys. Rep.* **182** (1989) n.6
7. C. Boccara et al., VIRGO internal note PJT93–018; C. Boccara and V. Loriette, VIRGO internal note PJT93–020
8. B. Caron et al., presented at "Frontier Detectors for Frontier Physics", La Biodola, May 1994, *Nucl. Instr. and Meth.* **A360** (1995) 379
9. C. N. Man et al., presented at the "7th Marcel Grossman Meeting", San Francisco, July 1994
10. J.–Y. Vinet and V. Brisson, VIRGO internal note PJT00–011
11. Pisa group, VIRGO internal note PMP 00–26; Orsay group, VIRGO internal notes PMP 00–21, 22
12. S. Braccini et al., *Rev. Sci. Instr.* **67** (1996) 2899
13. S. A. Hughes and K. S. Thorne, to be submitted to Phys. Rev. D

14. S. Braccini et al., *Phys. Lett.* **A199** (1995) 307
15. F. Marchesoni et al., *Phys. Lett.* **A187** (1994) 359
16. B. Caron et al., presented at "Frontier Detectors for Frontier Physics", La Biodola, May 1994, *Nucl. Instr. and Meth.* **A360** (1995) 375
17. B. Caron et al., presented at the "7th Marcel Grossman Meeting", San Francisco, July 1994 preprint LAPP–EXP–94–14

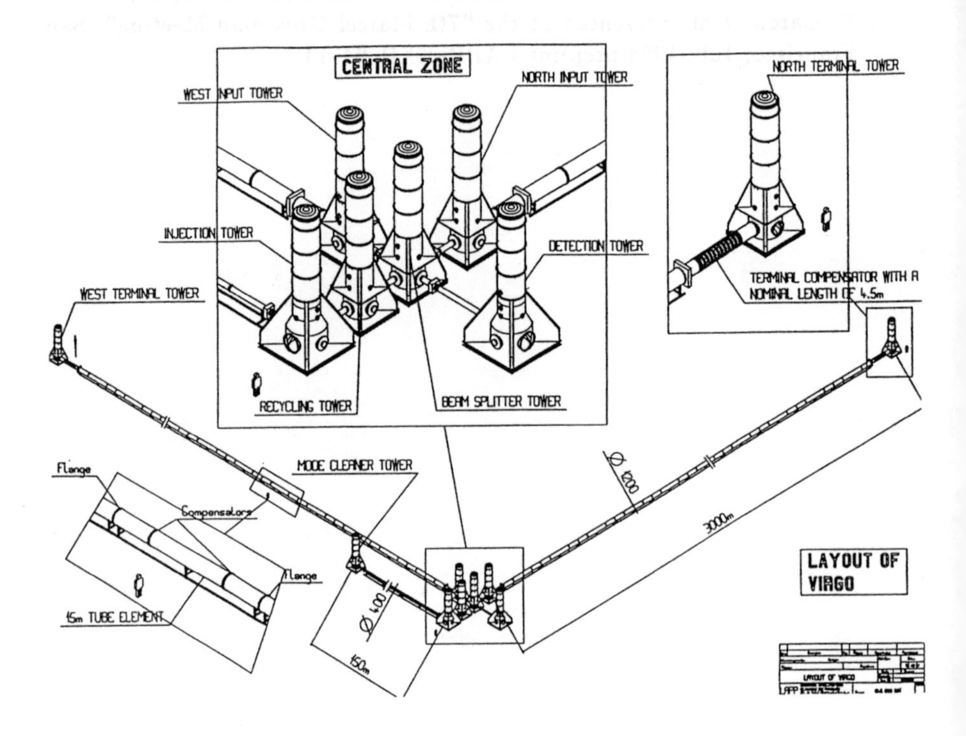

Figure 1: The layout of VIRGO with details of the towers for the optical elements

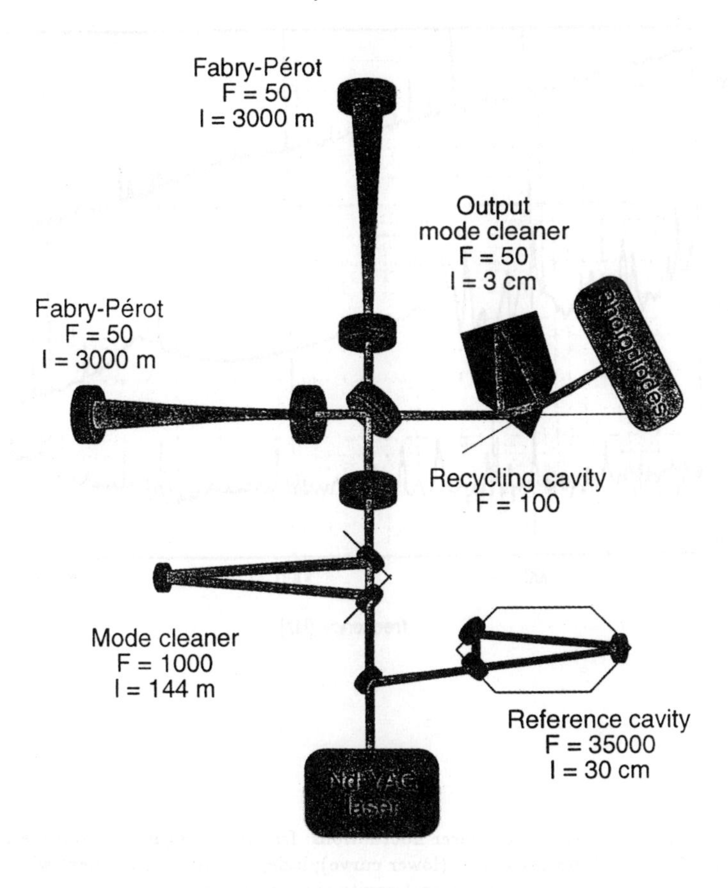

Figure 2: The conceptual scheme of the VIRGO detector

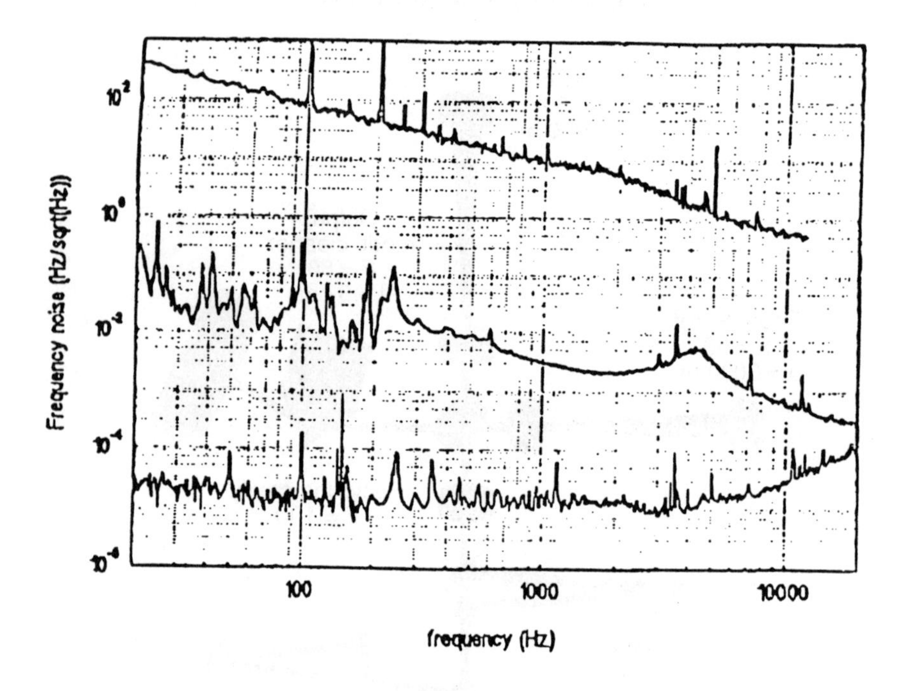

Figure 3: Spectral density of the laser fluctuations: free running (upper curve); error signal when locked to a high finesse cavity (lower curve); independent measurement with a second Fabry–Perot cavity (intermediate)

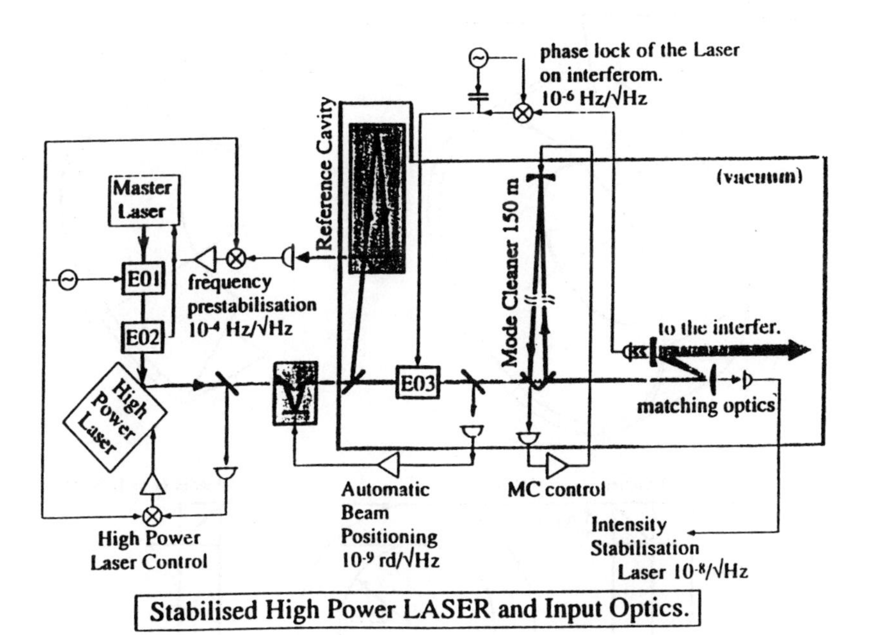

Figure 4: Scheme of the stabilized high power laser and input optics

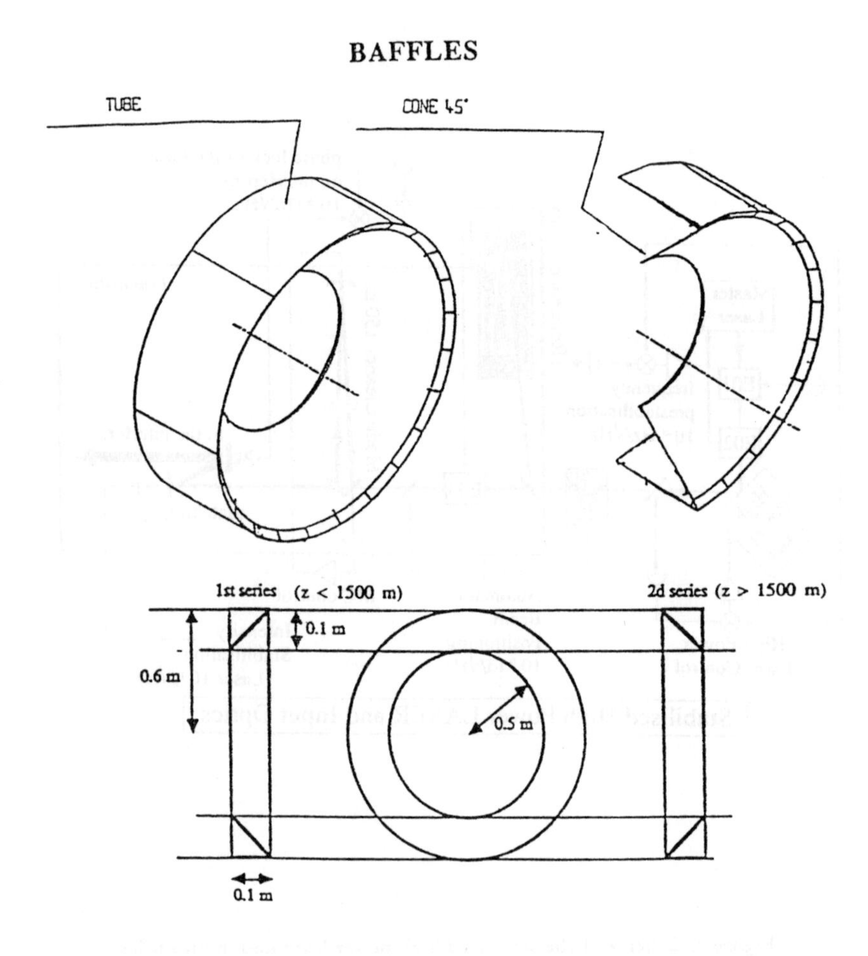

Figure 5: Details of the baffles used to reduce the contribution of scattered light

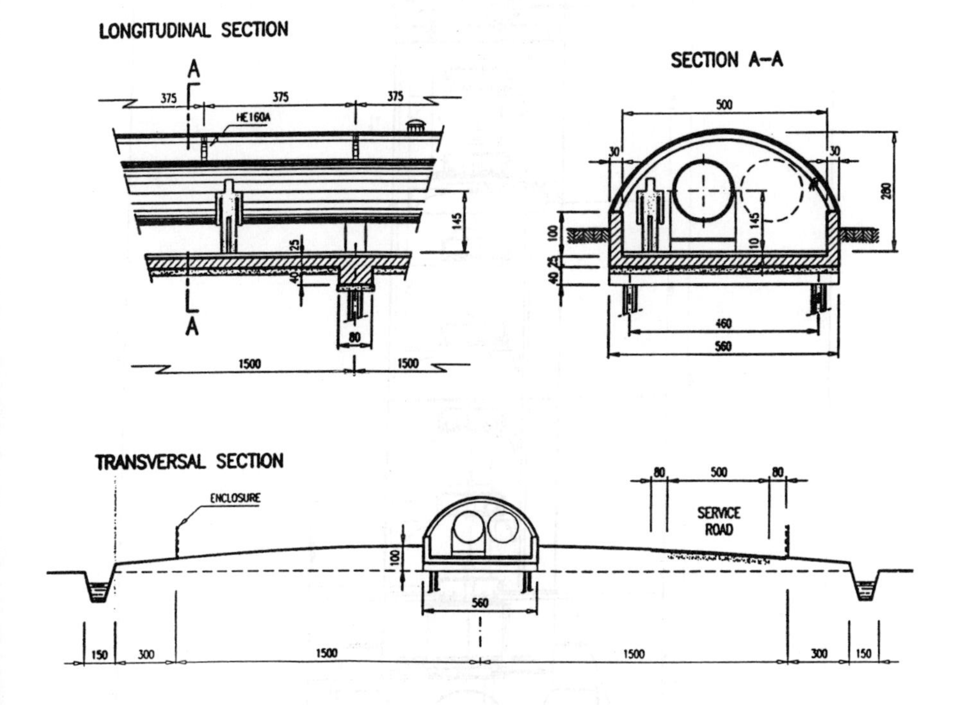

Figure 6: A preliminary version of the VIRGO tunnel

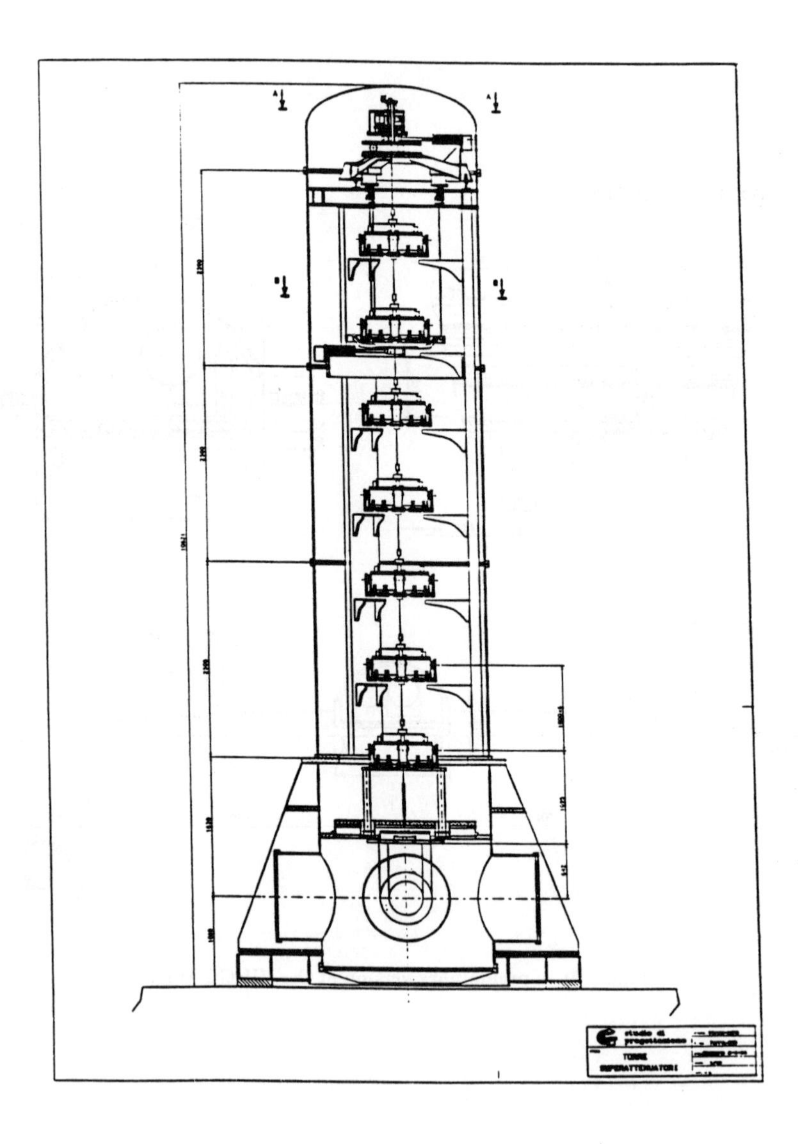

Figure 7: The vacuum tower with the SA cascade

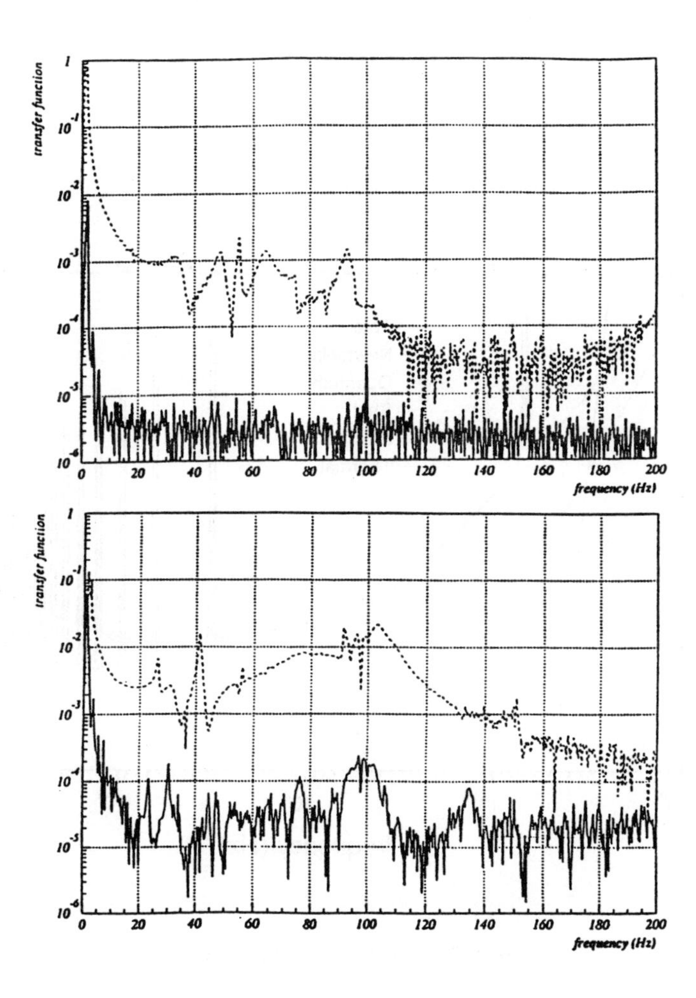

Figure 8: Attenuation achieved with one (dashed line) and two stages of the SA: a) horizontal direction; b) vertical direction

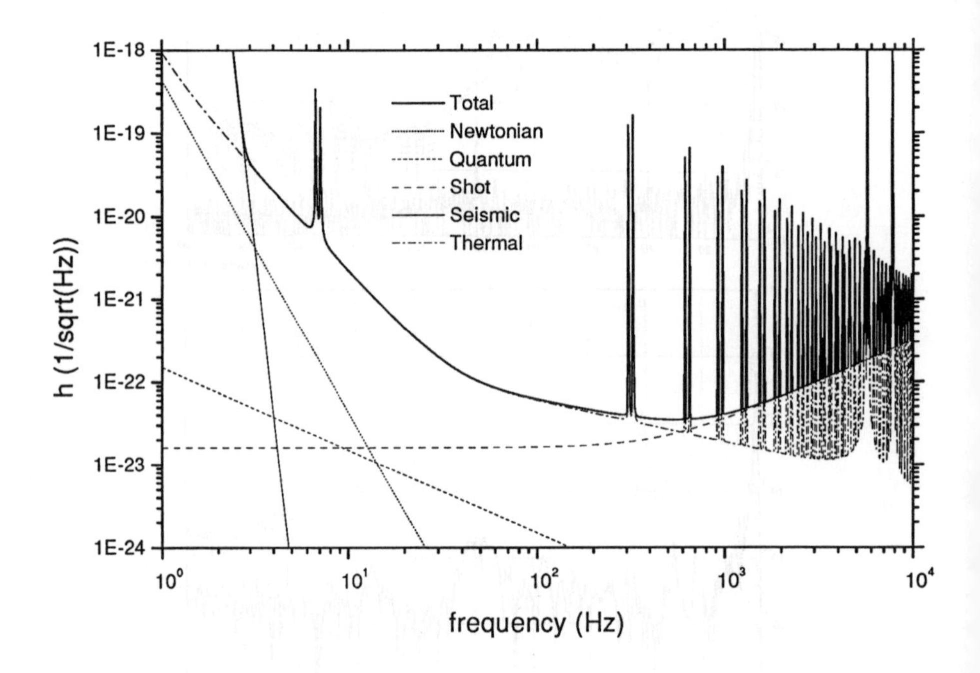

Figure 9: The VIRGO expected sensitivity

THE STATUS OF LIGO

MARK W. COLES
LIGO Project, California Institute of Technology, Pasadena, CA 91125, USA

The LIGO observatories at Hanford Washington and Livingston Louisiana are presently under construction. Earth work, concrete placement, beam tube enclosures, and vacuum component fabrication are all simultaneously in progress. This paper provides a brief overview of the LIGO facilities, a review of progress in the last year, and future plans for civil construction, vacuum system and detector development. References to LIGO vendors and specifications which may be of interest to others working in this field are also provided.

1 Overview

The LIGO Project is building two broad band gravity wave interferometer facilities, located in Hanford, Washington and Livingston, Louisiana, which will be operated as a single observing system. The Hanford observatory will initially house two Nd:YAG (1064 nm) laser interferometers having 2 km and 4 km long arms sharing a common vacuum space. The observatory to be constructed in Livingston will initially house a single 4 km interferometer. Initial strain sensitivity is expected to be $h_{\rm rms} \sim 10^{-21}$. A technical description of the design of the interferometer apparatus and its principles of operation are given in reference 1.

Figure 1 shows a plan view of the Hanford, Washington observatory interior. The observatory facilities have been designed in consultation with the Ralph M. Parsons Company (Pasadena, CA). The design[2] of the LIGO buildings includes consideration of the effects of potential external and operational sources of vibrational "noise" that might degrade the performance of the interferometer if not carefully addressed. Some of the design features implemented which reduce the level of vibration in the environment in which the interferometers must operate are:

- the vacuum equipment, laser, input/output optics, vertex test masses, and beam splitters are located on a 30 inch thick concrete slab (designated "Laser Vacuum Equipment Area" (LVEA) in figure 1) which is isolated from the building foundation.
- Vibrating equipment has been placed as far away as practical from the vacuum envelope containing the interferometers. Vacuum support equipment (such as roughing pumps and ion pump transformers) are located in a separate slab in the "Mechanical Room" (see figure 1). Rotating

and reciprocating equipment (such as compressors, chillers, and cooling water pumps) are located on vibration isolators having a resonant frequency of 3 Hz (the lowest practical frequency available) at least 300 feet from the vacuum envelope. The operating status of this apparatus will be continuously monitored and recorded and will be integrated with the interferometer output data.

- Oversized round HVAC ducting supported from spring hangers has been used throughout to prevent "oil canning" as an acoustic source. Special Q blocks have been placed in the HVAC[3] air plenums to further reduce acoustic vibration.

- Vehicular traffic is routed at least 200 feet away from the vacuum envelope.

- Liquid nitrogen transfer lines connecting the cryopumps to the storage dewars are being designed to minimize acoustic noise due to boil-off.

The Hanford observatory's LVEA is designed to accommodate three interferometers although only two are initially planned. This allows an additional interferometer to be added without requiring major civil construction.The Livingston observatory will accommodate two 4 km interferometers although initially only one is planned. Other interesting features incorporated into the design of both observatories are: clean rooms for optics and vacuum equipment preparation adjacent to the LVEA, service of the LVEA by 26.5 foot hook height, 5 ton bridge cranes, and control room, office, and conference room space for approximately 20-30 scientific, technical, and administrative staff and visitors.

2 Status

Major earth work at Hanford was completed in 1994 so that ample time for ground settlement could occur. This year, concrete slabs supporting the interferometer arms were placed in May and June. Fabrication of 2600 arched concrete enclosures[4] needed to cover the interferometer arms is in progress. These will be installed simultaneously with the beam tube so that it can be protected from damage.

Rough grading of the Livingston site is currently underway. Berms, drainage, and access roads are presently being built and will be completed in August. Following completion, ground settlement will be allowed to occur for about six months prior to commencement of building erection and placement of concrete for the interferometer arms. Process Systems International (Westborough, Massachusetts) has been selected to provide vacuum pumps, valves, vacuum instrumentation, and vacuum chambers for both LIGO observatories. The final design[5] of this equipment has been completed. Prototype chambers are now being fabricated for mechanical design validation and cleaning tests. These chambers will be completed and evaluated this year. Fabrication of the remaining chambers will commence following evaluation with completion scheduled for Fall 1997. First articles of pumps and 48 inch diameter gate valves have also been purchased. These items are now undergoing qualification tests.

CBI Services, Inc. (Chicago, Illinois) has been selected to fabricate and install the 16 km of spiral welded beam tube[6], expansion joints, and supports needed for both observatory sites. The beam tube has a diameter of 1.2 meters and is made from 3 mm thick spiral welded 304L stainless steel coils having widths of 24 and 36 inches. The beam tube will be baked out at 150 C prior to interferometer operation to remove water from its interior surface. A 0.100 inch thick expansion joint[7], manufactured by Hyspan (Chula Vista, CA), is welded between alternate beam tube joints to accommodate thermal expansion. Auxiliary pump ports, located every 238 meters along the beam tube, will be used to maintain vacuum during bake out, allow the attachment of RGA's for leak localization during commissioning, and can be used to install additional pumping capacity if necessary. Initial vacuum performance is required to be approximately $< 10^{-6}$ torr but pressures as low as 10^{-9} torr may be needed by future interferometers located within the vacuum space.

At present, CBI is outfitting a manufacturing facility in Pasco, WA, (approximately 30 miles from the LIGO site) to make the spiral tube using low hydrogen hot rolled stainless steel coils[8]. A Pacific Roller Die (Hayward, CA.) mill will be used to manufacture 20 meter spiral weld tube sections[9] within the facility. Stiffening rings and pump out ports will then be added and each tube will be helium leaked checked to 10^{-10} torr-liters/sec.

Many quality control tests are included in the fabrication of the beam tube to insure that it is leak free and has an acceptably low outgassing rate. For example, the hydrogen outgassing rates of coupons from each lot of baked steel are measured prior to fabrication. Also, cutoff sections from tube ends will be studied to insure complete weld penetration. After vacuum check and immediately prior to field installation, each tube will be cleaned[10] at the factory using a combination detergent and steam wash. Samples of a solvent rinse will be evaluated using FTIR spectroscopy to monitor hydrocarbon content as an indicator of the effectiveness of the wash process. Tubes will be girth welded together on site within a mobile clean room and the girth welds will be individually helium leak checked. As a final check, two kilometer lengths of beam tube will be independently evacuated and leak checked before and after the 150 C bake out.

Optical baffles[11] will be field installed into the interior of the tube as required to intercept light which might otherwise scatter off the walls into the main beam. These are presently being fabricated by Capital Industries (Seattle, WA) and Meyer Tool and Mfg. Co. (Oak Lawn, IL) and coated at West Coast Porcelain (Corona, CA).

The LIGO interferometers will use Nd:YAG (1064 nm) lasers. It was originally planned to use argon ion lasers, but was changed to allow an increase in optical power ($\approx$ 10 watts), relaxed mirror requirements, and anticipated higher operating reliability. LIGO has established a contract with Lightwave Inc. for commercial development of this laser[12]. A pathfinder program[13] is presently underway to identify polishing companies capable of meeting the LIGO figure requirement of $\lambda/800$. Organizations participating in the pathfinder program are CSIRO (Sydney, Australia), HDOS (Danbury, CT), General Optics (Moorpark, CA), and REO (Boulder, CO). Initial results look promising. LIGO is also examining uniformity results obtained from measurements of the coating thickness and plans to choose suppliers of glass, polishing, and coating by the end of this summer.

LIGO is working with Hytech Inc. (Los Alamos, NM) to optimize the design of the interferometer seismic isolation[14]. This work (based on the design presently used in the 40 meter prototype laser interferometer located on the Caltech campus) is a 4 layer viton/metal stack. Single loop suspension of the core optic elements is presently being pursued. Q tests of these test masses are underway and a single loop suspension system is being installed in the 40 meter prototype this summer. Mechanical design and finite element analysis of the suspension structure are in progress.

3 Conclusion

The LIGO project is now well into fabrication and construction. Specifications have been developed and suppliers have been contracted to provide the beam tube, vacuum chambers, pumps, valves, earth and concrete work, beam tube enclosures, and building erection. Important future milestones and dates are: completion of the buildings in Washington and Louisiana in November 1997 and March 1998 respectively; completion of the vacuum equipment installation approximately six months later at each site; interferometer installation will begin in July 1998; and coincident operation between the two sites will occur in 2000.

Acknowledgements

This work is supported under NSF Cooperative Agreement No. PHY-9210038 between the National Science Foundation, Washington, DC 20550 and the California Institute of Technology, Pasadena, CA 91125.

References

LIGO Project documents listed below are available from the LIGO Project Document Control Center, Caltech, Mail Code 51-33, Pasadena, CA 91125.

1. A. Abramovici, W.E. Althouse, R.W.P. Drever, Y. Gursel, S. Kawamura, F.J. Raab, D. Shoemaker, L. Sievers, R.E. Spero, K.S. Thorne, R.E. Vogt, R. Weiss, S.E. Whitcomb, and M.E. Zucker, *Science* **256**, 325 (1992).
2. LIGO document IFB EJ-239.
3. Construction of the Laser Vacuum Equipment Area is consistent with class 50,000 operation and is overpressured to prevent contamination influx. Only three air changes per hour are planned which reduces air velocity in the HVAC duct work to approximately 3 ft/sec. This approach was also found to be more cost effective - Class 100 portable clean rooms will be placed over the vacuum chambers when open to avoid contamination of internal components so that the entire high bay area need not be designed to this stringent standard.
4. For a description of the beam tube enclosure design, see LIGO document IFB No. JLT222. In addition to the parabolic arched sections, service entrances into the enclosure every 238 meters provide access to pump ports along the beam tube. Also, emergency access doors between the service entrances provide personnel access.
5. See "Vacuum Equipment Specification", LIGO-E940002.
6. See "Beam Tube Modules Detailed Design, Master Document List", LIGO-C950496 which lists all beam tube drawings, specifications, and procedures developed by CBI.
7. See "LIGO Beam Tube Expansion Joints", C-EJ-CO, LIGO-E950027
8. The steel coils are baked at 432 C for 36 hours to reduce hydrogen outgassing. See coil material bake specification C-CMBSI, LIGO-E950023.
9. The tube mill must be able to fabricate tube sections complying with the specification "LIGO Beam Tube Sections" C-BT-CO, LIGO-E950025.
10. The cleaning procedure is described in "Cleaning of Beam Tube Can Sections", CLA, LIGO-E950062.

11. About 1000 baffles are required. The baffle design and fabrication process is described in the following documents: "Specification for Porcelain Coating of Beam Tube Baffles" LIGO-E960028; "Specification for Mechanical Fabrication of Beam Tube Baffles" LIGO-E960037; "Beam Tube Baffle, Full Serration, Fabrication and Porcelain Coverage Detail" LIGO-D960045; "Beam Tube Baffle, Non-serrated, Fabrication and Porcelain Coverage Detail", LIGO-D960046; "Welding Procedure for Resistance Spot Welding of LIGO Beam Tube Baffles" LIGO-E960089. The baffle coating is a clayless black frit glass available as item number L034792 from Fero Frit Division, Fero-Corp., Cleveland, OH.
12. Technical requirements of the laser are detailed in "Nd^{3+} Laser Specifications", LIGO-E950081 and "Contractor Proposal for Design and Fabrication of Nd^{3+} Laser/LIGO", LIGO-C960599.
13. Requirements for this program are detailed in "Core Optics Components Requirements (1064 nm)", LIGO-E950099.
14. See "Seismic Isolation Design Requirements Document", LIGO-T960065.

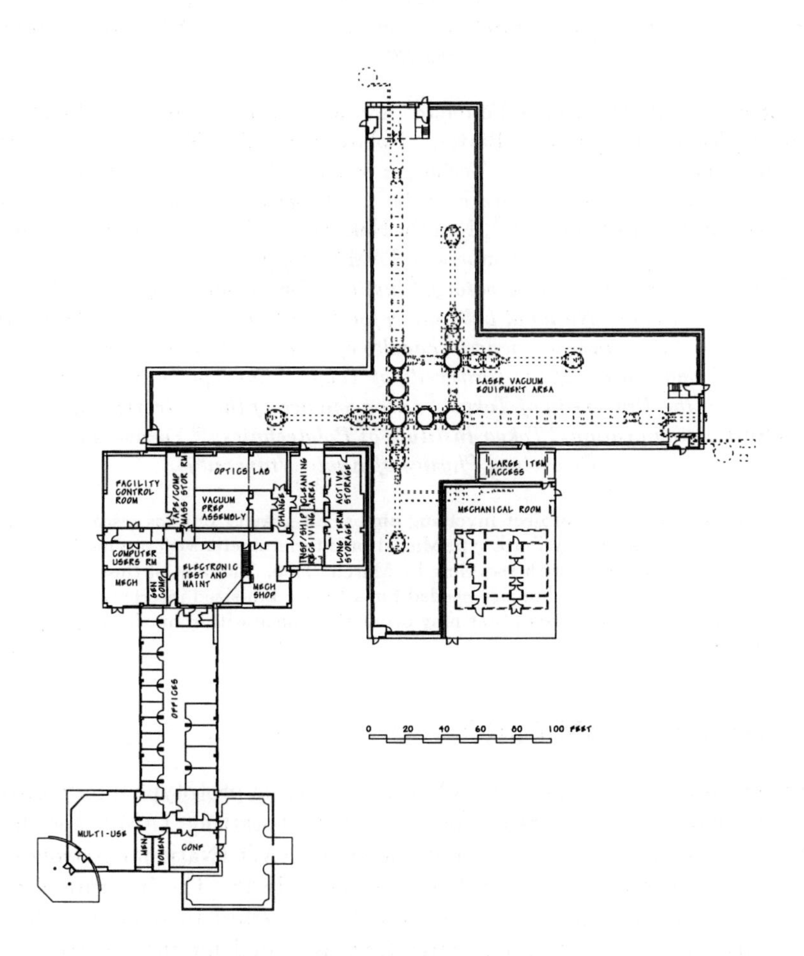

The LIGO observatory at Hanford, Washington

Status of TAMA

Kazuaki Kuroda
Institute for Cosmic Ray Research, University of Tokyo, 3-2-1 Midoricho, Tanashi, Tokyo 188, Japan

Yoshihide Kozai[1], Masa-katsu Fujimoto[1], Masatake Ohashi[1], Ryutaro Takahashi[1], Toshitaka Yamazaki[1], Mark A. Barton[1], Nobuyuki Kanda[2], Yoshio Saito[3], Norihiko Kamikubota[3], Yujiro Ogawa[3], Toshikazu Suzuki[3], Nobuki Kawashima[4], Ei-ichi Mizuno[4], Kimio Tsubono[5], Keita Kawabe[5], Norikatsu Mio[6], Shigenori Moriwaki[6], Akito Araya[7], Kenichi Ueda[8], Kenichi Nakagawa[9], Takashi Nakamura[10] and Members of TAMA group
[1] *National Astronomical Observatory,* [2] *Institute for Cosmic Ray Research of the University of Tokyo,* [3] *National Laboratory for High Energy Physics,* [4] *The Institute of Space and Astronautical Science,* [5] *Faculty of Science of the University of Tokyo,* [6] *Faculty of Engineering of the University of Tokyo,* [7] *Earthquake Research Institute of the University of Tokyo,* [8] *Laser Science of the University of Electro-Communication,* [9] *Tokyo Institute of Polytechnics,* [10] *Yukawa Institute for Theoretical Physics of Kyoto University*

TAMA is a five year project involving almost all gravity physics researchers in Japan. It adopts a Fabry-Perot type Michelson interferometer with pre-modulation and will be completed with recycling by March, 2000. The aim of this project is to develop advanced techniques needed for a future km-sized interferometer and to catch gravitational waves that may occur by chance within our local group of galaxies.

1 Overview of TAMA project

Members of the TAMA project had been developing prototype interferometric gravitational wave detectors for four years before starting TAMA, specifically the 20 m Fabry-Perot type interferometer at NAO (National Astronomical Observatory) and a 100 m delay-line type one at ISAS (The Institute of Space and Astronautical Science). We need a little more effort to attain the ultimate sensitivities in these detectors. After funding ended for these detectors, we started the TAMA project aiming to establish gravitational wave astronomy. In the lead up to TAMA, the 20 m Fabry-Perot detector at NAO has served as a work bench for the development of techniques necessary for a km-sized detector and the 100 m delay-line detector has provided information on the practicalities of an observational system. The main object of TAMA is to construct a 300 m baseline Fabry-Perot-Michelson interferometer at NAO, at Mitaka, a suburb of Tokyo. Needless to say, this interferometer will not be the final detector in Japan. After the end of funding for TAMA we plan to use the results towards a km-sized detector. Notwithstanding this long term plan, we are doing our best to attain the ultimate sensitivity with this 300 m base line system, incorporating new technologies from around the world, and we are hopeful of catching events occurring within the local group of galaxies including Andromeda.

TAMA started in the April of 1995 and will end in March of 2000. It is supported by seven organizations as listed in the author affiliation above and has a staff of about thirty people. Our schedule is to construct the body of the interferometer without implementing the recycling technique by the end of March, 1998. After this, the recycling mirror will be installed and the optical system will be changed appropriately. This improvement should be finished in two years. The budget of the first year was $5M: $3M for building with civil engineering and $2M for part of the vacuum system and the R&D expenditure.

2 Scientific object and interferometer design

The scientific object is the detection of gravitational wave bursts produced in star collapses and coalescence of binary neutron stars. On average, supernovae are estimated to occur once every ten years and coalescences of binary nuetron stars every three centuries in galaxies such as ours. Simulations of the outcome of supernova explosion give no more than 0.6% of the total mass energy of the system as gravitational wave energy, so, the sensitivity should be 2.6×10^{-21} in h with S/N (signal to noise ratio) of 10 to catch supernova events in Andromeda which is 640kpc away. Although more than ten galaxies are counted in the "local group", of these, only five (LMC, SMC, M31, M33 and ours) are expected to produce supernova explosions of types II and Ib, those which might radiate gravitational waves as mentioned above. The estimated frequency is 2.5 per century [1]. Optimistic theory predicts the emission of gravitational wave in the form of a chirp during the coalescence of binary neutron stars, of a strength sufficient to be detected by a detector with a sensitivity of 3×10^{-21} with a S/N of 10 as far away as Andromeda. This sensitivity of 3×10^{-21} can be attained by a 300 m baseline system as shown below. Such coalescences are only expected in spiral galaxies such as ours and Andromeda, and thus the frequency does not exceed once per century. Since the Virgo cluster, which hold about four thousand galaxies, lies 10 times further away than Andromeda, we need to attain a sensitivity of 10^{-22} to probe it, and this is impossible with TAMA. Accordingly, putting emphasis on the development of techniques for future detectors, we nontheless anticipate to have a chance of catching events in the local group of galaxies.

Table 1: Important parameters of TAMA

Parameter	Value
Sensitivity	3×10^{-21} (BW 300 Hz at 300 Hz)
Interferometer	Recombined Fabry-Perot Michelson
Baseline length	300 m
Cavity finesse	516
Laser	10 W LD pumped Nd:YAG 1064 nm
Recycling gain	10
Vacuum pressure	10^{-6}Pa

Table 1 summarizes important parameters of TAMA.

Fabry-Perot cavities are formed by flat near mirrors of 10 cm in diameter and 6 cm in thickness, and end mirrors of the same size but with a 450 m radius of curvature. The distances from the near mirrors to the beam splitter are different by 50 cm to allow pre-modulation with a frequency of 15.25 MHz. As shown in Fig. 1, pre-stabilized laser light is introduced with a mode matching telescope through appropriate adjusting mirrors to a 10 m ring mode cleaner. After the mode cleaner the light beam enters through a second mode-matching optical system into a recycling mirror. At present an optical circulator is planned to be inserted between the matching system and the recycling mirror. Optical alignment control is done by the method of wave front sensing, where light beams are picked off in front of near mirrors. The radius of curvature of the recycling mirror is 9 km. Commercial mirror manufacturers are unwilling to polish such a mirror so we decided to make it ourselves. This will be valuable experience for the future km-sized interferometer.

3 Design and development of key technologies for TAMA

Factors limiting the sensitivity are seismic isolation, suspension thermal noise, mirror internal thermal noise and laser shot noise. We have fairly concrete designs for a laser source with pre-stabilization and for an anti-vibration system and we have begun to design a practical suspension sytem. The following overviews of activities by all members of TAMA necessarily omit some of the many indispensable techniques that will be required for TAMA and its successor.

We have contracted with SONY corporation for making a LD pumped Nd:YAG laser source. Basically it is an injection-locked high-power laser with an input to change the laser frequency for stabilization. We have developed

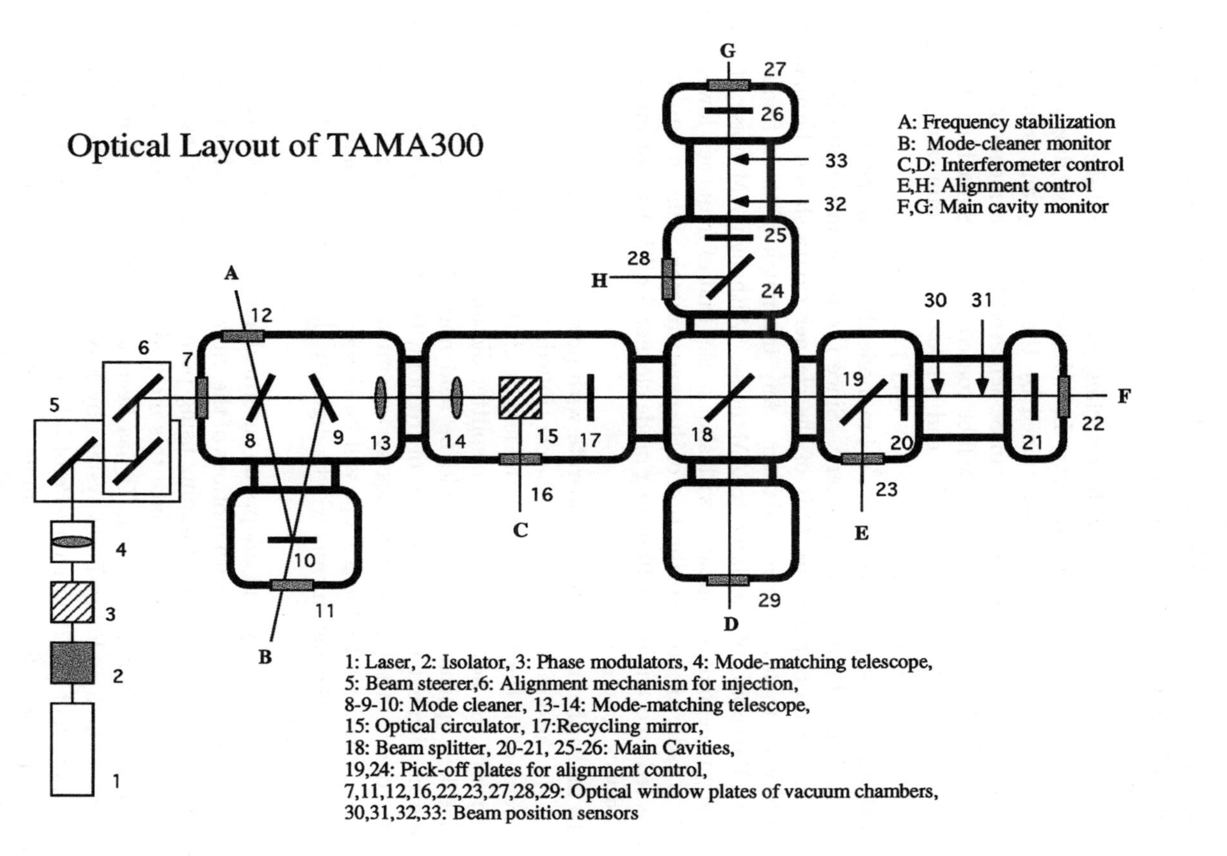
Optical Layout of TAMA300
A: Frequency stabilization
B: Mode-cleaner monitor
C,D: Interferometer control
E,H: Alignment control
F,G: Main cavity monitor
1: Laser, 2: Isolator, 3: Phase modulators, 4: Mode-matching telescope,
5: Beam steerer,6: Alignment mechanism for injection,
8-9-10: Mode cleaner, 13-14: Mode-matching telescope,
15: Optical circulator, 17:Recycling mirror,
18: Beam splitter, 20-21, 25-26: Main Cavities,
19,24: Pick-off plates for alignment control,
7,11,12,16,22,23,27,28,29: Optical window plates of vacuum chambers,
30,31,32,33: Beam position sensors

Figure 1: Optical layout of TAMA300

basic techniques for this stabilization [2]. An output of 10 W has been achieved and the integrated version of this laser will be delivered by the end of this June, 1996.

The main mirrors except for the recycling one will be polished and coated by Japan Aviation Electronics Industry Ltd (JAE). A very loss due to absorption and scattering, 6 ppm, was attained by the company and much better quality is expected. Such a coating can be achieved by refinements of sputtering technique. The recycling mirror will be coated by an IBS machine (Oxford) at NAO. As a first step towards this end we coated a 20 mm sample mirror and achieved a loss of 170 ppm, which is very promising.

Taking over the previous work [3], a 4 m mode cleaner is being developed for the 20 m Fabry-Perot interferometer [4]. Since the phase modulator is placed before the mode cleaner, the FSR of the cleaner cavity should be made to coincide with that of the modulation. Any mismatch of these frequencies converts FM noise of the laser to AM noise. Although AM noise of high frequency is negligible compared with shot noise, that of frequency lower than 300 Hz is badly affected by vibration of the cavity mirrors. Since a similar effect is expected to occur in the 10 m ring mode cleaner for TAMA, optimization of the feedback loop is under study.

To make the interferometer operate with power recycling, four degrees of freedom in the optical path have to be controlled: the sum and difference of the displacement of the main cavities, the length of the recycling cavity, and the difference of the near-mirror to beamsplitter distances. In addition to these, the five key optical elements should be held normal to the incident beam. Since each element has two degrees of freedom, ten degrees of freedom must be controlled. In total, the feedback system must control fourteen degrees of freedom. The main feedback loops will be closed by analog signals and those loops involving the end mirrors will be digitally closed. This control system is being designed.

The suspension system adopts an intermediate mass between the mirror and the suspension point, with motions of the intermediate mass being strongly suppressed by eddy current damping using permanent magnets, as originally developed by Tsubono *et al.* [5] Since this method is passive, it is easier to use than an active damping system. It has been tested using the direct recombination of the 3 m Fabry-Perot system by Kawabe *et al* [6] and by the direct recombination of the 20 m Fabry-Perot interferometer [4] at NAO. It introduces a strong magnetic field near the main mirror, which may induce noise from ambient magnetic fields, but the field can be reduced by shielding. In a preliminary test of this double pendulum system together with vertical isolation springs, many mechanical resonances appeared in the observational frequency range and degraded the performance of the suspension from the design level. However, since we cannot omit the function of mirror control from this suspension system, we will have to accept this degraded isolation at some level.

The suspension system is mounted on an isolation stack through an X-pendulum vibration isolation table. The stack consists of three legs, each containing three layers of rubber and heavy stainless blocks. We plan to cover the rubber with vacuum bellows. Since this type of stack is used in each of eight vacuum chambers, the total number of bellows amounts to 216. The air inside the bellows is evacuated before installation. The X-pendulum was invented by Barton and has been developed in ICRR[7]. The basic idea had been used in a field of geophysics as a tiltmeter but we were the first to apply it as a long period vibration isolation system[8]. The basic X-pendulum behaves as a pendulum of one degree of freedom. We extended this idea to two dimensions and designed a prototype system for TAMA with designed period of 10s. From results with a one-dimensional system, we expect a reduction ratio of 100 at 1 Hz[9]. Since this mechanism is made only of metal, it is suitable to be put into ultra-high vacuum system. The overall performance of the combined system with the X-pendulum on the isolation stack will be tested by this summer, 1996.

The data acquisition system covers a) data acquisition for interferometer signals with 20 kHz sampling rate, b) remote control, monitoring and data logging of interferometer components, and c) online pre-analysis for gravitational wave candidate signals[10]. An optical fiber will be installed for data transfer between the center and end chambers.

The diameter of the vacuum tube, 11 m in length, is 40 cm and the diameter of the beam splitter chamber is 1.2 m. The chambers for the near mirrors, end mirrors and the end of the mode cleaner have a diameter of 1.0 m. The vacuum tubes are connected via bellows with metal gaskets. By the end of the 1995 financial year (March 1996), vacuum tubes of 150 m for one arm and chambers for the mode cleaner had been completed. Electro-Chemical Buffing and TiN coating for non-baking system[11] were studied and will be applied for TAMA.

The vacuum system will be placed in 3 m depth below ground at NAO. The arm sections of the vacuum tubes are enclosed by precast concrete box culverts connected with sealing material. Putting the tubes underground gives us several benefits such as stability of temperature, lower seismic noise and so on. The facility building was finished this March, 1996.

4 Development and research

A laser power of 70 W has been attained using virtual point-source multipass-pump method[12] and 300 W is planned for the future. Radially mounted laser diodes emit light towards the center, where a Nd:YAG resonator rod is mounted along the axis. It lases in a single mode with a Gaussian power distribution. This is very promising for the future detector.

Although thermal noise of the suspension pendulum is important around 100 Hz according to our structure damping model, the internal vibration modes of mirrors turn out to be even more important in the middle frequency range, where all the design sensitivity curves of existing projects have an optimum point. Preliminary experiments to measure Q of various internal modes of a sample mirror suspended by two wires as in the TAMA design show that there are a few modes with very poor Q (of the order of 1000) which decreases the estimated sensitivity by more than one order in the middle frequency range. Since mirrors are necessarily suspended by thin wires and are very likely controlled by magnets, much more research on this point is needed.

A table top experiment using simple Michelson interferometer with suspended mirrors was conducted and a recycling gain of 60 was achieved. This is reported by Moriwaki in these proceedings. A new method to separate control signals clearly for recycling has been proposed and testing is planned.

5 Conclusion

The facility housing the vacuum chamber was completed this March, 1996. All of the vacuum system has been designed. Some parts have been ordered and the remainder will be finished by the year after next. The laser source will be delivered by this June. The seismic isolation system has been designed and its prototype has been partially tested. The suspension with control is being designed and the polishing and coating of the main mirrors have been ordered to be in time for the installation in 1997. Development of key techniques such as the mode cleaner, mirror coatings, mirror alignment and recycling are in progress. We are confident of achieving the object of TAMA.

6 Acknowledgments

TAMA is funded by a Grant-in-Aid for Creative Basic Research of the Minsitry of Education.

References

1. G.A. Tammann in *Frontiers of Neutrino Astrophysics*, ed. Y. Suzuki and K. Nakamura (Universal Academy Press, Tokyo, 1993) p255.
2. Noboru Uehara *et al.*, Optics Letters **20**, 530 (1995).
3. Akito Araya, Docter Thesis, University of Tokyo, 1995.
4. Masatake Ohashi, Doctor Thesis, University of Tokyo, 1994.
5. Kimio Tsubono *et al.*, Rev. Sci. Instrum. **64**, 2237 (1993).
6. Keita Kawabe *et al.*, Applied Optics **33**, 5498 (1994).
7. Mark A. Barton and K. Kuroda, Rev. Sci. Instrum. **65**, 3775 (1994).
8. Nobuyuki Kanda *et al.*, Rev. Sci. Instrum. **65**, 3780 (1994).
9. Mark A. Barton *et al.*, Rev. Sci. Instrum. submitted.
10. Norihiko Kamikubota *et al.*, ICRR report 342-95-8: KEK Preprint 95-141, 1995.
11. Yoshio Saito *et al*, Vacuum, in press (1996).
12. Noboru Uehara *et al.*, Opticis Letters **20**, 1707 (1995).

Effort of Stable Operation and Noise Reduction of 100m DL Laser Interferometer [TENKO-100] for Gravitational Wave Detection

E. MIZUNO, N. KAWASHIMA, S. MIYOKI, E. G. HEFLIN, K. WADA, W. NAITO, S. NAGANO, K. ARAKAWA
The Institute of Space and Astronautical Science (ISAS), 3-1-1 Yoshinodai, Sagamihara City, 229 Japan

Abstract

The construction of TENKO-100, a 100m delay-line laser interferometer gravitational-wave detector, has finished in 1994, and an effort to realize its stable operation has been going on since then. In July–August 1995, we performed a 100 hours run of the antenna including 50 hours of data recording using newly developed data acquisition system. The purpose of this experiment is to verify and increase the reliability of the antenna including the data collection system. In December 1995, we undertook a second data taking run, for 30 hours again. The analysis of these data tells us the typical time scale of this interferometer over which it is stable.

1 Purpose

To operate a laser interferometer as a gravitational wave antenna and get reliable data, the following points are important:

- the condition of the interferometer is stable during an operation
- all the signals that may affect the interferometer signal are recorded

In fact, recent large-scale laser interferometer projects, for example, LIGO, VIRGO and TAMA, are preparing data recorders with more than a few hundred channels [1,2,3]. There, all the data relevant to the operation of the interferometer are considered to be part of the interferometer data [1,3]. TENKO-100 is a prototype for TAMA 300m and a future km-class interferometers, and an effort has been devoted for improving the sensitivity [4]. To know how stable the system is, and understand the relationship between the interferometer output and other internal states, such as laser frequency, intensity, seismic noise, DL mirror alignment, we need to record all those data for sufficiently long time period.

2 Experiment and Analysis

Table 1 shows the signals recorded. Total number of channels was 38. Still more channels are under preparation now. In the summer run (July,

August), 10 hours of continuous data, and in the winter (December) run, 2 to 3 hours of data, were recorded every day.

Table 1: TENKO-100 operation: recorded signals

10kHz sampling: Main signal, Laser frequency(AC), Laser intensity (AC), Center tank acceleration (x, y, z), End tank acceleration (inside/outside),
25Hz sampling: h (1kHz narrow, 500Hz–2kHz), Main servo gain, Dark fringe level (DC), Absolute control fringe, Laser frequency (DC), Laser intensity (DC), Averaged acceleration power (Center tank, End tank/room), Accoustic noise power (Center room, Center/end outside), Vacuum pressure (center tank, two arms), Temperature (room, outside), Atmospheric pressure, Mirror DC offset (4 DL mirrors)

Fig. 1 shows an example of the linear spectral density, expressed by the brightness, of the interferometer output signal. There, unlocking of the servo occurs at time #1 and #2. For the time period plotted (1868 seconds), about 5 percent of the data correspond to the unlocking condition, which is typical for TENKO-100 during the experiment. The rest portion (95 percent) is valid, however, we can see the shape of the spectrum changes with time, especially at high frequencies close to the unity gain frequency of 10 kHz. From the DC dark fringe signal simultaneously recorded, it has become clear that decrease of the open-loop gain $G_{\mathrm{O.L.}}$ due to deterioration of dark fringe contrast is responsible to the amplitude decrease, since the signal represents $\frac{HG_{\mathrm{O.L.}}}{1+G_{\mathrm{O.L.}}}\Delta L$, not pure $H\Delta L$ (ΔL: displacement noise, H: pockels-cell efficiency). By comparing the seismic motion (Table 1) with the main signal for a long period, we now understand the timing of the servo unlock is in most cases synchronized to external seismic disturbances when the dark fringe contrast is small, implying servo becomes 'weak' then. We have also monitored the DC mirror offset in the direction of the arms for the four DL mirrors. Dark fringe contrast has strong correlation with position of the DL mirrors used to lock the fringe in low frequencies (below 100 Hz), implying the mirror actuator in this direction is coupled with mirror orientation. Based on these facts, we are going to modify the mirror local control system. Such a fact is difficult to know without taking and analyzing data for a sufficiently long time interval.

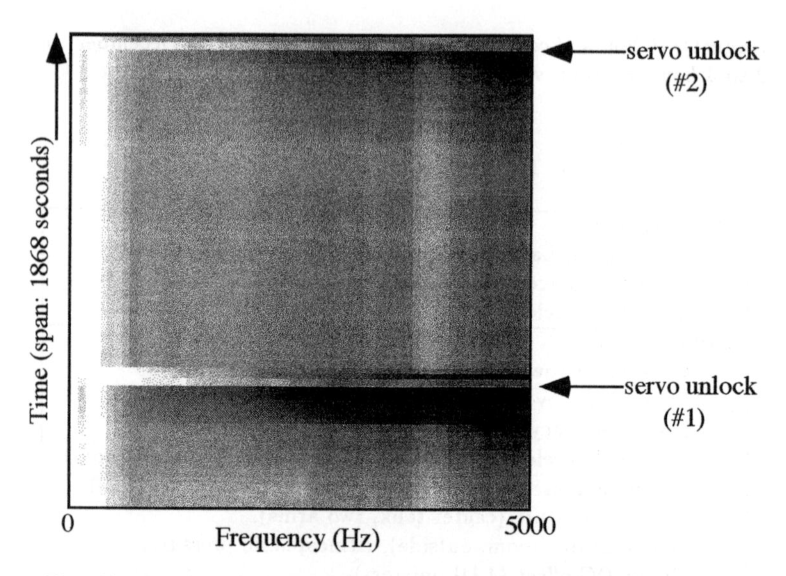

Figure 1: Time history of the linear spectral density of the pockels-cell feedback signal in the interferometer servo on December 19, 1995. It represents the higher frequency portion (above 300 Hz – 5 kHz) of the mirror displacement noise, or strain sensitivity of the interferometer. Amplitude is expressed by brightness. The ratio of the maximum (white) to the minimum (black) is 30 dB. Sharp horizontal lines (#1 and #2) corresponds to the periods when the servo is unlocked.

3 Conclusion

From Fig. 1, we can clearly see the signal is approximately stationary, or the system is stable for ~1000 seconds or less, however, for longer time scales, it becomes unstationary, and the cause is discussed above. The next step for us is to increase the time scale up to at lease several hours without decreasing the sensitivity so that the data may be used for various analysis.

References

1. http://www.ligo.caltech.edu/LIGO_web/overview/CDS.ps
2. *VIRGO Final Conceptual Design*, 1992
3. Kamikubota *et al.*, "Design for the Control and Data-Acquisition System for the TAMA300 300m Laser Interferometer", *Proc. ICALEPCS'95*, 1995
4. Miyoki, "Development of a 100-meter Delay-Line Laser Interferometer", Doctoral Dissertation, The University of Tokyo, 1996

VIRGO MIRROR METROLOGY

A. C. Boccara, Ph. Gleyzes, V. Loriette and J. P. Roger
Laboratoire d'Optique Physique, UPR A0005 CNRS
ESPCI, 10 rue Vauquelin 75005 Paris, France

1 Motivations for the development of an efficient metrology for Virgo

Most of interferometric antennas for gravity wave detection such as VIRGO or LIGO use a dark-fringe Michelson interferometer with Fabry-Pérot (FP) cavities in each arm and a power recycling mirror. This optical scheme can only reach its optimum sensitivity (from a purely optical point of view) when the optical elements exhibit characteristics which are at the limit of nowadays technology.

The required quantitative values are based on numerical simulations of the antenna showing the degradation of the shot-noise limited sensitivity S with regard to optical specifications. For instance a residual absorption of the coatings inside the FP cavities (15 kW of stored power) of 10 ppm will reduce S by 3 %. This reduction is far from being dramatic; nevertheless the next parameter to be fixed is the balance of losses between the two arms: a 10 % difference (1 ppm) will reduce the sensitivity by 40 % ! To relax this value we must go into the ppm losses range, where 10 % asymmetry reduces S by only 10 %. This kind of behaviour holds for many characteristics of the mirrors: we have to reach very good geometrical, optical, mechanical performances and (for most of them) have them as symmetric as possible. In order to achieve the fabrication process, to characterize the preliminary and final mirrors and to insert realistic values in the simulation of the antenna that calculate the ultimate sensitivity we need an efficient and almost permanent metrology (see table 1).

2 Static path difference: Geometrical Parameters

In our opinion these parameters are close to the limits of modern instrumentation especially for specific spatial frequency ranges. Indeed the first simulations of the Virgo antenna have fixed the wavefront distortions from the ideal shape to less than 10 nm ($\lambda/100$) p-p over a range of 12 cm for the end mirrors.

Table 1: Metrology tools used to test the Virgo mirrors

TARGET	INSTRUMENT	SENSITIVITY
Absorption losses (1 ppm)	"Mirage" Bench collinear geometry	10 ppb @ 1 W
Roughness (substrate) Equiv. roughness (coating) (100 pm or 1 ppm)	Polarization based profilometer	(TIS) 1 pm or 410^{-5} ppm
Total losses	1.06 Fabry Pérot confocal cavity (pulses)	Easy for low losses 1 ppm
Wave front control $\lambda/100$ p-p	a) Laboratory differential slope measurement b) Zygo mark IV xp	10 nm (reproducibility)
Birefringence	Pol. modul. Bench	$\lambda/10^5$ (Local)
Reflection coef equal to 10^{-4}	Differential reflection bench	$6 \cdot 10^{-5}$
Ellipsometry	Extension in the Near IR of a SOPRA Laboratory Ellipsometer	

Moreover the scattering level requirements being in the sub ppm range told us that high spatial frequencies wavefront distortions should be in the 0.1 nm range.

For these two extremes limits commercial interferometers (optimized Fizeau) used to investigate the low spatial frequency region — and scatterometers or interferometric microscopes dedicated to the measurement of high frequency roughness — are close to give the required numbers; but in between what do we need ?

Here again simulations can help us to get realistic values : for surfaces whose topography $h(x, y)$ follows a power law $h \propto f^a$ (e.g.: most of the optical and mechanical surfaces are found to be self affine) the loss of s/n ratio of the reduced antenna (Virgo 98) shows a dramatic dependence on a. The values that we have observed experimentally (a=1.67) are relatively favorable from the s/n point of view nevertheless they imply that we control the wavefront with a precision of 10 nm over 12 cm, about 0.2 nm over 1 cm, etc...

In order to enhance the high frequency wavefront distortion detection we have built a slope-based detection system (fig. 1) capable of such a sensitivity.

Other set-ups are used to control path differences at various scales in transmission (fig. 2 shows the bench which takes care of the birefringence and of the wavefront distortion induced by the beamsplitter) or in reflection (fig. 3 shows the photoelastic phase modulated Nomarsky microscope used to check the roughness with a few picometers differential sensitivity).

3 Coating and Bulk Absorption and Transmission

The dynamic behaviour simulation of the Virgo antenna takes into account the heating effects which are particularly strong in the Fabry-Perot. The absorption of coatings prepared by the SMA group in Lyon is decreasing since the beginning of the project and losses as small as 0.3 ppm have been obtained. These losses are measured by using a photothermal technique called Mirage Detection in which a focused modulated heating beam induces a thermal lens whose edges (maximum gradient) deflect periodically an HeNe probe beam. With such device a noise equivalent signal of $6 \cdot 10^{-3}\,\mathrm{ppm}\sqrt{\mathrm{Hz}}$ has been reached which is enough to perform the absorption losses measurement required (homogeneity is checked by moving the sample).

The same technique can be applied to the bulk absorption measurement, with slight changes: indeed the heating and probe beam must overlap over a few mm to insure a good sensitivity (typically $10^{-7}/\,\mathrm{cm}$ in SiO_2). The best silica measured by N. Man *et al.* exhibited $10^{-6}/\,\mathrm{cm}$ absorption losses.

Finally a differential set-up has been used to probe the equality and the homogeneity of mirrors reflection coefficients with a sensitivity better than 10^{-4}.

In conclusion we have shown through various examples how the coupling between simulation and metrology leads us to a better optimisation of optical components of Virgo and a realistic determination of the sensitivity of the antenna.

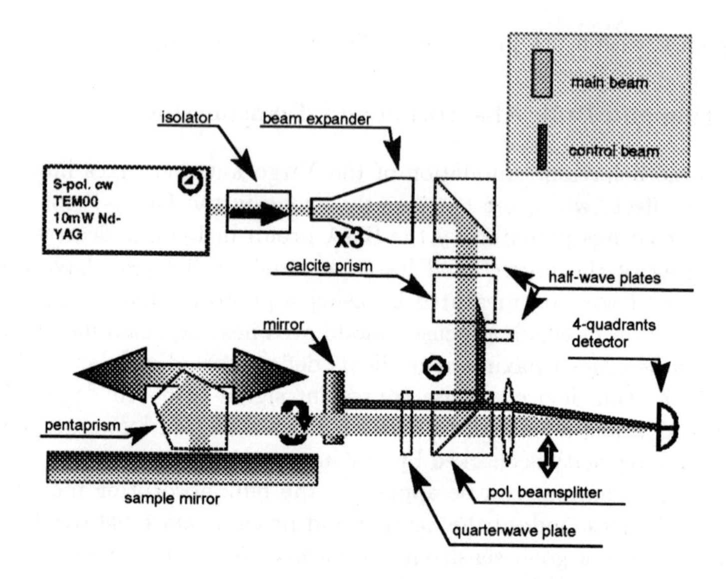

Figure 1: Optical profilometer based on absolute slope measurement.

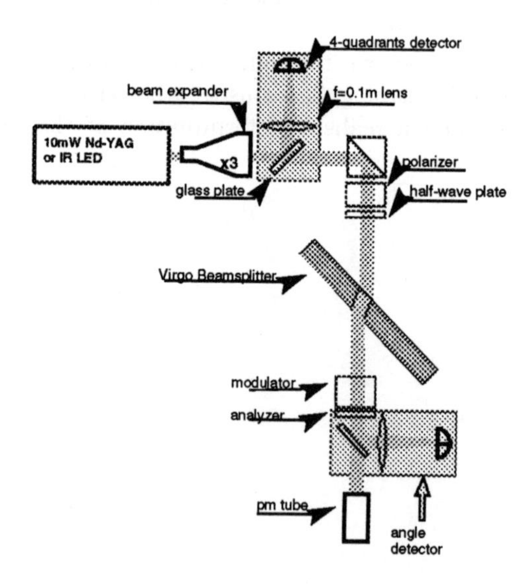

Figure 2: Birefringence and optical thickness bench.

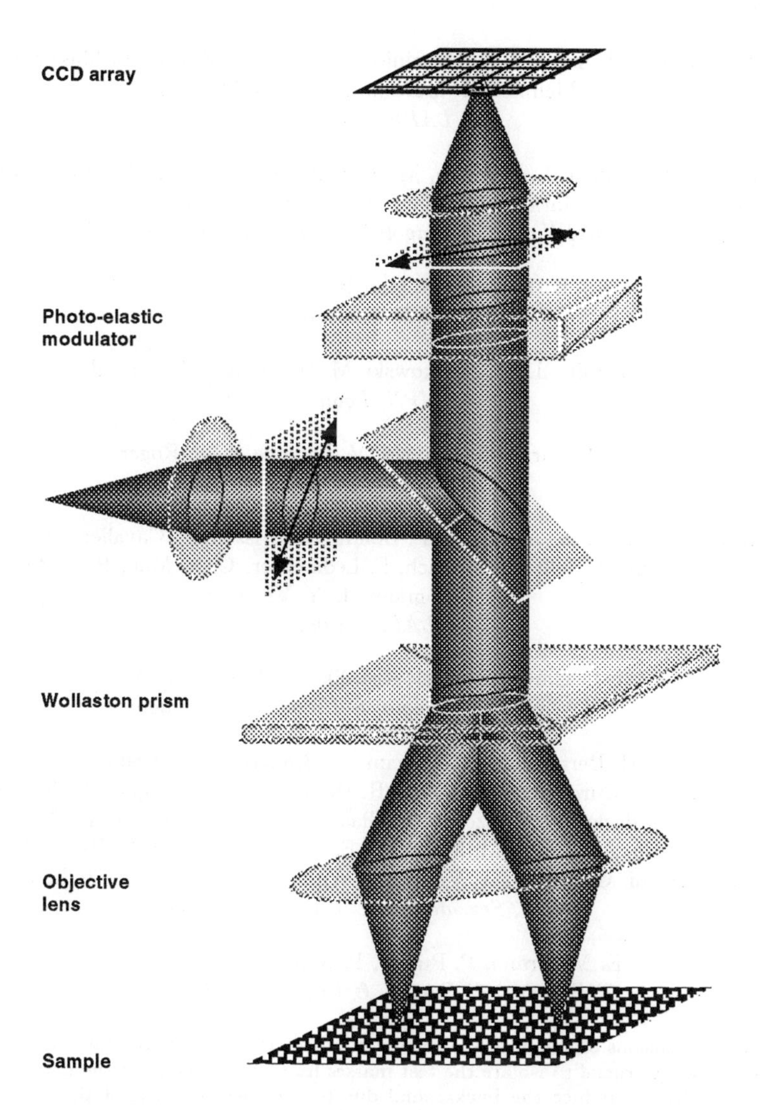

Figure 3: Photoelastic phase modulated Nomarski microscope.

THE VIRGO SUSPENSION SYSTEM

B. Caron, A. Dominjon, R. Flaminio, X. Grave, F. Marion, L. Massonet, C. Mehmel, R. Morand, B. Mours, M. Yvert
LAPP, Annecy

F. Barone, E. Calloni, L. Di Fiore, M. Flagiello, A. Grado, M. Longo, M. Lops, S. Marano, L. Milano, G. Russo, S. Solimeno
INFN, Sez. di Napoli & Università - Napoli

D. Babusci, G. Giordano, G. Matone
LNF, Frascati

L. Dognin, J.–M. Mackowski, M. Napolitano, L. Pinard
IPN, Lyon

C. Boccara, Ph. Gleyzes, V. Loriette, J.–P. Roger
ESPCI, Paris

Y. Acker, M. Barsuglia, A. Brillet, F. Bondu, V. Brisson, F. Cavalier, M. Davier, H. Heitmann, P. Hello, L. Latrach, F. Lediberder, C. N. Man, P.T. Manh, M. Taubmann, J.–Y. Vinet
LAL, Orsay

G. Cagnoli, L. Gammaitoni, J. Kovalik, F. Marchesoni, M. Punturo
INFN, Sez. di Perugia, & Università - Perugia

M. Beccaria, M. Bernardini, S. Braccini, C. Bradaschia, G. Cella, A. Ciampa, E. Cuoco, G. Curci, R. Del Fabbro, R. De Salvo, A. Di Virgilio, D. Enard, I. Ferrante, F. Fidecaro, A. Giazotto, A. Giassi, G. Gorini, L. Holloway, P. Lapenna, G. Losurdo, A. Luiten, M. Morganti, F. Palla, H.–B. Pan, D. Passuello, R. Poggiani, G. Torelli, A. Vicere', J. Winterflood, R. Woode, Z. Zhang
INFN, Sez. di Pisa & Università - Pisa

E. Majorana, P. Puppo, P. Rapagnani, F. Ricci
INFN, Sez. di Roma1, & Università - Roma 1

The suspensions of the optical elements of an interferometric gravitational wave antenna are crucial to isolate the test masses from seismic noise in the detection bandwidth, to reduce the background due to the thermal noise of the system and to keep the interferometer on its working point. We give here an overview of the suspension and control system of the VIRGO antenna, which has been designed in order to have a bandwidth starting from frequencies as low as 4 Hz and extending up to 10 kHz with a sensitivity which will permit to detect low frequency gravitational waves produced by rotating neutron stars and coalescing binaries.

1 Introduction

The interferometric gravitational wave antenna VIRGO has been designed with the aim to detect gravitational radiation signals in a frequency bandwidth starting at values at least as low as 4 Hz and extending up to a few kHz.

This feature of VIRGO is essential to observe signals coming from a wide range of sources as rotating neutron stars in our galaxy, coalescing binaries, SNe and matter accreted by massive objects [1].

Currently, pulsars are the most reliable and well known sources of gravitational radiation: the emitted radio signal gives with high precision the rotation frequency of the star, and a strict upper limit on the emitted wave amplitude can be inferred from the variation of the rotation period.

Nowadays one of the most energetic sources of gravitational waves available for g.w. detection is thought to be the coalescing process of binary systems of compact objects. The outgoing signal has the additional advantage of a strong *chirp* signature due to the increasing frequency of the binary orbit as the coalescing occurs.

SNe explosions are still thought to be the most frequent high energy sources of g.w.: due to extended observations, SN statistics is currently well known, and indicates a frequency of events of about 5 SNe/year in the Virgo Cluster [2]. SNe explosions are currently well tracked down by wide sky surveys, allowing a strict temporal correlation between the g.w. signals and the optical e.m. observations. In this case, the amplitude and spectral shape of the signal can give strong hints on the star structure and on collapse dynamics.

Signal at low frequency are also emitted by massive black holes ($M > 100M_\odot$) perturbed by infalling matter [3].

2 Expected Sensitivity and Seismic Noise

The short list of possible sources recalled in the preceding section shows clearly that the low frequency sensitivity of VIRGO is essential to open up gravitational wave astronomy in a range very promising and not accessible to other detection instruments. However at these low frequencies the sensitivity can be severely limited by the seismic noise.

Typically, in the 10 Hz range, the spectrum of the seismic noise is well represented by the empirical expression:

$$x_s = \frac{a}{\nu^2}\frac{m}{\sqrt{Hz}} \tag{1}$$

with $a \simeq 10^{-7}(m/s^2)/\sqrt{Hz}$.

If this vibration amplitude is totally transmitted to the test mass the minimum amplitude which can be detected is $h > x_s/L$, being L the interferometer arm. We point out that this value cannot be improved increasing the interferometer sensitivity, but only increasing its length.

It can be shown [5] that in order to reach a sensitivity of at least $h \simeq 10^{-21}/\sqrt{Hz}$ at 10 Hz, the test masses must be isolated from seismic noise down to a level of $10^{-18}m/\sqrt{Hz}$, i.e., according to (1), a mechanical attenuation of 200 dB is needed.

3 The Suspension System

The usual method to isolate a test mass from mechanical noise is to connect it to the environment through a mechanical oscillator of frequency ω_f much lower than the detection frequency ω_d. It can be easily shown that in this way the external disturbances are reduced by a factor ${\omega_f}^2/{\omega_d}^2$. This technique has been extensively developed and is currently used for instance in ultracryogenic resonant gravitational wave antennas [4]. As an example, in the Nautilus detector, operating at 100 mK, an overall attenuation from external disturbances of about 200 dB at the detection frequency of 916 Hz is provided. However, the necessity to have a low detection frequency in VIRGO requires frequencies of the filtering apparatus lower than 1 Hz, which are extremely difficult to implement in practice.

Moreover all six degree of freedom should be isolated because of the unavoidable coupling between rotations and displacements, and a further constraint is given by the necessity to control the relative r.m.s. position of the test masses within $\simeq 1\mu m$ in order to lock the interferometer. Hence, in order to decrease the vibration amplitude at resonance, the quality factor must be reduced using passive or active damping systems. The development of the seismic isolation system has started as soon as VIRGO was born.

It was soon realised that to reach the goal sensitivity and bandwidth an eight stage suspension had to be constructed each stage having longitudinal and transverse mode frequencies at 0.5 Hz.

A first attempt to develop a suspension stage used a gas springs system which had very good attenuation characteristics, but had the problem of large drifts of the working position [6].

The current solution is shown in fig. 1. The vertical attenuation is given by a spring system made by hard steel blades working in a cantilever configuration. Each suspension stage is a pendulum of frequency 0.5 Hz in the transverse direction as well. Because of the large moment of inertia of the filter, low torsional frequencies are achieved.

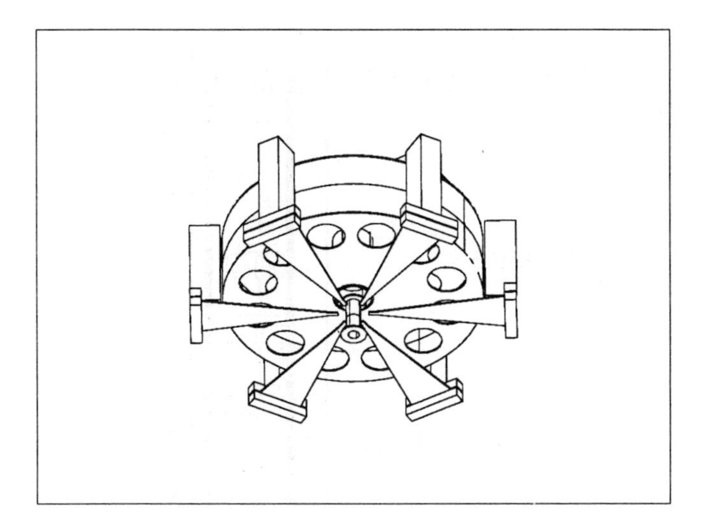

Figure 1: Scheme of a Stage of the Suspension System for VIRGO Interferometer

The longitudinal frequency is further decreased by magnetic antisprings[7] The absolute position of the apparatus is regulated by means of pz actuators acting on the blades and on the suspension point of the whole filter column.

A damping of the modes of the suspension is currently achieved by reducing the quality factor at each resonance. An active system monitoring the accelerations on all six degree of freedom to compute the correction forces to be applied is currently under study.

In fig. 2 a scheme of the complete filtering system is shown. In fig. 3 and 4 are preliminary measurements showing the results of tests on a double stage suspension system. The dotted lines are the transfer function curves for one stage, and the solid lines are the total attenuation, both in the vertical and the horizontal direction obtained with the two coupled stages system. The measured value is in good agreement with the expected one, showing that no unwanted effect due to a spurious coupling between stages is present.

The overall attenuation according to design is $\simeq 10^{-10}$ at 4 Hz, pushing the low frequency limit for VIRGO well below 10 Hz.

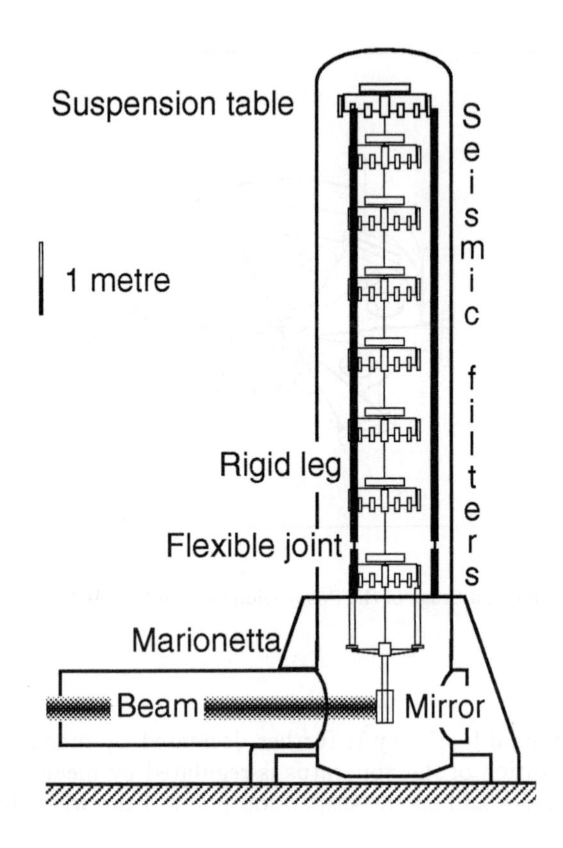

Figure 2: Suspension System for the optical elements of VIRGO Interferometer

4 The Last Stage Suspension System

The last stage suspension system is also used to control the orientation of the mirror in order to align the interferometer. The mirror is hanged to a special intermediate mass, called *marionetta*, which can be steered by e.m. actuators. However the displacement of the marionetta does not allow to control the position of the mirror in the entire detection bandwidth. As a consequence a reference mass is hanged to the marionetta as well, and acts directly on the mirror.

In fig. 5 a photo of a prototype system is shown.

5 The Thermal Noise

All test masses are in thermal equilibrium with the environment, at temperature T, and exhibit a random amplitude fluctuations with energy kT for each degree of freedom, k being Boltzmann's constant. The Brownian motion spectrum has resonances at each mode of the system, with a bandwidth determined by the mechanical dissipations. In order to have a good sensitivity in a wide range of frequencies it is essential to have these bandwidths as narrow as possible. This can be achieved by a careful design of the suspension of the mirror.

An extensive study for the optimisation of the dissipation behaviour of the mirror pendulum is currently being made. Recently, a $Q \simeq 7\ 10^5$ at 0.6 Hz has been obtained for a test system. Such a technology, implemented on the mirror suspension should permit to neglect the thermal noise at the goal sensitivity of VIRGO.

6 Conclusion

The suspension system has been carefully designed and the different parts are now being tested before starting the assembly of VIRGO interferometer in 1997. The experimental results obtained are in good agreement with the design requirements for VIRGO goal sensitivity.

References

1. B. F. Schutz, *Proceedings of 1995 Les Houches School on Astrophysical Sources of Gravitational Radiation*, Ed. J.-A. Marck and J.-P. Lasota (Springer, Berlin, 1996)
2. P. Rapagnani, Astronomy and Astrophysics **229**, 28 (1990)
3. V. Ferrari, *these Proceedings.*
4. P. Astone *et al.*, *Phys. Rev.* D **47**, 362 (1993)
5. A. Giazotto, *Physics Reports* **182**, 365 (1989)
6. R. Del Fabbro *et al.*, *Phys. Lett.* A **132**, 237 (1988)
7. S. Braccini *et al.*, *Rev. Sci. Instr.* **64**, 310 (1993).

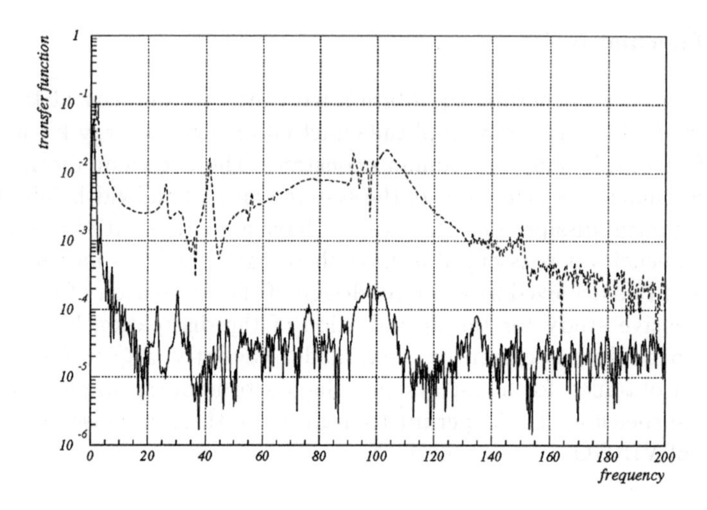

Figure 3: Transfer Function for vertical motion of a suspension test system with two stages (solid line). The dotted line is the transfer function of a single stage.

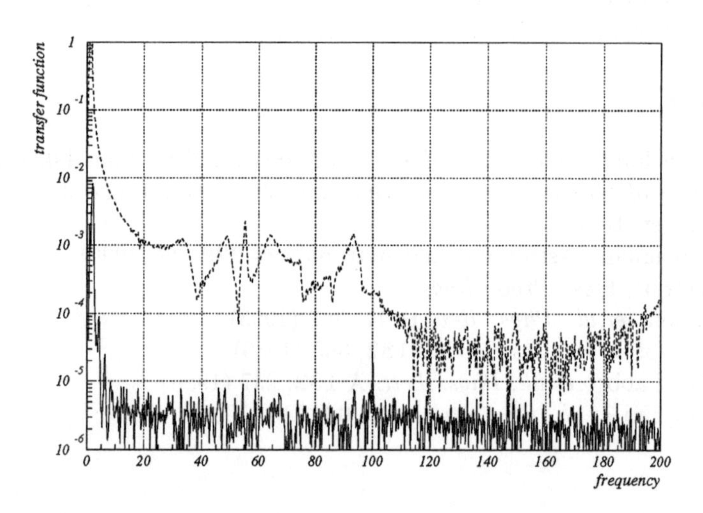

Figure 4: Transfer Function for the horizontal motion of a suspension test system with two stages (solid line). The dotted line is the transfer function of a single stage.

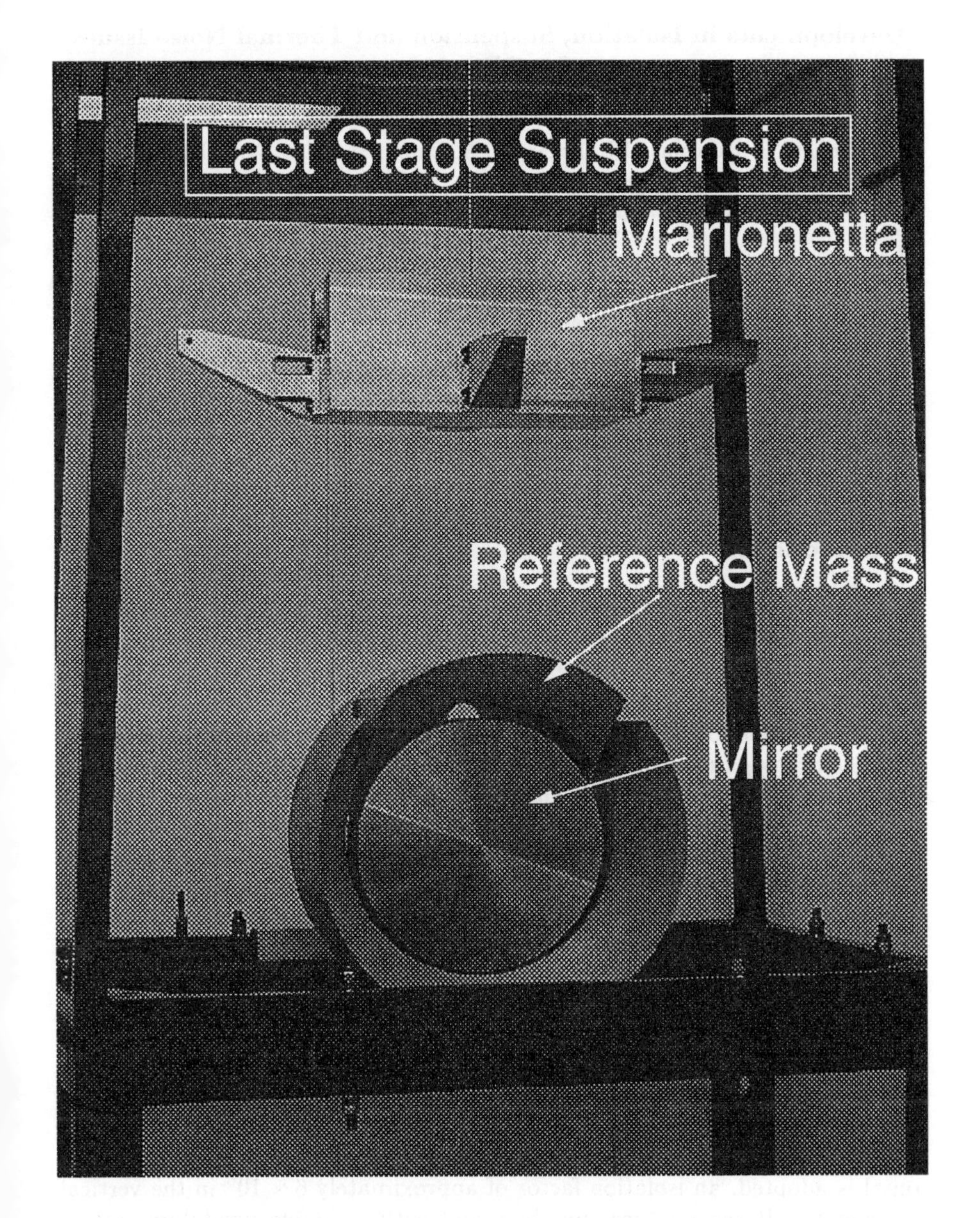

Figure 5: Prototype of the Last Stage Suspension System for VIRGO Interferometer

Developments in Isolation, Suspension and Thermal Noise Issues for GEO 600

J. Hough, R. Hutchins, J.E. Logan, A. McLaren, M. Plissi, N.A. Robertson, S. Rowan, K.A. Strain, S.M. Twyford
Department of Physics and Astronomy, University of Glasgow, Glasgow, G12 8QQ, UK

A revised specification for GEO 600 calls for the system to achieve a sensitivity of $2\times10^{-22}/\sqrt{\mathrm{Hz}}$ at 50 Hz, the main limitation at this frequency being thermal noise. To allow this it is proposed that seismic isolation is provided by two layer stacks encapsulated in stainless steel bellows in combination with two vertical spring stages and a double pendulum for suspending each test mass. The lower stage of each pendulum will incorporate a fused silica test mass suspended with fused silica fibres. Significant progress has been made on stack development and on evaluating material losses in fused silica.

1 Introduction

The specification for GEO 600 was originally defined to allow operation at frequencies above 100 Hz with an rms sensitivity of $10^{-22}/\sqrt{\mathrm{Hz}}$, this being determined by the expected levels of thermal noise in the system[1]. Seismic isolation was designed such that the seismic 'wall' lies just below this frequency. However in order to encompass a larger range of possible sources it now seems desirable to design the system such that it remains thermal noise limited down to 50 Hz, at which frequency the sensitivity limit from thermal noise is expected to be $2\times10^{-22}/\sqrt{\mathrm{Hz}}$. For the optical system envisaged with four light beams in each arm this corresponds to a test mass motion of approximately $7\times10^{-20}\,\mathrm{m}/\sqrt{\mathrm{Hz}}$ at 50 Hz. In order that seismic noise is not significant at this frequency the design goal for the seismic isolation is to achieve a seismic noise level at each test mass of $7\times10^{-21}\,\mathrm{m}/\sqrt{\mathrm{Hz}}$.

2 Seismic Isolation

Assuming a typical seismic noise level of $10^{-7}/f^2\,\mathrm{m}/\sqrt{\mathrm{Hz}}$ in all dimensions an isolation factor of approximately 6×10^{9} at 50 Hz is required in the principal horizontal dimension, and if a coupling factor of 0.1 % of vertical into horizontal is adopted, an isolation factor of approximately 6×10^{6} in the vertical is adequate. Based on these numbers and with a double pendulum system as outlined previously[1] it appears that the required seismic isolation can be achieved using a stack consisting of two layers of stainless steel blocks and

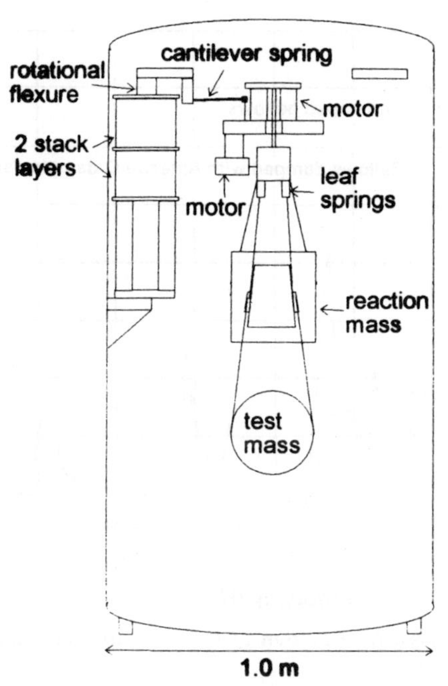

Figure 1: *Possible Vibration Isolation and Suspension System for GEO 600. Only one leg of the three stack legs is shown for clarity. Note the reaction pendulum associated with the intermediate mass in the double pendulum. This will act as a mounting platform for control transducers.*

silicone rubber with two stages of vertical springs as shown in figure 1.

There are three legs in the stack and in each leg each silicone rubber stage consists of three cylinders of diameter approximately 30 mm and height approximately 40 mm encased in a stainless steel bellows unit which is evacuated for operation. The spring constant for each bellows/rubber combination is approximately $6 \times 10^4\,\mathrm{Nm}^{-1}$ in the horizontal and approximately $7 \times 10^4\,\mathrm{Nm}^{-1}$ in the vertical. The stainless steel mass between the first and second rubber stage is of 15 kg mass and the top plate is equivalent to a mass of the order of 30 kg at the top of each leg. In order to reduce the rotational stiffness of the system, the top plate may be attached to the tops of the bellows by cross spring hinges.

From the top of the stack a plate is suspended by cantilever springs and the double pendulum is suspended by a further set of leaf springs. The vertical resonance of each spring/mass system is chosen to be approximately 3 Hz.

Motor actuators are provided to adjust the position of the double pendulum in three dimensions as shown.

Silicone rubber tends to be rather underdamped for use in seismic isola-

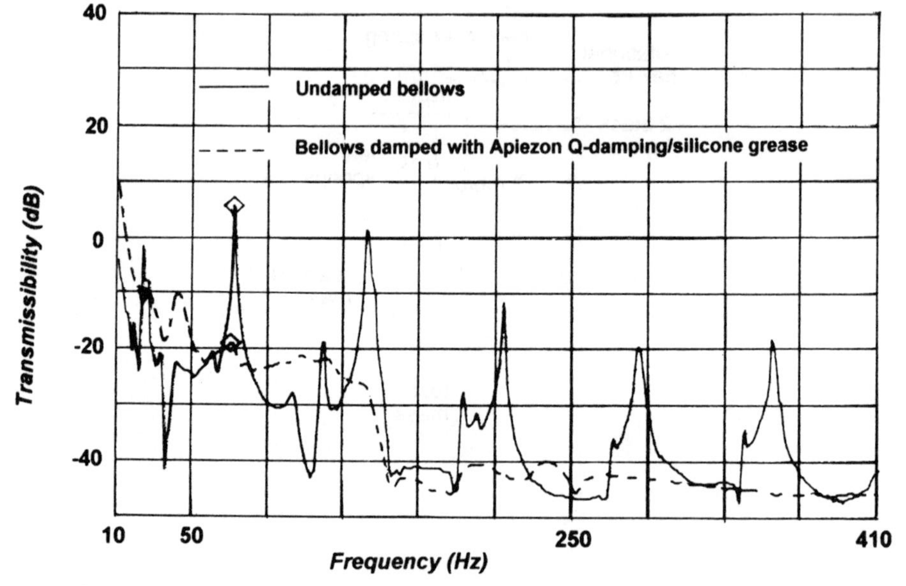

Figure 2: *Vertical Transfer Function of Bellows with and without Damping.*

tion stacks with the resonances being a problem. However it has been found that loading of the rubber with graphite can reduce the quality factor of the resonances to an acceptable value of approximately 5. Silicone rubber also tends to become stiffer under load[2] and this must be taken into account in the design of the rubber elements. There is a potential problem in using stainless steel bellows due to their mechanical resonances. However we have found, by coating the inside of the bellows with a layer of Apiezon Q compound mixed with silicone grease, that these resonances can be reduced to an acceptable level (see figure 2).

3 Thermal Noise Requirements

It is the likely thermal noise level which defines the performance specification for GEO 600. Fused silica is chosen as the test mass material and for the suspension fibres because of its known high Q properties[3]. Our analysis suggests that the thermal noise from the internal modes of the fused silica test masses will be dominant.

It is generally accepted that the damping in fused silica is structural in nature (i.e. Q essentially independent of frequency) and if the Q of each internal mode is taken as 5×10^6 this sets the noise level of the system at approximately $2 \times 10^{-22}/\sqrt{\mathrm{Hz}}$ at 50 Hz. For this to be the dominant limitation the thermal noise from the pendulum mode of each test mass has to be significantly lower. For the test masses used (16 kg) this requires having pendulum suspensions of quality factor approximately 10^7. For such a quality factor to be achieved the most suitable choice of the material for the fibres is fused silica. The GEO 600 design requires that each test is suspended by four fused silica fibres, with expanded ends of the fibres optically contacted on to the test masses.

3.1 Research into Fused Silica Fibres for Suspending Test Masses

Initial work with fused silica fibres has been carried out with a small glass test mass mass (200 g) suspended by two fibres. These fibres (of typical diameter 100 μm) have been formed by pulling fused silica rods in an oxy-hydrogen flame or in a radio-frequency induction furnace. Short lengths of rod at the ends of the fibres are attached by glueing, or by a combination of glueing and clamping, to the test mass and to a rigid supporting structure in a vacuum system. The Q of the pendulum is measured by observing its ringdown in a vacuum of approximately 3×10^{-7} mbar, the motion of the pendulum being monitored by the level of light transmitted by a flag attached to the test mass.

A typical ring down is shown in figure 3 where it can be seen that the Q starts at a high value (typically up to 8×10^6) and falls with time.

This fall in Q is currently under investigation and appears to result from charging up of the pendulum mass as a result of the effect on the system of ultra-violet light emitted by the ion pump used to evacuate the system and by the inverted magnetron vacuum gauge in use. The value of Q obtained at best is close to that required for GEO 600 but is not as high as expected from our measurements on the raw fused silica material and experiments continue in this area. In general for a pendulum suspension it is expected that the Q factor of the pendulum will exceed that of the raw material by a factor in excess of 100 due to most of the potential energy being stored in the lossless gravitational field rather than in the suspension fibre[4].

We have carried out measurements on the basic material by drawing ribbon fibres in an induction furnace and monitoring the ring-downs of the first four modes of these fibres, mounted horizontally, under high vacuum. Again a shadow technique was used for sensing the motion of the fibres.

Typical results are shown in figure 4 where it can be seen that material Q factors of approximately 10^6 are being obtained. The use of this material should thus allow pendulum Q factors of approximately 10^8.

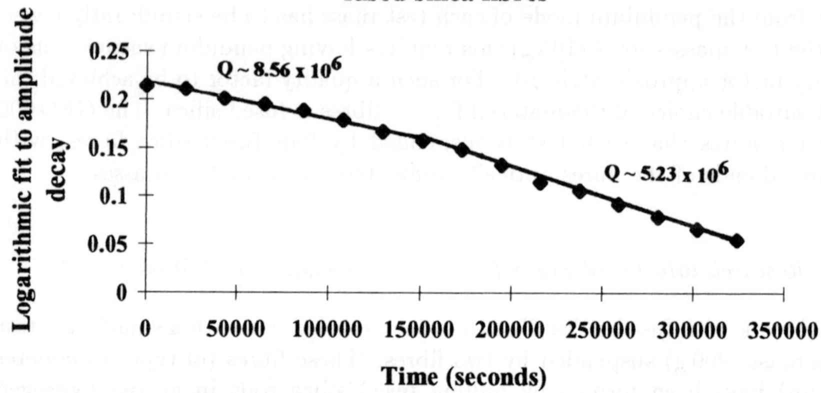

Figure 3: *Typical Logarithmic fit to Ringdown of Pendulum indicating decrease in Q as a function of time.*

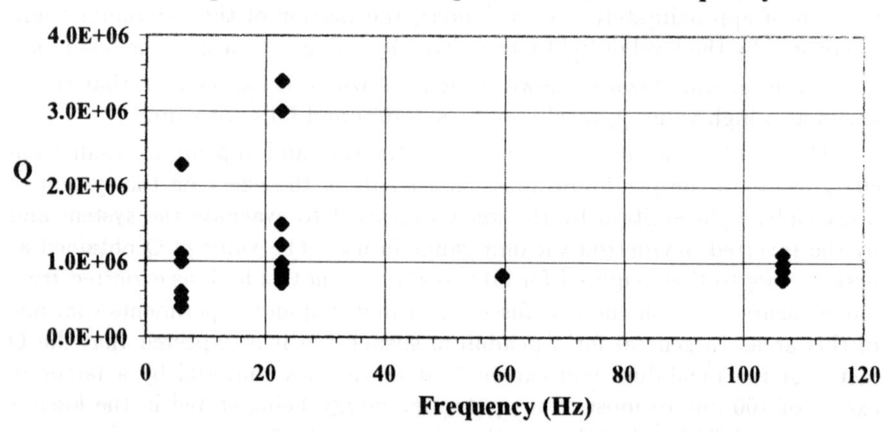

Figure 4: *Q Factors of the First Four Modes of a Fused Silica Ribbon Fibre. Fibre is 12.5 cm long, 0.3 cm wide and 54 μm thick and is drawn from a silica slide, part of which is left attached to allow firm clamping. Note the spread of results - thought to be due to mode coupling which varied as a function of temperature. The single point at 59 Hz was obtained when the room temperature was very low. No further points were possible due to very strong mode coupling being present at higher temperatures.*

4 Future Work on Fused Silica Fibres

This will be focussed on a number of areas: understanding and improving the Q factors of the pendulums; testing optical contacting for jointing the fibres to the test masses and monitoring the effects of such jointing on the internal Q of fused silica masses; scaling the present systems to the full size systems required for GEO 600.

Acknowledgments

This research was supported by PPARC and the University of Glasgow. We should like to thank our colleagues in Glasgow, Garching and Hannover for many useful discussions. We would also like to thank Dr. Jim Faller of JILA, Boulder, for his stimulating advice.

References

1. K. Danzmann *et al.*, *Proc. First Edoardo Amaldi Conf.*, eds. E. Coccia, G. Pizzella, F. Ronga, (World Scientific, Singapore, **1995**), p100.
2. G.W. Painter, *Rubber Age*, (74), (5) 701, **1954**.
3. V.B. Braginskii, V.P. Mitrofanov and S.P. Vyatchanin, *Rev. Sci. Instrum.*, (65), (12), 3771, **1994**.
4. P.R. Saulson, *Phys. Rev. D*, (42), 2437, **1990**.

Cosmology

STRING COSMOLOGY AND RELIC GRAVITATIONAL RADIATION

G. VENEZIANO

Theoretical Physics Division, CERN
CH - 1211 Geneva 23

String theory counterparts to Einstein's gravity, cosmology and inflation are described. A very tight upper bound on the Cosmic Gravitational Radiation Background (CGRB) of standard inflation is shown to be evaded in string cosmology, while an interesting signal in the phenomenologically interesting frequency range is all but excluded. The generic features of such a stringy CGRB are presented.

1 Introduction

In this talk I will first explain why string theory offers an interesting alternative to Einstein's gravity and cosmology. The standard post-big-bang picture emerges as just the *late-time* history of a Universe which, in a prehistoric (*pre-big-bang*) era, underwent an inflationary expansion driven by the growth of the universal coupling of the theory.

I will then turn to describing one of the most interesting physical consequences of this new scenario: the production of a Cosmic Gravitational Radiation Background (CGRB), which could by far exceed, in the relevant frequency range, the one predicted by ordinary inflationary models.

I will leave the detailed discussion of the near-future prospects for observability of our CGRB to the following talk by R. Brustein and refer you, for more details on the scenario and the computations, to the collection of papers on string cosmology appearing on WWW under:

http:/www.to.infn.it/teorici/gasperini/

The precious collaboration of Maurizio Gasperini throughout the development of the pre-big-bang scenario, and the additional one of Ramy Brustein, Massimo Giovannini and Slava Mukhanov in working out its consequences for gravitational perturbations, are gratefully acknowledged.

2 Einstein Gravity and Standard Cosmology

In order to introduce string gravity and a cosmological model based on it, I will first recall a few known facts about Einstein gravity and standard cosmology (see for instance [1]).

The well-known Einstein equations:

$$R_{\mu\nu} - 1/2 g_{\mu\nu} R + \Lambda g_{\mu\nu} = -8\pi G T_{\mu\nu} \tag{1}$$

follow from setting to zero the variation of the Einstein–Hilbert action (I will use $c = \hbar = 1$ throughout):

$$S = -\frac{1}{16\pi G} \int d^4x \sqrt{-g}\ [R - 2\Lambda] + S_{matter}\ . \tag{2}$$

Einstein cosmology follows from Einstein's equations upon insertion of a homogeneous (and, for simplicity, spatially flat) ansatz for the metric:

$$ds^2 \equiv g_{\mu\nu} dx^\mu dx^\nu = dt^2 - a(t)^2 dx^i dx^i \tag{3}$$

and after assuming that also matter is homogeneously distributed. The Einstein–Friedman equations (of which only two are independent) then follow:

$$\begin{aligned} H^2 \equiv (\dot{a}/a)^2 &= \frac{8\pi G}{3}\rho + \frac{\Lambda}{3} \\ \dot{H} + H^2 \equiv (\ddot{a}/a) &= -\frac{4\pi G}{3}(\rho + 3p) + \frac{\Lambda}{3} \\ \dot{\rho} &= -3H(\rho + p) \end{aligned} \tag{4}$$

where the matter energy density ρ and pressure p are defined in terms of $T_{\mu\nu}$ by $T_\mu^\nu = \text{diag}\ (\rho, -p, -p, -p)$. Notice that the effect of a non-vanishing cosmological constant Λ is equivalent to that of a special kind of matter, the "vacuum", with $\rho_{vac} = -p_{vac} = \frac{\Lambda}{8\pi G}$. Normal matter has $(\rho + 3p) > 0$ and therefore leads to a decelerated expansion of the Universe: in particular, a matter-dominated Universe $(p/\rho \sim 0)$ expands like $t^{2/3}$, while a radiation-dominated Universe $(p/\rho \sim 1/3)$ expands like $t^{1/2}$.

A trivial but important remark for the following discussion: if we regard (as we should) the first of eqs. (4) as expressing the vanishing of the total energy of the matter-plus-gravity system, we see that the expansion of the Universe contributes with a *negative* kinetic energy to such an equation.

Inflation, i.e. a long phase of accelerated expansion of the Universe ($\dot{a}, \ddot{a} > 0$), is badly needed in order to solve the outstanding problems of the standard cosmological model[2]. Unlike ordinary matter, a cosmological constant can easily do the job. The same is true of potential energy originating from a scalar field (the so-called inflaton) which, during some cosmic epoch, was approximately frozen away from the minimum of its potential and thus provided an effective (positive) cosmological constant $\Lambda_{eff} = 8\pi GV$. In this case a (quasi) de Sitter exponential expansion of the Universe takes place:

$$a(t) \sim \exp(Ht)\ , \quad H^2 = \frac{8\pi}{3} GV = \Lambda_{eff}/3\ . \tag{5}$$

3 The disappointing CGRB of standard inflation

In standard potential-energy-driven inflation, while the inflaton slowly rolls down to the true minimum of the potential (where, by assumption, the potential energy is very small), the Hubble parameter H stays constant or decreases slowly. The Hubble radius H^{-1} thus remains constant (or increases slowly) during the inflationary epoch and then starts to grow like cosmic time t during the radiation- and matter-dominated eras.

In Fig. 1 this behaviour of the Hubble radius is plotted together with the behaviour of different physical scales which, by definition, grow like the scale factor $a(t)$ itself. It is easily found that scales cross the "horizon" outward (exit) during inflation and cross it again inward (re-enter) during the matter- or radiation-dominated epochs. Larger scales exit earlier and re-enter later than shorter scales. In order to solve the homogeneity problem of standard cosmology, it is necessary that the scale corresponding to the present horizon, $O(H_0^{-1})$, once upon a time, was inside the horizon. For this to happen a total red-shift of about

$$z_{infl} \equiv \frac{a_{end}}{a_{beg}} > 10^{30} \tag{6}$$

during inflation is needed[2].

One of the celebrated bonuses of inflation [2] is a natural explanation of the origin of large-scale structure. Let us assume that, initially, there were no inhomogeneities other than the minimal ones due to quantum mechanics. In other words, let us ask ourselves whether the origin of a structure in the Universe can be found in the initial vacuum quantum fluctuations. Vacuum fluctuations of the metric (defined as usual by $g_{\mu\nu} = \eta_{\mu\nu} + h_{\mu\nu}$) with wavelength

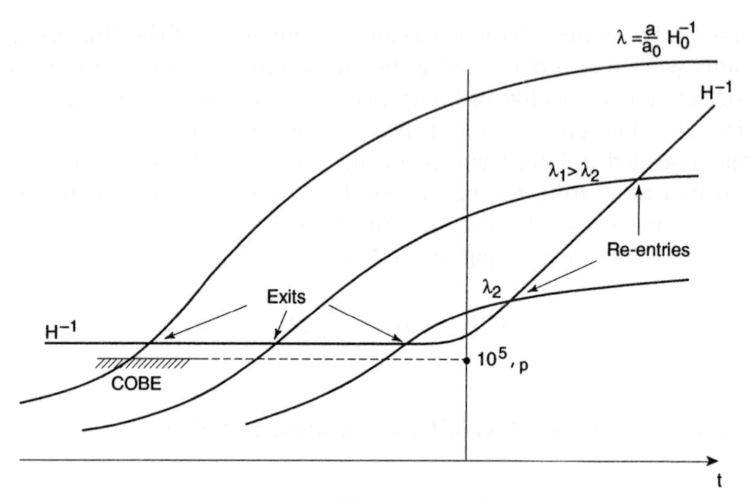

Figure 1

λ have a typical magnitude

$$\delta h(\lambda) \sim \frac{\ell_P}{\lambda} \,, \tag{7}$$

where $\ell_P = \sqrt{G}$ is the Planck length whose magnitude controls the size of quantum gravity effects. We see that the shorter the wavelength the larger the quantum fluctuation.

It is not hard to show that these original perturbations (inhomogeneities) are adiabatically damped (i.e. they follow eq. (7) with $\lambda \sim a$) as long as their physical wavelength stays inside the horizon, while they freeze-out (stay constant) after going outside. Since larger wavelengths spend a longer time outside the horizon (Fig.1) they stay frozen for a longer time. In other words there is a competition between two effects: quantum mechanics favours short scales while classical freeze-out favours large scales.

Combining the two effects leads to a simple and suggestive formula [2] for the present magnitude of tensor metric perturbations i.e of gravitational waves (GW):

$$\rho_\gamma^{-1} \frac{d\rho}{d\log\omega} \equiv \frac{\Omega_{GW}(\omega)}{\Omega_\gamma} \sim (\ell_P H)^2|_{ex} \,, \tag{8}$$

where $\Omega_\gamma = \frac{\rho_\gamma}{\rho_{cr}} \sim 10^{-4}$ and $\Omega_{GW}(\omega) = \rho_{cr}^{-1} \frac{d\rho_{GW}}{d\log\omega}$ and the label ex indicates that $l_P H$ has to be evaluated, for each scale λ, at the time of its exit. This is the crucial quantity for the GW yield at any given frequency. As we have explained, H is constant or slowly decreasing during inflation, hence the same is true of the GW spectrum as a function of ω. This is the celebrated (quasi) scale-invariant Harrison–Zeldovich spectrum, which appears to be quite efficient for generating the observed large-scale structure (if combined with an appropriate model for dark matter).

Unfortunately, for the purpose of this talk, the above result is bad news, i.e. represents a disappointing spectrum of GW in the relevant frequency region. Indeed, COBE's observation [3] of a $\frac{\Delta T}{T}$ of order 10^{-5} at large angular scales implies $H^{-1} > 10^5 \ell_P$ when scales of the order of the present Hubble radius went out of the horizon, and an even smaller value when shorter scales did (see again Fig.1). Inserting such a limit in eq. (8), we immediately arrive at:

$$\Omega_{GW}(\omega) < 10^{-14} \text{ to } 10^{-15} \tag{9}$$

in the interesting (Hz to MHz) frequency range. This upper limit makes the CGRB produced by ordinary inflation an unobservable signal for some time to come ...

4 String Gravity

Being a theory of extended objects, string theory contains a fundamental length scale λ_s, a built-in ultraviolet (short-distance) cut-off [4]. As a result, string gravity differs from Einstein gravity in a subtle and essential way. Instead of the action (1.1), string theory gives [5]:

$$\begin{aligned} \Gamma_{eff} &= \frac{1}{2}\int d^4x\sqrt{-g}\, e^{-\phi}\left[\lambda_s^{-2}(R+\partial_\mu\phi\partial^\mu\phi)+F_{\mu\nu}^2+\bar{\psi}D\!\!\!/\psi\right] \\ &\quad + \left[\text{higher orders in } \lambda_s^2\cdot\partial^2\right] + \left[\text{higher orders in } e^{\phi}\right] . \end{aligned} \tag{10}$$

As indicated in (10), string gravity has (actually needs!) a new particle/field, the so-called dilaton ϕ, a scalar particle. It enters Γ_{eff} as a Jordan–Brans–Dicke [6] scalar with a "small" negative ω_{BD} parameter, $\omega_{BD} = -1$. Bounds on the present rate of variation of α and G imply that, today, $\dot{\phi} < H_0$, while precision tests [7] of the equivalence principle put an upper (lower) limit on the range of the dilaton-exchange force (on the dilaton mass) [8]:

$$m_\phi > 10^{-4} \text{ eV} . \tag{11}$$

Both problems are solved by assuming that a non-perturbative dilaton potential has to be added to (10). Such a potential will freeze the dilaton to its present value ϕ_0 and make us recover Einstein's theory (and its experimental successes) at late times.

The value ϕ_0 provides[9] today's unified value of the gauge and gravitational couplings at energy scales of $O(\lambda_s^{-1})$. In formulae:

$$\begin{aligned} \ell_P^2 &\equiv 8\pi G_N = e^{\phi_0}\lambda_s^2 \,, \\ \alpha_{GUT}(\lambda_s^{-1}) &\simeq \frac{e^{\phi_0}}{4\pi} = \frac{\ell_P^2}{\lambda_s^2}, \end{aligned} \tag{12}$$

implying (from $\alpha_{GUT} \approx 1/20$) that the string-length parameter λ_s is about 10^{-32} cm. Note, however, that the above formulae, in a cosmological context in which ϕ evolves in time, can only be taken as giving the *present* values of α and ℓ_P/λ_s. In the scenario we will advocate, both quantities were much smaller in the very early Universe!

Equation (10) contains two dimensionless expansion parameters. One of them, the above-mentioned $g^2 \equiv e^\phi$, controls the analogue of loop corrections in quantum field theory (QFT), while the other, $\lambda^2 \equiv \lambda_s^2 \cdot \partial^2$, controls string-size effects, which are of course absent from QFT. Obviously, the expansion in λ^2 is reliable at small curvatures (derivatives), i.e. at energies smaller than the string scale λ_s^{-1}, while higher orders in g^2 will be negligible at weak coupling.

The first and main assumption of our scenario is that the Universe started its evolution in a regime that was perturbative with respect to both expansions, i.e. in a region of weak coupling and small curvatures (derivatives). During that phase the string-gravity equations take the simple form:

$$\begin{aligned} R_{\mu\nu} + \nabla_\mu \nabla_\nu \phi = -\lambda_s^2 \, e^\phi T_{\mu\nu} \\ R - \nabla_\mu \phi \nabla^\mu \phi + 2\nabla^2 \phi + 2\Lambda = 0 \,, \end{aligned} \tag{13}$$

which are similar to Einstein's equations, yet substantially different. As already stressed, we wish to recover general relativity at late times; nonetheless, we want to take advantage of the difference for the prehistory of the Universe.

Before closing this section I would like to briefly comment on a point that appears to be the source of much confusion: it is the dilemma between working in the so-called string frame and working in the more conventional Einstein frame. The two frames are not to be confused with different coordinate systems: they are instead related by a local field redefinition, a conformal, dilaton-dependent rescaling of the metric, to be precise. All physical quantities are independent of the frame one is using. The question is: What should we call the metric? Although, to a large extent, this is a question of taste, one's intuition may work better with one definition than with another. Note also that, since the dilaton is time-independent today, the two frames now coincide.

Let us compare the virtues and problems inherent in each frame.

A) **String Frame.** This is the metric appearing in the original (σ-model) action for the string. Classical, weakly coupled strings sweep geodesic surfaces with respect to this metric [10]. Also, the dilaton dependence of the low-energy effective action takes the simple form indicated in (10) only in the string frame. The advantage of this frame is that the string cut-off is fixed and the same is true of the value of the curvature at which higher orders in the σ-model coupling λ become relevant. The main disadvantage is that the gravitational action is not so easy to work with.

B) **Einstein Frame.** In this frame the pure gravitational action takes the standard Einstein–Hilbert form. Consequently, this is the most convenient frame for studying the cosmological evolution of metric perturbations. The Planck length is fixed in this frame, while the string length is dilaton- (hence generally time-) dependent. In the Einstein frame, Γ_{eff} takes the form:

$$\begin{aligned}\Gamma_{eff} &= \int \frac{d^4x\sqrt{-g}}{16\pi G_N}\left[R + \partial_\mu\phi\partial^\mu\phi + e^{-\phi}F^2_{\mu\nu} + \partial_\mu A\partial^\mu A + e^{\phi}m^2A^2\right] \\ &\quad + \left[G_N e^{-\phi}R^2 + \ldots\right], \end{aligned} \tag{14}$$

showing that the constancy of G in this frame is only apparent, since masses are dilaton-dependent (even at tree level). The same is true of the value of R at which higher order stringy corrections become important.

For the above reasons I will choose to base the discussion (although not always the calculations) in the string frame.

5 String cosmology

There is an exact (all-order) vacuum solution for (critical) superstring theory. Unfortunately, it corresponds to a free theory ($g = 0$ or $\phi = -\infty$) in flat,

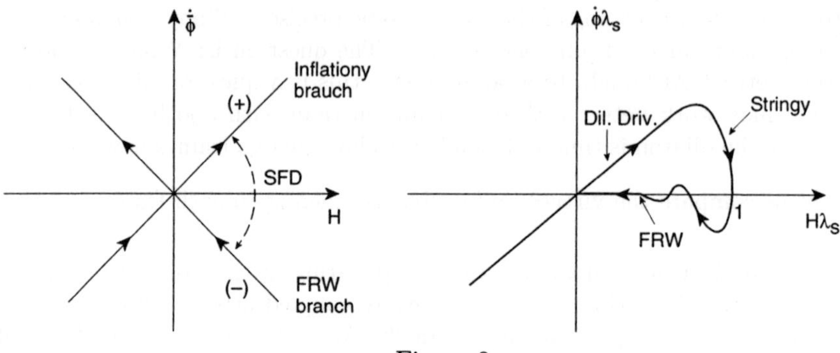

Figure 2

ten-dimensional, Minkowski space-time, nothing like the world we are living in today! Could this instead have been the *original* state of the Universe? The very basic postulate of the pre-big-bang scenario [11, 12] is that this is indeed the case.

Such a postulate is supported by the observation that, in the space of homogeneous (and, for simplicity, spatially-flat) perturbative solutions to the field equations, the trivial vacuum is a very special, *unstable* solution. This is depicted in Fig. 2a for the simplest case of a ten-dimensional cosmology in which three spatial dimensions evolve isotropically while six "internal" dimensions are static (it is easy to generalize the discussion to the case of dynamical internal dimensions, but then the picture becomes multidimensional).

The straight lines in the $H, \dot{\bar{\phi}}$ plane (where $\dot{\bar{\phi}} \equiv \dot{\phi} - 3H$) represent the evolution of the scale factor and of the coupling constant as a function of the cosmic time parameter (arrows along the lines show the direction of the time evolution). As a consequence of a stringy symmetry, known [11, 13] as Scale Factor Duality (SFD), there are two branches (two straight lines). Furthermore, each branch is split by the origin in two time-reversal-related parts (time reversal changes the sign of both H and $\dot{\bar{\phi}}$).

As mentioned, the origin (the trivial vacuum) is an "unstable" fixed point: a small perturbation in the direction of positive $\dot{\bar{\phi}}$ makes the system evolve further and further from the origin, meaning larger and larger coupling and absolute value of the Hubble parameter. This means an accelerated expansion (inflation) or an accelerated contraction. It is tempting to assume that those

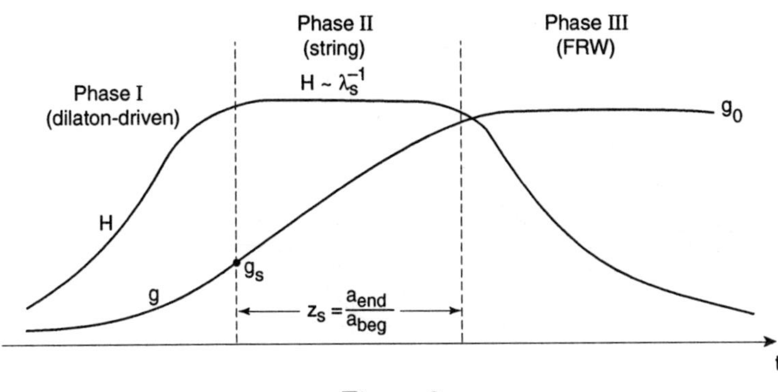

Figure 3

patches of the original Universe that had the right kind of initial fluctuation have grown up to become (by far) the largest fraction of the Universe today.

In order to arrive at a physically interesting scenario, however, we have to connect somehow the top-right inflationary branch to the bottom-right branch, since the latter is nothing but the standard FRW cosmology, which has presumably prevailed for the last few billion years or so. Here the so-called *exit problem* of string cosmology arises. At lowest order in λ^2 (small curvatures in string units) the two branches do not talk to each other. The inflationary (also called $+$) branch has a singularity in the future (it takes a finite cosmic time to reach ∞ in our gragh if one starts from anywhere but the origin) while the FRW ($-$) branch has a singularity in the past (the usual big-bang singularity).

It is widely believed that QST has a way to avoid the usual singularities of classical general relativity or at least a way to reinterpret them [14], [15]. It thus looks reasonable to assume that the inflationary branch, instead of leading to a non-sensical singularity, will evolve into the FRW branch at values of λ^2 of order unity. This is schematically shown in Fig. 2b, where we have gone back from $\dot{\bar{\phi}}$ to $\dot{\phi}$ and we have implicitly taken into account the effects of a non-vanishing dilaton potential at small ϕ in order to freeze the dilaton at its present value. The need for the branch change to occur at large λ^2, first argued for in [16], has recently been proved [17].

There is a rather simple way to parametrize a class of scenarios of the kind defined above. They contain (roughly) three phases and two parameters, which can be easily visualized in Fig. 3.

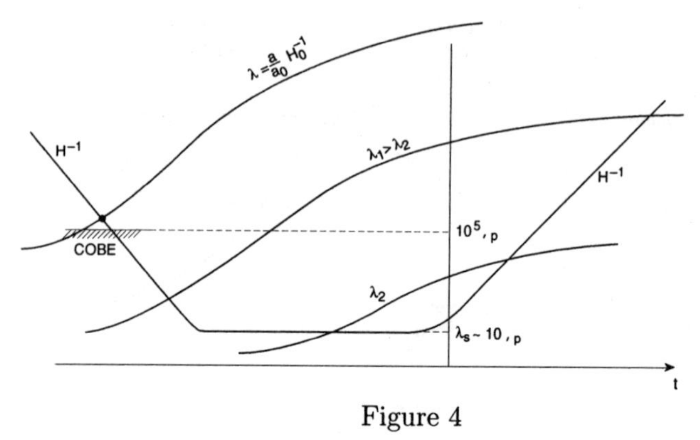

Figure 4

In phase I the Universe evolves at $g^2, \lambda^2 \ll 1$ and is thus close to the trivial vacuum. This phase can be studied using the tree-level low-energy effective action (10); it is characterized by a long period of dilaton-driven inflation. The accelerated expansion of the Universe, instead of originating from the potential energy of an inflaton field, is driven by the growth of the coupling constant (i.e. by the dilaton's kinetic energy) with $\dot{\phi} = 2\dot{g}/g \sim H$ during the whole phase. Notice that, as for ordinary inflation, the negative value of the kinetic energy associated with an expanding Universe is crucial.

Phase I supposedly ends when the coupling λ^2 reaches values of $O(1)$, so that higher-derivative terms in the effective action become relevant. Assuming that this happens while g^2 is still small (and thus the potential is still negligible), the value g_s of g at the end of phase I (the beginning of phase II) is an arbitrary parameter (a modulus of the solution).

During phase II, the stringy version of the big bang, the curvature as well as $\dot{\phi}$ are assumed to remain fixed at their maximal value, given by the string scale (i.e. we expect $\lambda \sim 1$). The coupling g will instead continue to grow from the value g_s until, in turn, it reaches values $O(1)$. At that point, thanks to a non-perturbative effect in g, the string phase will come to an end and the dilaton will be attracted to the true non-perturbative minimum of its potential; the standard FRW cosmology can then start, provided the Universe was by then heated-up and filled with radiation (see below). The second important parameter of this scenario is the duration of phase II or better the total red-shift, $z_s \equiv a_{end}/a_{beg}$, which has occurred from the beginning to the end of the stringy phase.

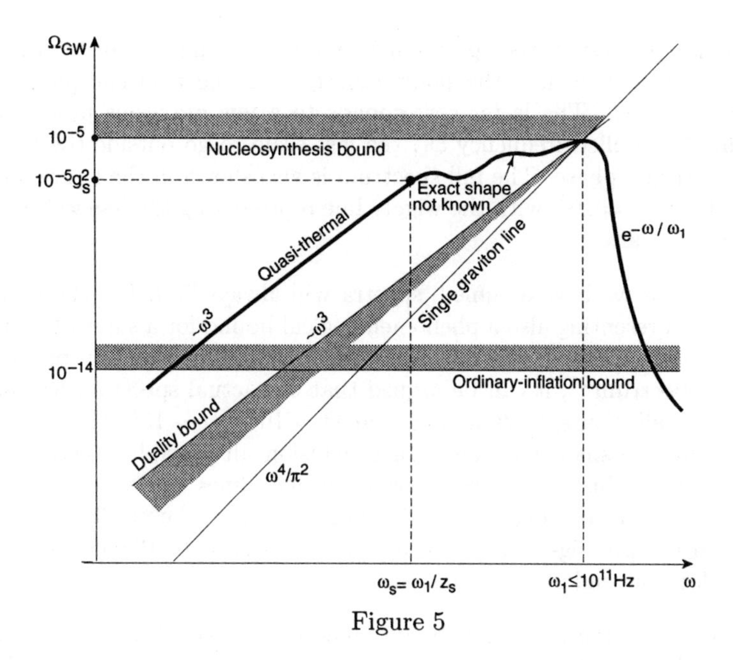

Figure 5

6 Perturbations and CGRB in String Cosmology

Starting from Fig. 3 let us draw the analogue of Fig. 1 for string cosmology, i.e. the behaviour of the horizon and the way different scales cross it. This is depicted in Fig. 4. Recalling eq. (8) we can easily understand why the bound (9) is now easily avoided. All we need in order to satisfy COBE's constraint is that $\ell_P H$ had been small enough at the time when the present horizon's scale crossed the Hubble radius during inflation (see hatched region in Fig. 4). Since the horizon is shrinking during superinflation, this does not prevent $\ell_P H$ from having been much smaller when scales of interest for GW detection (say above 1 Hz) crossed the horizon.

The final outcome for the GW spectrum [18, 19] is shown schematically in Fig. 5 (leaving a more detailed description to Ramy Brustein's talk). For a given pair g_s, z_s one identifies a point in the ω, Ω_ω plane as illustrated explicitly in the case of $g_s = 10^{-3}, z_s = 10^6$. The resulting point (indicated by a large dot) represents the end-point $(\omega_s, \Omega_{\omega_s})$ of the ω^3 spectrum corresponding to scales having crossed the horizon during the dilatonic era.

Although the rest of the spectrum is more uncertain, it can be argued[19] that it must smoothly join the point $(\omega_s, \Omega_{\omega_s})$ to the true end-point $\Omega \leq 10^{-5}, \omega \sim 10^{11}$ Hz. The latter corresponds to a few gravitons produced at the maximal amplified frequency ω_1, the last scale to go outside the horizon during the stringy phase. The full spectrum is also shown in the figure for the case $g_s = 10^{-3}$, $z_s = 10^6$, with the wiggly line representing the less well-known high-frequency part.

If $g_s < 1$, as we have assumed, spectra will always lie below the $\Omega_{GW} = 10^{-5}$, a line representing also a phenomenological bound for a successful nucleosynthesis to take place [20]. On the other hand, by invoking duality properties of the GW spectrum [21], it can be argued that the actual spectrum will never lie below the self-dual spectrum ending at $\Omega \sim 10^{-5}, \omega \sim 10^{11}$ Hz (the thick line bordering the shaded region). In conclusion all possible spectra sweep the angular wedge inside the two above-mentioned lines and a signal close to the NS bound is all but excluded. The large signal can be attributed to the fact that, in the pre-big-bang scenario, curvatures close to Planck's scale are reached before the end of inflation.

Having left to the next talk the discussion of further details on the GW spectrum and on the future prospects of detecting them, I will use the remaining time to mention a few more encouraging consequences of the pre-big-bang scenario. Like the generation of GW, they have something to do with the well-known phenomenon [22] of amplification of vacuum quantum fluctuations in cosmological backgrounds.

The first concerns scalar perturbations: Do they remain small enough during the pre-big-bang not to destroy the quasi-homogeneity of the Universe? The answer to this question turns out to be yes! This is not a priori evident since, in commonly used gauges (see e.g. [23]) for scalar perturbations of the metric (e.g. the so-called longitudinal gauge in which the metric remains diagonal), such perturbations appear to grow very large during the inflationary phase and to destroy homogeneity or, at least, to prevent the use of linear perturbation theory. In ref. [24] it was shown that, by a suitable choice of gauge (an "off-diagonal" gauge), the growing mode of the perturbation can be tamed. This can be double-checked by using the so-called gauge-invariant variables of Bruni and Ellis [25]. The bottom line is that scalar perturbations in string cosmology behave no worse than tensor perturbations. An interesting question arises here, in connection with the detectability of scalar perturbations of this type by using spherical antennas.

The second point that I wish to mention concerns a rather unique prediction of our scenario: the amplification of EM perturbations. Because of the scale-invariant coupling of gauge fields in four dimensions, electromagnetic (EM) perturbations are *not* amplified in a conformally flat cosmological background (even if inflationary). In string cosmology, the presence of a time-dependent dilaton in front of the gauge-field kinetic term allows the amplification of EM perturbations. Seeds for generating the galactic magnetic fields through the so-called cosmic-dynamo mechanism [26] can thus be obtained.

The final outcome can be expressed [27, 28] in terms of the fraction of electromagnetic energy stored in a unit of logarithmic interval of ω normalized to the one in the CMB, ρ_γ. One finds:

$$r(\omega) = \frac{\omega}{\rho_\gamma} \frac{d\rho_B}{d\omega} \simeq \frac{\omega^4}{\rho_\gamma} |c_-(\omega)|^2 \equiv \frac{\omega^4}{\rho_\gamma} (g_{re}/g_{ex})^2 \ . \tag{15}$$

where g_{ex} (g_{re}) refer to the value of the coupling at exit (re-entry) of the scale ω under consideration.

In terms of $r(\omega)$ the condition for seeding the galactic magnetic field through ordinary mechanisms of plasma physics is [29]

$$r(\omega_G) \geq 10^{-34} \ , \tag{16}$$

where $\omega_G \simeq (1 \text{ Mpc})^{-1} \simeq 10^{-14}$ Hz is the galactic scale. Using the known value of ρ_γ, we thus find, from (15) and (16):

$$g_{ex} < 10^{-33} \ , \tag{17}$$

i.e. a very tiny coupling at the time of exit of the galactic scale.

The conclusion is that string cosmology stands a unique chance to explain the origin of the galactic magnetic fields. Indeed, if the seeds of the magnetic fields are to be attributed to the amplification of vacuum fluctuations, their present magnitude can be interpreted as prime evidence that the fine structure constant has evolved to its present value from a tiny one during inflation. The fact that the needed variation of the coupling constant ($\sim 10^{30}$) is of the same order as the variation of the scale factor needed to solve the standard cosmological problems, can be seen as further evidence for scenarios in which coupling and scale factor grow roughly at the same rate during inflation.

Finally, I would like to mention a more theoretical bonus following from the pre-big-bang picture: a possible explanation of standard cosmology's hot initial state.

The question is: Can one arrive at the hot big bang of the SCM starting from our "cold" initial conditions? The reason why a hot Universe can emerge at the end of our inflationary epochs (phases I and II) goes back to an idea of L. Parker [30], according to which amplified quantum fluctuations can give origin to the CMB itself if Planckian scales are reached.

Rephrasing Parker's idea in our context amounts to solving the following bootstrap-like condition: At which moment, if any, will the energy stored in the perturbations reach the critical density? The total energy density ρ_{qf} stored in the amplified vacuum quantum fluctuations is roughly given by:

$$\rho_{qf} \sim N_{eff} \frac{M_s^4}{4\pi^2} (a_1/a)^4 \ , \tag{18}$$

where N_{eff} is the number of effective (relativistic) species, which get produced (whose energy density decreases like a^{-4}) and a_1 is the scale factor at the (supposed) moment of branch change. The critical density (in the same units) is given by:

$$\rho_{cr} = e^{-\phi} M_s^2 H^2 \ . \tag{19}$$

At the beginning, with $e^{\phi} \ll 1$, $\rho_{qf} \ll \rho_{cr}$; but, in the $(-)$ branch solution, ρ_{cr} decreases faster than ρ_{qf} so that, at some moment, ρ_{qf} will become the dominant source of energy while the dilaton kinetic term will become negligible. It would be interesting to find out what sort of initial temperatures for the radiation era will come out of this assumption.

7 Conclusions

- All cosmological inflationary models lead to the prediction of a stochastic CGRB that should surround us today very much like its electromagnetic analogue.

- Standard inflationary models must unfortunately satisfy the constraint $\Omega_{GW} < 10^{-10}\Omega_\gamma$ in the interesting frequency range.

- Inflationary models, such as those suggested by string theory, in which the Hubble parameter grows during inflation and eventually reaches values $O(\lambda_s^{-1}, \ell_P^{-1})$, evade the above constraint and (may) naturally lead to $\Omega_{GW} \leq 0.1\Omega_\gamma$ in the interesting frequency range.

- Observation of such a CGRB would open a unique window on the very early Universe and thus on fundamental physics at the Planck (string) scale.

- Last but not least, as emphasized to me by Emilio Picasso, trying to detect a stochastic CGRB is not just relying on getting a gift from the sky!

References

1. S. Weinberg, *Gravitation and Cosmology*, John Wiley & Sons, Inc., New York (1972).
2. L.F. Abbott and So-Young Pi (eds.), *Inflationary Cosmology*, World Scientific, Singapore (1986);
   E. Kolb and M. Turner, *The Early Universe*, Addison-Wesley, New York (1990).
3. G. Smoot et al., *Astrophys. J.* **396** (1992) L1.
4. G. Veneziano, "Quantum strings and the constants of Nature", in *The Challenging Questions*, Erice, 1989, ed. A. Zichichi, Plenum Press, New York (1990).
5. C. Lovelace, *Phys. Lett.* **B135** (1984) 75;
   C.G. Callan, D. Friedan, E.J. Martinec and M.J. Perry, *Nucl. Phys.* **B262** (1985) 593;
   E.S. Fradkin and A.A. Tseytlin, *Nucl. Phys.* **B261** (1985) 1.
6. P. Jordan, *Z. Phys.* **157** (1959) 112;
   C. Brans and R.H. Dicke, *Phys. Rev.* **124** (1961) 925.
7. See, for instance, E. Fischbach and C. Talmadge, *Nature* **356** (1992) 207.
8. T.R. Taylor and G. Veneziano, *Phys. Lett.* **B213** (1988) 459.
9. E. Witten, *Phys. Lett.* **B149** (1984) 351.
10. N. Sanchez and G. Veneziano, *Nucl. Phys.* **B333** (1990) 253;
11. G. Veneziano, *Phys. Lett.* **B265** (1991) 287.
12. M. Gasperini and G. Veneziano, *Astropart. Phys.* **1** (1993) 317; *Mod. Phys. Lett.* **A8** (1993) 3701; *Phys. Rev.* **D50** (1994) 2519.
13. A.A. Tseytlin, *Mod. Phys. Lett.* **A6** (1991) 1721;
    A.A. Tseytlin and C. Vafa, *Nucl. Phys.* **B372** (1992) 443.
14. E. Kiritsis and C. Kounnas, *Phys. Lett.* **B331** (1994) 51;
    A.A. Tseytlin, *Phys. Lett.* **B334** (1994) 315.
15. P. Aspinwall, B. Greene and D. Morrison, *Phys. Lett.* **B303** (1993) 249;
    E. Witten, *Nucl. Phys.* **B403** (1993) 159;
    A. Strominger, *Massless black holes and conifolds in string theory*, hep-

th/9504090.
16. R. Brustein and G. Veneziano, *Phys. Lett.* **B329** (1994) 429.
17. N. Kaloper, R. Madden and K. A. Olive, *Towards a singularity-free inflationary universe?*, Univ. Minnesota preprint UMN-TH-1333/95 (June 1995).
18. R. Brustein, M. Gasperini, M. Giovannini and G. Veneziano, *Phys. Lett.* **B361** (1995) 45;
see also M. Gasperini and M. Giovannini, *Phys. Rev.* **D47** (1992) 1529.
19. R. Brustein, M. Gasperini and G. Veneziano, *Peak and end point of the relic graviton background in string cosmology*, hep-th/9604084;
A. Buonanno, M. Maggiore and C. Ungarelli, *Spectrum of relic gravitational waves in string cosmology*, Pisa preprint IFUP-TH 25/96.
20. N. Hata et al., *Phys. Rev. Lett.* **75** (1995) 3977;
C. Copi et al., *Phys. Rev. Lett.* **75** (1995) 3981; and references therein.
21. R. Brustein, M. Gasperini and G. Veneziano, in preparation.
22. L.P. Grishchuk, *Sov. Phys. JEPT* **40** (1975) 409;
A.A. Starobinski, *JEPT Lett.* **30** (1979) 682;
V.A. Rubakov, M. Sazhin and A. Veryaskin, *Phys. Lett.* **B115** (1982) 189;
R. Fabbri and M. Pollock, *Phys. Lett.* **B125** (1983) 445.
23. See, for instance, V. Mukhanov, H.A. Feldman and R. Brandenberger, *Phys. Rep.* **215** (1992) 203.
24. R. Brustein, M. Gasperini, M. Giovannini, V. Mukhanov and G. Veneziano, *Phys. Rev.* **D51** (1995) 6744.
25. G. F. R. Ellis and M. Bruni, *Phys. Rev.* **D40** (1989) 1804;
M. Bruni, G. F. R. Ellis and P. K. S. Dunsby, *Class. Quant. Grav.* **9** (1992) 921.
26. E. N. Parker, *Cosmical Magnetic Fields*, Clarendon, Oxford (1979);
Y. B. Zeldovich, A. A. Ruzmaikin and D. D. Sokoloff, *Magnetic fields in astrophysics*, Gordon and Breach, New York (1983).
27. M. Gasperini, M. Giovannini and G. Veneziano, *Phys. Rev. Lett.* **75** (1995) 3796; *Phys. Rev.* **D52** (1995) 6651.
28. D. Lemoine and M. Lemoine, *Primordial magnetic fields in string cosmology*, Inst. d'Astrophysique de Paris preprint (April 1995).
29. M. S. Turner and L. M. Widrow, *Phys. Rev.* **D37** (1988) 2743.
30. L. Parker, *Nature* **261** (1976) 20.

SPECTRUM OF COSMIC GRAVITATIONAL WAVE BACKGROUND

RAM BRUSTEIN

Department of Physics, Ben-Gurion University, Beer-Sheva, 84105, ISRAEL

Models of string cosmology predict a stochastic background of gravitational waves with a spectrum that is strongly tilted towards high frequencies. I give simple approximate expressions for spectral densities of the cosmic background which can be directly compared with sensitivities of gravitational wave detectors.

A class of string cosmology models in which the evolution of the Universe starts with a dilaton-driven inflationary phase, followed by a high curvature (or "string") phase and then by standard radiation and matter dominated evolution was presented [1], and shown to predict a cosmic gravitational wave (GW) background of characteristic type. Using general arguments, numerical estimates of spectral parameters were improved [2]. The purpose of this talk is to provide simple approximate formulae for the spectrum of the cosmic GW background which could be used for direct comparison with measurements of present and planned experiments. More details about the general framework, the models and the computation of the spectrum as well as previous relevant work can be found in [3-10].

The energy density today in GW at frequency f in a bandwidth equal to f, $\rho_G(f) = \dfrac{dE_G}{d^3x\, d\ln(f)}$, produced during a dilaton-driven inflationary phase was computed in [1] and is given, approximately, by

$$\rho_G(f) = \rho_G^S(y_S)\left(\frac{f}{f_S(z_S)}\right)^3 \qquad f \le f_S \tag{1}$$

The two parameters of the model are 1) y_S, the ratio of the string coupling parameter at the beginning and at the end of the high curvature phase, which is taken to be $y_S \lesssim 1$, and 2) z_S, the ratio of the scale factor of the Universe at the end and at the beginning of the string phase. The parameter $z_S \ge 1$ may take very large values. The two parameters y_S, z_S may be traded for the two parameters $\rho_G^S(y_S)$ and $f_S(z_S)$.

The energy density in GW produced during the string phase cannot be computed at present because of our inadequate understanding of high curvature dynamics. However, we may boldly extrapolate the spectrum using a single power to obtain the following spectrum $\rho_G(f) = \rho_G^S + (\rho_G^{max} - \rho_G^S)\left(\frac{f-f_S}{f_1-f_S}\right)^\beta$

for $f_S \leq f \leq f_1$ where $\beta = -\frac{\ln y_S}{\ln z_S} > 0$ and f_1 is the end-point frequency, today, of the amplified spectrum. Note that $\rho_G(f_1)$ is the maximal energy density ρ_G^{max}.

We turn now to discuss numerical estimates for the pairs (f_1, ρ_G^{max}) and (f_S, ρ_G^S). To that end it is useful to consider the "minimal spectrum", in which the the dilaton-driven inflationary phase connects almost immediately to FRW radiation dominated evolution. For the minimal spectrum $z_S = 1$, $y_S = 1$, $f_1 = f_S$ and $\rho_G^{max} = \rho_G^S$. The end-point frequency today $f_1(t_0)$, was red-shifted from its value at the onset of the radiation era $f_1(t_r)$ due to the expansion of the Universe. Using entropy balance consideration we may evaluate the amount of red-shift in terms of ratios of temperatures and effective numbers of degrees of freedom. To evaluate $f_1(t_r)$ we use energy balance considerations at $t = t_r$ to relate $T(t_r)$ and the Hubble parameter at $H(t_r)$ and a geometrical relation between the end-point frequency and $H(t_r)$, $f_1 = H(t_r)/2\pi$. To determine $\rho_G(f_1)$ for the minimal spectrum we may use the "one-graviton" criterion, identifying the crossover from accelerated to decelerated evolution when only one graviton per phase space cell is produced $\rho_G(f_1) = 16\pi^2 f_1^4$. The result is the following range, depending on various parameters [2]

$$\begin{aligned} f_1 = f_S &= 0.6 - 2 \times 10^{10} Hz \\ \rho_G^{max} = \rho_G^S &= .03 - 3 \times 10^{-6} \rho_c h_{50}^{-2}. \end{aligned} \tag{2}$$

In the last equation the critical energy density ρ_c and the Hubble parameter in units of 50 *km/sec/mpc* were introduced. Equations (2) define together the coordinates of the minimal spectrum. As will be shown elsewhere using S-duality symmetry arguments, the minimal spectrum is also a lower bound on the amount of GW energy density produced during the dilaton-driven phase. For more details on the determination of (f_1, ρ_G^{max}) see [2].

At the moment, the most restrictive (indirect) experimental bound on the spectral parameters comes from nucleosynthesis (NS) constraints [11] $\rho_G \lesssim 0.1 \rho_R$, $\rho_G(f_1) < 5 \times 10^{-5} \rho_c h_{50}^{-2}$. As can be seen from eqs.(2), the nucleosynthesis bound is satisfied quite well. Previous, and very recent analysis of relevant direct and indirect bounds on ρ_G can be found in [12].

For non-minimal spectra $f_S = f_1/z_S$ and $\rho_G^S = \rho_G^{max} y_S^{-2}$, and the spectrum has a break at f_S as in figure 1. The position of the break (f_S, ρ_G^S) covers a wedge in the (f, ρ_G) plane. We obtain estimates similar to eq.(2) on ρ_G^S and ρ_G^{max} and summarize the results in figure 1, describing the interesting wedge in the (f, ρ_G) plane, bounded in between the NS bound and the minimal spectrum.

Most GW experiments present their sensitivity in terms of the spectral density $\sqrt{S_h(f)}$ of a metric perturbation h rather than that of $\rho_G(f)$. We will

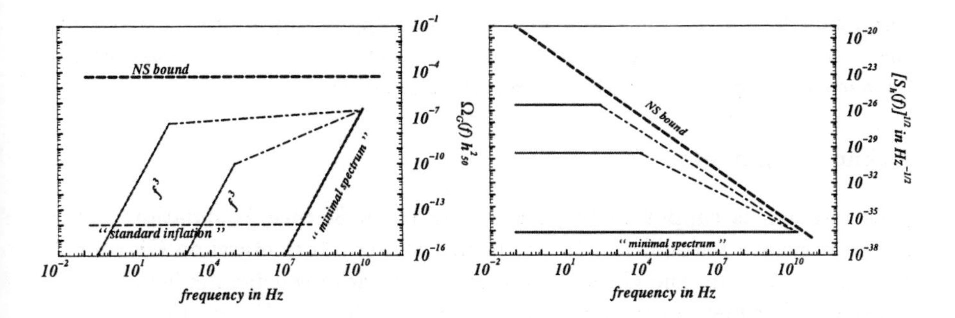

Figure 1: Energy density $\Omega_G(f) = \rho_G(f)/\rho_c$ (left) and spectral amplitude (right) of gravitational waves (GW). The solid lines are possible spectra of GW produced during the dilaton-driven phase. The minimal spectrum provides a lower bound on energy density in GW produced durng the dilaton-driven phase and the thick dashed line marks a nucleosynthesis upper bound. The dot-dashed lines are extrapolations of the spectrum into the string phase. The spectrum expected from slow-roll inflation is shown (left) for comparison.

translate the previous results and present the corresponding $S_h(f)$. The two-sided spectral density $S_h(f)$ is given by $\langle \widetilde{h}(f_1)\widetilde{h}^*(-f_2)\rangle = \frac{1}{2}\delta(f_1+f_2)S_h(f_1)$ where $\widetilde{h}(x,f)$ is Fourier transform of each of the two physical, transverse-traceless metric perturbations and $\langle\cdots\rangle$ denotes time or ensemble averages. The average GW energy density for isotropic and homogeneous background such as ours is given by $\frac{dE_G}{d^3x} = \frac{1}{8\pi G_N}\langle \dot{h}^2\rangle$ (G_N is Newton's constant), which can be expressed as $\frac{dE_G}{d^3x} = \frac{\pi}{2G_N}\int_0^\infty df\, f^2 S_h(f)$. Therefore $\rho_G(f) = \frac{\pi}{2G_N} f^3 S_h(f)$ and finally $\sqrt{S_h(f)} = \sqrt{\frac{2G_N}{\pi}}\frac{\sqrt{\rho_G(f)}}{f^{3/2}}$. For our class of models the dilaton-driven GW spectrum is given in eq.(1) and therefore $\sqrt{S_h(f)}$ for that part of the spectrum is constant

$$\sqrt{S_h(f)} = \sqrt{\frac{2G_N}{\pi}}\frac{\sqrt{\rho_G^S}}{f_S^{3/2}}. \tag{3}$$

Now, all we need to do is to translate the coordinates of the end-point of the minimal spectrum from the (f, ρ_G) plane to the $\left(f, \sqrt{S_h(f)}\right)$. We summarize the results in figure 1 showing the interesting region in the $(f, \sqrt{S_h(f)})$ plane.

The cryogenic resonant detectors EXPLORER and NAUTILUS provide the best direct upper limit on the existence of a relic graviton background, $S_h^{1/2} < 6\times 10^{-22}\mathrm{Hz}^{-1/2}$, at $f = 920$ Hz [13]. This limit is still too high to be significant for our background. An important way of improving sensitivity for the detection of our background is to perform cross-correlation measurements between as many working GW detectors as possible.

Acknowledgments

This research is supported in part by the Israel Science Foundation and by an Alon grant. I would like to thank my collaborators Massimo Giovannini, Maurizio Gasperini and Gabriele Veneziano and acknowledge the help of many experts, experimentalists and theorists, in the field of GW detection.

References

1. R. Brustein et al., *Phys. Lett.* B361, 45 (1995).
2. R. Brustein, M. Gasperini and G. Veneziano, preprint hep-th/9604084.
3. G. Veneziano, *Phys. Lett.* B265, 287 (1991).
4. M. Gasperini and G. Veneziano, *Astropart. Phys.* 1, 317 (1993).
5. M. Gasperini and M. Giovannini, *Phys. Lett.* B282, 36 (1992); *Phys. Rev.* D47, 1519 (1993).
6. M. Gasperini and G. Veneziano, Mod. Phys. Lett. A8, 3701 (1993); *Phys. Rev.* D50, 2519 (1994).
7. K. A. Meissner and G. Veneziano, *Phys. Lett.* B267, 33 (1991); Mod. Phys. Lett. A6, 3397 (1991); M. Gasperini and G. Veneziano, *Phys. Lett.* B277, 256 (1992); A. A. Tseytlin and C. Vafa, *Nucl. Phys.* B372, 443 (1992).
8. R. Brustein, et al., *Phys. Rev.* D51, 6744 (1995).
9. R. Brustein and G. Veneziano, *Phys. Lett.* B329, 429 (1994); N. Kaloper, R. Madden and K. Olive, *Nucl. Phys.* B452, 677 (1995); M. Gasperini, J. Maharana and G. Veneziano, preprint hep-th/9602087.
10. M. Gasperini, M. Giovannini and G. Veneziano, *Phys. Rev. Lett.* 75, 3796 (1995); *Phys. Rev.* D52, 6651 (1995); D. Lemoine and M. Lemoine, *Phys. Rev.* D52, 1955 (1995).
11. V. F. Schwartzmann, *JEPT Lett.* 9, 184 (1969); T. Walker et al., *Ap. J.* 376 (1991) 51.
12. R. L. Zimmerman and R. W. Hellings, *Ap. J.* 241, 475 (1980); B. Allen, preprint gr-qc/9604033.
13. P. Astone et al., *Upper limit for a gravitational wave stochastic background measured with the EXPLORER and NAUTILUS gravitational wave resonant detectors*, (Rome, February 1996), to appear.

PROPAGATION OF GRAVITATIONAL WAVES IN MATTER

A. Degasperis
Dipartimento di Fisica, Università di Roma "La Sapienza"
Istituto Nazionale di Fisica Nucleare, Sezione di Roma
P.le A.Moro 2, 00185 Roma, Italy

V. Ricci
Dipartimento di Matematica, Università di Roma "La Sapienza"
P.le A.Moro 7, 00185 Roma, Italy

This is a very short report on a project which is still in progress, and is aimed to investigate on the current wisdom according to which gravitational waves propagate from the source region to terrestrial detectors (for instance, a distance of about 10^7 pc from Virgo cluster to Virgo detector) linearly and without dispersion. In a plane-wave model, if x is the coordinate in the propagation direction, the metric is $g_{\mu\nu} = \eta_{\mu\nu} + h_{\mu\nu}(x - ct)$, where the nonvanishing components of the linear field $h_{\mu\nu}$ are $h_{22} = -h_{33}$ and $h_{23} = h_{32}$ (the index μ runs from 0 to 3 and $\eta_{\mu\nu} = diag(+,-,-,-)$). It is true, however, that waves propagate in nonempty space, with matter density $\rho_0 \sim 10^{-30} gr/cm^3$, and they experience self-interaction due to the nonlinearity of Einstein equation. Sound arguments in favor of dispersionless and linear propagation are the very weak interaction with matter, say $\frac{8\pi G\rho_0}{\omega^2} \sim 10^{-44}$ (ω is the wave angular frequency, $\omega \sim 10^4 Hz$), and the very small wave amplitude, $h \sim 10^{-20}$. It is therefore extremely unlikely that a linear and nondispersive model should be replaced, for any practical purpose, by a nonlinear and dispersive one. With different emphasis, and by means of various mathematical techniques, several estimates of nonlinear and dispersive effects [1] have been computed, and support this claim.

Despite all this, we still give our attention to this propagation problem for two reasons. First, our approach is based on a perturbation theory, the multiple-scale method, which has been very successfully applied to nonlinear optics [2] (leading to the prediction of soliton propagation in fibers), but never applied, as far as we know, to gravitational waves. Second, the results published on this matter are not always in agreement with each other [3], and their derivation is not always mathematically transparent. Therefore, in view of the many efforts and resources devoted to the detection of gravitational waves, we deem it appropriate to bring further support to our current belief by providing firm estimates of nonlinear and dispersive effects on wave propagation.

Here below we outline the strategy of our method, with few comments. The starting point is, of course, the Einstein field equation

$$R_{\mu\nu} - \frac{1}{2}Rg_{\mu\nu} = \frac{8\pi G}{c^2}T_{\mu\nu}, \tag{1}$$

together with the integrability conditions (Bianchi identities)

$$T^{\mu\nu}{}_{;\nu} = 0 \tag{2}$$

which provide the matter equation of motion. In order to build up a dispersion theory, we introduce first a model of the medium through which waves propagate. The medium, together with its associated gravitational field, has to be a sufficiently well known solution of equations (1) and (2), which will be referred to as the background, or zero-order (no wave) solution. A standard, if very idealized, model of the medium is provided by the perfect fluid, whose energy-momentum tensor reads

$$T_{\mu\nu} = (\rho + \frac{p}{c^2})g_{\mu\alpha}g_{\nu\beta}u^\alpha u^\beta - \frac{p}{c^2}g_{\mu\nu}, \tag{3}$$

where ρ and p are, respectively, the mass density and pressure, while u^μ is the fluid velocity field with $g_{\mu\nu}u^\mu u^\nu = 1$. Moreover, we assume, for the sake of simplicity, that the pressure p be proportional to the density

$$p = c_s^2\rho\ , \tag{4}$$

c_s being the sound velocity. Our background solution $g_{\mu\nu} = b_{\mu\nu}, \rho = \rho_b$ and $u^\mu = u_b^\mu$ is static as a result of pressure-gravity balance. Next, we introduce the following symmetry assumption, which is well suited to our purposes, namely we ask that the background solution depends only on one space coordinate, say x. It turns out that this condition is strong enough to allow us to find the exact solution of Einstein equations [4]. Thus, our medium is like a standing wall, whose density and gravitational field have a definite profile in the x-direction, while being constant in the transverse y and z-directions. In the present context, however, we need only the Newtonian limit of our exact solution, which obtains at the first order in the parameter $\alpha = c_s^2/c^2 \sim 10^{-12}$, and has the following expression

$$b_{\mu\nu} = \eta_{\mu\nu} + 4\alpha\, log\, \cosh\xi\ \delta_{\mu\nu} - 4\alpha\xi diag\{0,1,1,1\}, \tag{5a}$$

$$\rho_b = \rho_0 \cosh^{-2}\xi\ ,\ u_b^\mu = (1 - 2\alpha\, log\, \cosh\xi)(1,0,0,0), \tag{5b}$$

where ξ is the dimensionless coordinate $\xi = x/(2L_s)$, with $8\pi G\rho_0/c^2 = L^{-2}$, $L_s = \sqrt{\alpha}L$, and note that $L \sim 10^{28} cm$ so that the wall thickness is around 10^{22} cm.

Next we look for a second solution of the same equations (1) and (2), with (3) and (4), of the form

$$g_{\mu\nu} = b_{\mu\nu}(x) + h_{\mu\nu}(x,t) \tag{6}$$

where $h_{\mu\nu}$ is supposed to describe a gravitational wave propagating in the x direction. Since the wave-length $\lambda \sim 10^7$ cm is much smaller than the wall thickness, we let the wave propagate around $\xi = 0$, and take all contributions of the background as evaluated at $\xi = 0$. In this way, we obtain the dispersion law for gravitational waves by linearizing the Einstein equations with respect to $h_{\mu\nu}$. This takes the simple form

$$\omega(k) = c\sqrt{k^2 - L^{-2}}, \tag{7}$$

in agreement with other authors [3]. The modes corresponding to (7) are the two polarizations h_{22} and h_{23} in the transverse-traceless gauge, namely, at this order, all entries of $h_{\mu\nu}$ vanish with the exception of $h_{22} = -h_{33}$ and $h_{23} = h_{32}$. This analysis yields, of course, also the acoustic mode with the dispersion relation $\omega = c_s\sqrt{k^2 - \frac{1}{2}L_s^{-2}}$, that implies the well-known Jeans instability.

Since the exact equations for $h_{\mu\nu}, \rho_h$ and u_h^μ, which obtain by inserting (6), together with $\rho = \rho_b + \rho_h$ and $u^\mu = u_b^\mu + u_h^\mu$, into the full equations (1) and (2), are highly nonlinear and involve many field variables, the only way to compute nonlinear effects is via a suitable perturbation scheme. The one we adopt here, the multiple-scale method, applies to gravitational pulses which are sufficiently far from the source (so as to neglect source gravitational field effects), and are quasi-monochromatic plane waves. This amounts to ask that the first order approximation of a polarization field h (that may be a linear combination of h_{22} and h_{23}) reads

$$h(x,t) = \epsilon(A \exp[i(kx - \omega t)] + A^* \exp[-i(kx - \omega t)]) + 0(\epsilon^2), \tag{8}$$

where the smallness parameter ϵ is the relative width of the Fourier spectrum, i.e. $\epsilon = \Delta k/k$ (or, equivalently, $1/\epsilon$ is the number of wavelengths in the pulse), and k and ω are the wave-number and angular frequency of the gravitational wave (according to reasonable estimates [5], $\epsilon \sim 10^{-4}$ so that $|A| \sim 10^{-16}$). In analogy with the anharmonic oscillator, the nonlinearity affects an expansion of h in powers of ϵ in two ways: it introduces all other harmonics (say $\exp[in(kx -$

$\omega t)]$ with $n = 0, \pm 2, \pm 3, ...)$, and it generates an unwanted secular growth of the amplitudes. In order to get rid of secularities, one let the amplitude A in (8) depend on slow coordinates according to the following scaling

$$A = A(x_1, t_1, t_2)\ ,\ x_1 = \epsilon x,\ t_1 = \epsilon t,\ t_2 = \epsilon^2 t, \tag{9}$$

and imposes appropriate evolution equations with respect to the slow times t_1 and t_2. Thus, at $t = 0$, the slowly varying function $A(x_1, 0, 0)$ modulates the amplitude of the carrier wave $\exp[i(kx - \omega t)]$, with a profile which contains $N \sim 1/\epsilon$ wavelengths. As for the slow time t_1, the wave propagates according to the equation

$$\partial A/\partial t_1 + \omega_1 \partial A/\partial x_1 = 0, \tag{10}$$

where ω_1 is the group velocity $d\omega/dk$ ($\omega_1 \sim c$). However, with respect to the slower time t_2 (see (9)), the amplitude profile propagates according to a nonlinear equation. Similarly to the propagation of electromagnetic waves in a homogeneous nonlinear medium, we expect that the two polarizations amplitudes A_{22} and A_{23} satisfy the following system of nonlinear Schroedinger equations

$$i\partial A_{22}/\partial t_2 + \omega_2 \partial^2 A_{22}/\partial x_1^2 + \gamma(|A_{22}|^2 + |A_{23}|^2)A_{22} = 0 \tag{11a}$$

$$i\partial A_{23}/\partial t_2 + \omega_2 \partial^2 A_{23}/\partial x_1^2 + \gamma(|A_{22}|^2 + |A_{23}|^2)A_{23} = 0, \tag{11b}$$

which, if only one polarization state is present, reduces to

$$i\partial A/\partial t_2 + \omega_2 \partial^2 A/\partial x_1^2 + \gamma |A|^2 A = 0. \tag{12}$$

Here $\omega_2 = \frac{1}{2} d^2\omega/dk^2$ is the dispersion constant, and γ is the nonlinearity parameter.

Although the main properties of the nonlinear Schroedinger equation (12) are well-known, since this equation shows up in several physical contexts, few observations are appropriate. First, we note that the wave behaviour crucially depends on the sign of the quantity $\omega_2\gamma$; if $\omega_2\gamma < 0$ (defocusing case), the dispersion effect cannot be compensated by the nonlinearity, and the wave, for very large time ($t_2 \to \infty$) disperse away as in the linear case. However, if $|\gamma A^2| >> |\omega_2| k^2$, the nonlinearity prevails up to the time $t_N \sim \ell^2 |\gamma| |A|^2 / (\omega_2^2 k^2)$ (here ℓ is the length of the initial pulse profile), and generates an amplitude-dependent modulation of the phase. More interesting is the opposite case, $\omega_2\gamma > 0$ (focusing case), because nonlinearity and dispersion can balance each other giving rise to nondispersive pulses (solitons), which propagate with definite velocity and amplitude profile.

Resonant Bars

THE CRYOGENIC GRAVITATIONAL WAVE ANTENNAS EXPLORER AND NAUTILUS

Giovanni Vittorio Pallottino
INFN and University of Rome "La Sapienza", P. A. Moro 2, 00185 Rome, ITALY

We report on the resonant gravitational wave detectors of the Rome group: the cryogenic antenna Explorer, operating at CERN since 1990, and the first ultracryogenic antenna Nautilus that recently started to operate in Frascati at 0.1 kelvin. A short description of these detectors - both using an aluminum bar of 2300 kg, a capacitive transducer and a SQUID amplifier - is followed by a discussion of their performance and of the experimental results obtained up to now. We remark that surely interesting perpectives for science are now opened by the operation of other resonant detectors, such as Allegro (USA) and Niobe (Australia), and by the next coming into operation of the ultracryogenic detector Auriga in Legnaro.

1 Introduction

After the pioneering work of Joseph Weber a new generation of cryogenic resonant detectors was conceived around 1970, built during the 80's and finally put in continuous operation starting from 1990. A further step was the development of ultracryogenic detectors, at temperatures well below 1 kelvin, the first of them only recently begun to take data. So we now have five resonant detectors in operation or very close to operate, as shown in fig.1, with experimental pulse sensitivity of $\sim 6 \times 10^{-19}$.

Overall, the task turned out much more difficult than originally thought, in spite of the strong effort of a number of people and research groups all over the world. We had, in fact, to push many technologies - mechanics, cryogenics, electronics, etc. - to their extreme limits and beyond. A major problem is pointed out in fig. 1, where we see large gaps in the actual operation of the detectors. This is the duty cycle problem, technically called availability [1], that is the probability that at any time an apparatus performs correctly which depends on both its reliability and its maintainability. The reliability, for those experiments, is very critical, as given by the product of the individual reliabilities of a number of devices. Even more critical is the maintainability, which depends on the average time required to correct any malfunction. We have in fact two barriers that hinder our corrective interventions: vacuum and low temperature. The latter, to be more specific, means several months for opening a cryostat, repairing the instrumentation and cooling it again, mostly due to the large thermal time constants involved.

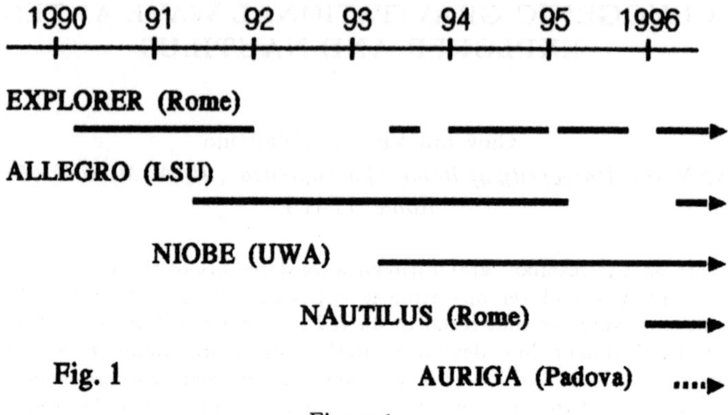

Figure 1:

Table 1: Present cryogenic detectors: resonant cylindrical bars (1996)

	mass	temp.	Sensit. theor.	Sensit. exper.
Allegro (1991)	2300 kg Al	4.2 K	3×10^{-19}	6×10^{-19}
Auriga	2.300 kg Al	100 mK	3×10^{-20}	
Explorer (1990)	2300 kg Al	2.5 K	3×10^{-19}	6×10^{-19}
Nautilus (1996)	2300 kg Al	100 mK	3×10^{-20}	6×10^{-19}
Niobe (1996)	1500 kg Nb	6 K	3×10^{-19}	6×10^{-19}

2 The Explorer and Nautilus detectors

The Explorer detector [2,3,4], installed at CERN, and the Nautilus detector [5,6,7], now operating in Frascati, INFN, are very similar as regards the resonant bar and the readout instrumentation. The important difference, is the cryogenic system and the operating temperature. Explorer is cooled between 2 and 2.5 K by a bath of helium brought in superfluid state. Nautilus is cooled at 100 mK and below using a dilution refrigerator $^3He -^4 He$.

Both Nautilus and Explorer interact with the gravitational field using a cylinder of aluminum with length of 3 m and mass of 2270 kg, whose vibrations in the first longitudinal mode are converted into electrical signals by a capacitive resonant transducer[8] (the gap is $63\mu m$ for Explorer, $49\mu m$ for Nautilus). Since the resonant frequency of the transducer is very close to that of the bar,

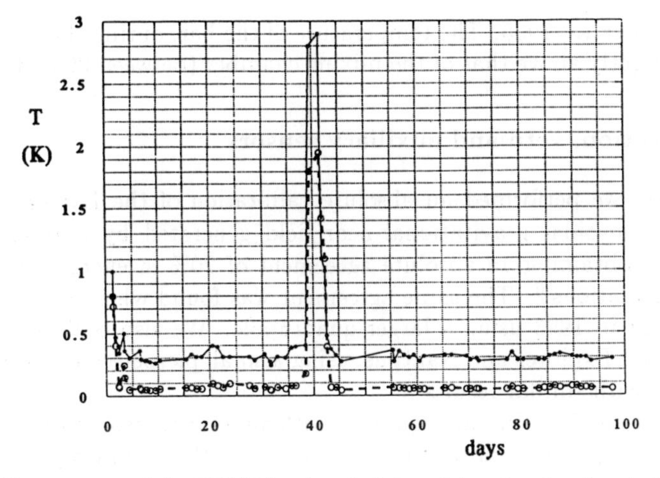

Figure 2: Temperature of the SQUID(–•–) and of the mixing chamber (–o–) of the Nautilus detector during 100 days starting from Dec 13, 1995.

the coupling of the two oscillators gives rise to two modes.

The electrical signal of the transducer is applied, through a superconducting transformer, to a DC SQUID amplifier[9] with flux noise of $2 \div 3 \times 10^{-6}$ flux quanta / $\sqrt{Hz}$. The data acquisition is performed both sampling the output of the SQUID amplifier at high speed (fast channel, $\Delta t = 4.54ms$) and sampling at low speed (slow channels, $\Delta t = 290ms$) the outputs of lock-in amplifiers, driven by the same signal, that are tuned at the frequencies of the modes.

The bar of Explorer is suspended with a titanium cable wrapped around its central section, thereby making up the last stage of a number of mechanical filters providing a total attenuation of 210 dB, at the frequencies of the modes, to the laboratory acoustic noise. The bar of Nautilus, instead, is suspended with a copper cable to a chain of thermal shields that are essential for reducing the thermal input. The copper cable sustaining the bar also provides the thermal contact with the mixing chamber, which is the coldest point of the refrigerator. The same thermal shields, and their suspensions, are the final stages of a mechanical filtering system that provides an overall attenuation of 220 dB to the laboratory noise. But here we also have the mechanical noise due to the refrigerator itself, that, on the other hand, requires a good thermal contact with the bar. The filtering of this noise along the thermal connection between the mixing chamber and the bar amounts to 90 dB. The temperatures of the mixing chamber and of the SQUID amplifier, which is located outside the innermost thermal shields, are shown in fig. 2 during 100 days starting from December 13, 1995. The temperature of the bar, during the same period, was only slightly above that of the mixing chamber (between 90 and 100 mK).

3 Calibration, veto and auxiliary sensors

During 1989 we performed an absolute calibration of the Explorer detector using the sinusoidal gravitational near field generated by a small rotating quadrupole, obtaining an absolute agreement of 20% in energy with the numerical calculations[10]. The same agreement was found when we repeated the measurement in 1993 using a larger quadrupole. We also performed calibrations, both on Explorer and Nautilus, using a small piezoelectric ceramic glued on the bar, with similar results.

Explorer and Nautilus are equipped with auxiliary sensors (accelerometers, search coil, etc.) that monitor the environment of the laboratory and allow to veto any event, observed by the detector, that occurs in the presence of external disturbances. Nautilus, in addition, since its design sensitivity is higher than Explorer, is equipped with a cosmic ray veto system [11] consisting of layers of streamer tubes placed above and below the cryostat. We expect, in fact, to observe about two cosmic ray events per day when the noise of this detector will reach the level of 1 mK, with an higher rate for further improvements of its sensitivity.

4 Temperature of the modes and filtering of the data

A basic and preliminary requirement is that the vibration modes of the oscillator exhibit a temperature corresponding to the thermodynamical temperature of the oscillator itself. As regards Explorer, this agreement has been obtained, for most of the time, since several years. As regards Nautilus, to obtain this agreement is more difficult, because its operating temperature is much lower and also for the presence of the refrigerator whose noise acts inside the cryostat, close to the bar. Only since the beginning of 1996, in fact, the temperature of the modes of Nautilus was close to about 100 mK, as shown in fig. 3.

The next step is to filter the data in order to improve the signal to noise ratio for short bursts of gravitational radiation. This is done using optimum filters, operating both on line and off line [12,13,14]. We filter both the "slow" data, representing the observations performed, separately, at the two modes with sampling time of 290 ms, and the "fast" data recorded at 4.54 ms. In the former case at each time instant we obtain two estimates (at the two modes) of the energy innovations in the bar, which we combine together to obtain a single estimate: here the distribution of the noise energy turns out to be exponential,

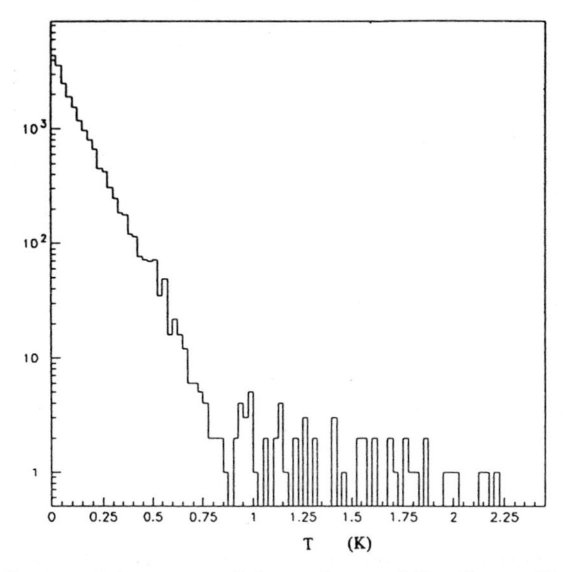

Figure 3: Distribution of the energy of the mode ν_- of Nautilus on March 4, 1996 with temperature of 111 mK.

and we call effective temperature its mean value. This quantity represents the energy sensitivity of the apparatus for pulse detection (for SNR=1). The "fast" filter, instead, directly provides overall estimates of the energy innovations, with a distribution of the noise energy which follows the Rayleigh law since its amplitude is Gaussian. Here again we call effective temperature the mean value of the noise energy. While the two filtering procedures should provide the same effective temperature, their actual performance is different. The effective temperature provided by the "fast" filter is, in fact, closer to the theoretical value. In addition, we had to recognize that the two filters operate on data with different sampling times (4.54 ms and 290 ms) so that their behavior is different for those bursts that are short for the "slow" filter, but long for the "fast" one. This is a rather delicate point for coincidence experiments between detectors whose data are sampled at different rate.

We remark, moreover, that more important than the effective temperature of a detector is the quality of its noise distribution, represented by the rate of samples that deviate from the statistics. The tails of the distributions determine, in fact, the background for any coincidence experiment. We consider very good, for example, the distribution shown in fig.4 with only one event (with energy of 578 mK) during 25 hours of continuous operation (all the samples deviating from the exponential law are in fact due to the same

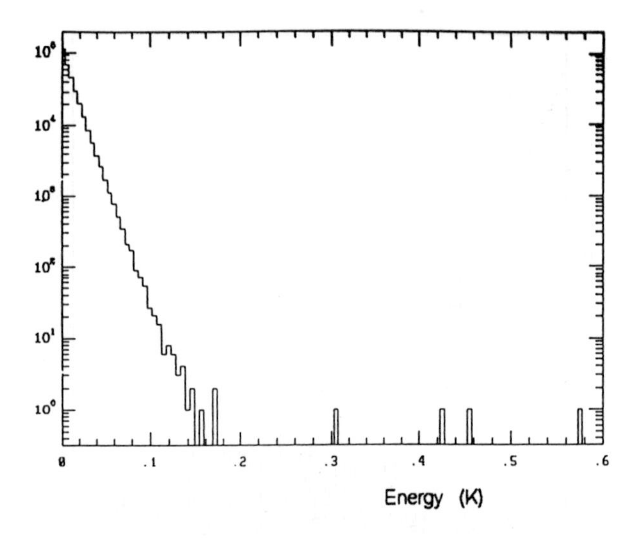

Figure 4: Distribution of the energy innovàtions of Explorer: 25 hours, October 6, 1995.

event, as originated from the response of the filter). We finally remark that, in addition to the search for short bursts, we are now using the data collected by our detectors to detect periodic waves over long time periods [15,16] as well as for giving upper limits for the stochastic background of cosmological origin [17,18]. In those cases the requirements are different than for pulse detection: here the amplifier noise (which determines the observation bandwidth) is less important, while it is very important to have high values of the quality factor of the modes and low values of the operating temperature. This point will be discussed in more detail in the paper of Pia Astone [19], so that here we only recall that the spectral strain sensitivity of the present detectors, even if in a small region of the spectrum, has already reached the remarkable figure of $6 \times 10^{-22}\sqrt{Hz}$.

5 Conclusions

The perspectives opened by the operation of a number of cryogenic resonant detectors are surely very interesting since it is clear that no detection has meaning if not obtained in coincidence by two or more antennas. While the results of the coincidence experiments reported in the past gave negative results[20,21,22], a more refined analysis of the data collected up to now and, above all, the analysis of the data the various detectors are now collecting is giving new chances to a task started so many years ago.

References

1. I. Bazovsky "Reliability Theory and Practice" (Prentice-Hall, 1961).
2. E. Amaldi e al. Nuovo Cimento 9C, 829 (1986)
3. E. Amaldi et al. Europhysics Letters 12, 5 (1990)
4. P. Astone et al. Phys. Rev. D, 47, 362 (1993)
5. P. Astone et al. Europhysics Letters 16, 231 (1991)
6. E.Coccia, V.Fafone, I.Modena Rev.Sci. Instr. 63 5432 (1992)
7. P. Astone et al. "The gravitational wave detector Nautilus: apparatus and preliminary experimental results" submitted to Phys. Rev. D (1996)
8. P.Rapagnani Nuovo Cimento 5C, 385 (1982)
9. C.Cosmelli, M.G.Castellano, P. Carelli in "SQUID 91" Proceedings of the 4th International Conference, Berlin, Germany, 1991, edited by H. Koch and H. Luebbig (Springer-Verlag, Berlin, 1992)
10. P. Astone et al. Z. Phys. C 50, 21 (1991)
11. E. Coccia et al. Nucl. Instr. and Meth. A 355, 624 (1995)
12. P. Bonifazi et al. Nuovo Cimento 1C, 465 (1978)
13. P. Astone et.al. Nuovo Cimento 17C, 713 (1994)
14. P. Astone et al. "The fast matched filter for gravitational wave data analysis: characteristics and applications" Dept. of Physics, Univ. of Rome La Sapienza, Internal Report n.1052 (1995)
15. G.V. Pallottino, G. Pizzella Nuovo Cimento 7C, 155 (1984)
16. S. Frasca, M. Gabellieri, G.V. Pallottino Nuovo Cimento 10C, 1 (1987)
17. R. Brustein et.al. Physics Lett. B 361, 45 (1995)
18. P. Astone et al. "Upper limit for gravitational wave stochastic background with the Explorer and Nautilus Detectors" submitted to Physics Lett. A (1996)
19. P. Astone et.al. Proceedings of this Conference
20. E. Amaldi et al. Astron. Astrophys. 216, 325 (1989)
21. P. Astone et al. "Results of a Preliminary Data Analysis in Coincidence between the LSU and Rome Gravitational Wave Antennas" in "General Relativity and Gravitational Physics" Proc. of the 10th Conference edited by M. Cerdonio et al. (World Scientific, Singapore, 1994)
22. Z.K. Geng et al. "Operation of the Allegro detector at LSU" in "Gravitational Wave Experiments" Proc. of the 1st Amaldi Conference, Frascati, 1994 edited by E.Coccia et al. (World Scientific, Singapore, 1995)

THE ULTRACRYOGENIC GRAVITATIONAL WAVE DETECTOR AURIGA

G.A.PRODI, J.P.ZENDRI, G.FONTANA, R.MEZZENA, S.VITALE
Department of Physics, University of Trento and I.N.F.N. Gruppo Coll. Trento Sezione di Padova, I-38050 Povo, Trento, Italy

L.TAFFARELLO, E.CAVALLINI, P.FORTINI
Department of Physics, University of Ferrara, and I.N.F.N., Sezione di Ferrara, Ferrara, Italy

P.FALFERI, M.BONALDI
Centro CeFSA, ITC-CNR, Trento and I.N.F.N. Gruppo Coll. Trento Sezione di Padova, I-38050 Povo, Trento, Italy

M.CERDONIO, D.CARLESSO, A.COLOMBO, D.PASCOLI
Department of Physics, University of Padova and I.N.F.N. Sezione di Padova, Via Marzolo 8, I-35131 Padova, Italy

G.VEDOVATO, A.ORTOLAN, S.CARUSO, L.CONTI, V.CRIVELLI VISCONTI, G.MARON
I.N.F.N. National Laboratories of Legnaro, via Romea 4, I-35020 Legnaro, Padova, Italy

We discuss the latest results of the AURIGA gravitational wave detector at the INFN National Laboratories of Legnaro, Italy. AURIGA is an ultracryogenic resonant bar detector, whose design is similar to the other INFN ultracryogenic detector NAUTILUS. The first cryogenic run of AURIGA is currently in progress. During these first months of operation we tested the performances of the cryogenics, of the mechanical suspensions and of the data acquisition system. The transducer output has been monitored through a room temperature port either to calibrate the detector or to measure the antenna noise. Here we present preliminary results on calibration and diagnostic of the antenna at liquid helium temperatures. In particular, the antenna noise performance has been thermal over a long time span at about 6K.

1 Introduction

AURIGA [1] is a resonant bar detector for gravitational waves, which has been set up at the Laboratori Nazionali di Legnaro of the Istituto Nazionale di Fisica Nucleare. The experimental activity on AURIGA started in 1990 and the first cryogenic run of the complete detector is currently in progress from May 1995. The first months of operation at liquid Helium temperature have been dedicated to test several components, to calibrate the detector and to check the antenna noise temperature. In the near future we plan to operate the antenna at $\sim 0.1K$; at that temperature the present configuration of the detector should allow a burst sensitivity of $h_{min} \sim 3\ 10^{-19}$ with a post detection bandwidth $\Delta\nu \sim 1Hz$. This performance is similar to that expected for the present configuration of NAUTILUS [2]. The perspectives for future improvements in sensitivity rely mainly on a lower noise d.c.SQUID amplifier, on a better coupling between transducer and SQUID by means of a tuned LC resonator and on an optimized transducer with higher mass and capacitance. In order to maximize the chances of detecting in coincidence, AURIGA is parallel to the other resonant antennae [3] NAUTILUS, ALLEGRO and EXPLORER and within a few degrees from NIOBE.

The general designs of the liquid Helium cryostat, of the mechanical suspensions and of the resonant displacement transducer are derived from the other INFN ultracryogenic antenna NAUTILUS [2]. The most important differences are in the set up of the internal mechanical suspensions, of the $^3He-^4He$ dilution refrigerator, of the ultracryogenic thermal links and of the signal amplification chain [1]. In each stage of the cryogenic suspensions the elastic rods have been equally tensioned, and the room temperature stacks of rubber disks has been loaded by additional lead masses. For what concerns the refrigerator, we implemented additional features in order to improve the availability of the detector, such as twin condenser lines, and to provide for vibration isolation, such as mechanically soft pipelines and thermal links. A cryogenic switch connects the signal leads from the transducer either directly to an external test port or to the internal d.c.SQUID amplification chain. Up to now the measurements have been made through the test port with room temperature electronics both for calibrating the detector and for monitoring the antenna noise. An external auxiliary dewar is also provided at the test port in order to house an alternative SQUID amplification chain outside the detector main cryostat.

2 Experimental results

The first cool down of the AURIGA detector to liquid helium temperatures took about 40 days, including about 10 days for testing at liquid nitrogen temperatures. About 10000 liters of liquid nitrogen and 3300 liters of liquid helium were needed. In the first 6 months of operation from June 1995, the antenna was cooled to $6 - 10\ K$ by exchange gas in the experimental chamber, with a tipical pressure of $2\ 10^{-5} mbar$. In the subsequent months the antenna has been cooled to $2\ K$ by pumping out the helium exchange gas and by operating the 1 K pot of the dilution refrigerator. The average liquid helium evaporation

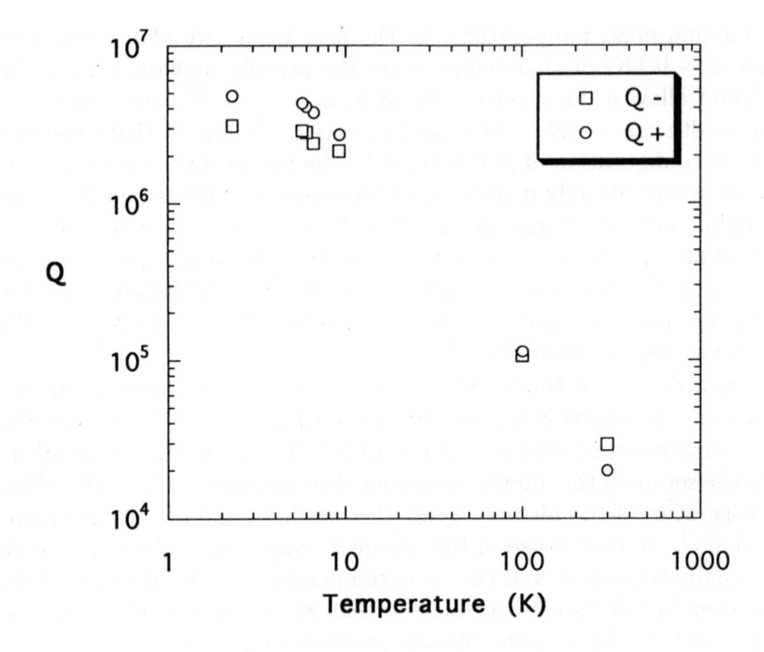

Figure 1: mechanical quality factors of mode - , $913Hz$, and mode +, $931Hz$, as a function of temperature.

rate was about 90 $liters/day$. The typical bias field of the transducer was $\sim 0.3 - 0.6\ MV/m$ and the electric charge showed a negligible leakage over 2 months. The quality factors of the antenna-transducer resonances at 2 K are 3 and 5 millions respectively for the 913 Hz and 931 Hz modes, see fig.1.

The vibration attenuation of the mechanical suspensions has been measured at both resonances of the detector by a monochromatic excitation applied to several positions on the building floor and on the external vacuum tank. We measured the acceleration of the external flanges, to which all the internal suspensions of the detector are hung, and we monitored the resulting oscillation amplitudes of the transducer output at steady state. The latter can be converted to the corresponding excitation at input of the antenna, since we calibrated by separate experiments the relationship between transducer output and thermal vibration of the antenna-transducer system. The resulting mechanical attenuation of the internal suspensions referred at antenna input is $\sim 245dB$, with an uncertainty of about $\pm 10dB$, including the repeatibility for different excitation positions and amplitudes. This performance should provide for sufficient vibration isolation of the present detector under normal ambient vibrational noise in the frequency bandwidth around resonances. Moreover, the overall attenuation is not far from the product of the attenuations of the single stages as measured at room temperature when the dilution refrigerator was not yet installed.

In all the measurements described the transducer output has been connected to the external test port. This allows an absolute calibration procedure for the noise of the detector. In fact, the two sharp and well separated mechanical resonances can be approximated by distinct oscillators close to each resonant frequency. From the test port, these oscillators can be modeled by two *parallel LCR* resonators, which can be fully determined by measuring the resonant frequencies $f_{0\pm}$, the quality factors $Q_\pm$ and the maxima of the real impedance at resonance $R_\pm = max\{Re[Z_\pm(f)]\}$. The Nyquist prediction of the thermal noise of each resonator gives a mean square value in potential of $\sigma_\pm^2 = 4\ K_B\ T\ R_\pm\ /\tau_\pm$, where $\tau_\pm$ is the decay time of the amplitude, T the thermodynamic temperature and K_B the Boltzmann constant. Typical values for our detector are $\tau_\pm \sim 10^3 s$ and $R_\pm \sim 10^6 - 10^7 \Omega$. In particular, since $R_\pm$ is proportional to $Q_\pm$ and to the square of the transducer bias field, we have chosen to operate with bias fields such that $R_\pm$ results in the region of lowest noise figure of the FET amplifier used to monitor the antenna noise.

The FET amplifier that we developed shows the following voltage and current bilateral noises : $V_n \sim 1.85\ nV/\sqrt{Hz}$ and $I_n \sim (0.8 \pm 0.4)\ fA/\sqrt{Hz}$. These correspond to a noise temperature $T_n \sim\ (0.10 \pm 0.05)\ K$ and to a noise resistance $R_n \sim\ (2.4 \pm 1.2)\ M\Omega$. While monitoring the antenna, the total noise in potential seen by the amplifier includes the additive contributions of V_n^2 and of $I_n^2\ \mid Z_\pm(f)\mid^2$, which are wideband and narrowband respectively. The input impedance of this amplifier does not affect the resonators since it was measured at about $1KHz$ to be $R_{in} = -12G\Omega$ in parallel with a $C_{in} \sim 30pF$; in particular, since $R_\pm \ll \mid R_{in} \mid$, the quality factor of the mechanical resonators are not affected by the presence of the FET amplifier.

The preamplified signal of the antenna is then demodulated at each resonance by means of lock-in with an averaging time $\tau_{li} = 100s$, close to optimum. The resulting mean square of a component of the lock-in output, σ_x^2, is determined from the record of the acquired data; in fact, $2\ \sigma_x^2$ is equal to the σ^2 of the distribution of the antenna mode energy, which is given by the sum of the squares of the two lock-in components. The data taking must be many hours long in order to get statistically significant results, because the energy correlation time is $\tau_\pm/2$.

Since the prediction for σ_x^2 is $\sigma_x^2 = 2K_B T_\pm R_\pm/(\tau_\pm + \tau_{li}) + \sigma^2_{ampli}$ where $\sigma^2_{ampli} = V_n^2 + I_n^2 R_\pm/(\tau_\pm + \tau_{li})$, we can determine the noise temperature of each mode from measured parameters:

$$T_\pm \;=\; \frac{\tau_\pm + \tau_{li}}{2K_B R_\pm}\,(\sigma_x^2 - \sigma^2_{ampli}). \tag{1}$$

The typical contribution of the amplifier in our experimental conditions has been $\sim 0.5\ K$ from the wideband and $\sim 0.1\ K$ from the narrowband noises respectively. Over a 31 day span while the antenna was kept at a thermodynamic temperature of $(6.5 \pm 0.5)\ K$, we could take data for 12.8 days. The acquisition was not continuous mostly because of the calibration measurements and cryogenic maintenance. The measured data give mean antenna noise temperatures $T_+ \sim 9.8\ K$ and $T_- \sim 8.1\ K$ for the + and - modes respectively, with a root mean square deviation of the data $\sim \pm 3\ K$. Therefore, we can conclude that the antenna noise was quite close to a thermal behaviour. We notice, however, that the fluctuations of the measured noise temperatures $T_\pm$ around the mean are greater than the contribution due to calibration accuracy, usually accounting for $\sim 5\%$ of $T_\pm$. These data have also been filtered with a zero order prediction filter, taking energy innovation over a time span equal to τ_{li}: the results are average effective temperatures $\sim 3.4\ K$ and $\sim 2.7\ K$ for the + and - modes respectively, with a r.m.s. deviation of $\sim \pm 2\ K$. The hystograms made with the counts of the energy levels follow the expected Boltzmann distribution, apart from a few events per day.

While running the antenna in the subsequent months at $2K$ by operating the 1 K pot of the dilution refrigerator, the mean noise temperature of the antenna over 1.5 days reached about 3 K. By applying the zero order prediction filter to these data we found effective noise temperature of $\sim 0.9\ K$ and $\sim 0.7\ K$ for the + and - modes respectively. However, in the last months we had problems with plugs in the 1 K pot refill line which caused the detector to loose the cryogenic working point several times.

References

1. M.Cerdonio *et al.* in *Proc. of the 1th E. Amaldi International Meeting on g.w. Experiments* (World Scientific, Singapore, 1995), p. 176.
2. E.Coccia *et al.* in *Proc. of the 1th E. Amaldi International Meeting on g.w. Experiments* (World Scientific, Singapore, 1995), p. 161.
3. see elsewhere in these Proceedings.

Detector Networks

BAR-INTERFEROMETER OBSERVING

K.A. COMPTON
Department of Physics and Astronomy
University of Wales College of Cardiff, Cardiff, U.K.

B.F. SCHUTZ
Department of Physics and Astronomy
University of Wales College of Cardiff, Cardiff, U.K.
and
Max Planck Institute for Gravitational Physics
The Albert Einstein Institute
Potsdam, Germany

We examine the opportunities for joint observations between interferometric and bar (more generally, solid-mass) gravitational wave detectors in the near and more distant future. We give simple formulas to estimate the sensitivity of joint searches, and we present results of more detailed calculations for certain combinations. Bars and interferometers can do searches for pulsars with competitive sensitivity, and can therefore confirm each other's observations. Cross-correlation of two detectors permits searches for a stochastic background. Restricted to higher frequencies, bar-interferometer pairs are less sensitive than two interferometers for stochastic backgrounds of constant energy density (the Harrison-Zel'dovich spectrum), but they are competitive for backgrounds that have constant spectral density, as are predicted by recent calculations in superstring cosmologies. Particularly interesting are the NAUTILUS detector with the VIRGO or GEO600 interferometers, and even more sensitive would be a spherical solid-mass detector built at the site of VIRGO, LIGO or GEO.

1 Introduction

Gravitational wave observations require joint observing by a network of detectors, both to increase confidence in a detection and to provide information on the direction and polarisation of the waves. There have already been coordinated observations using networks of bar detectors [1] and interferometers [2], but coordinated observing between bars and interferometers has not been discussed much until recently [3,4,5]. Given the difficulties inherent in gravitational radiation detection, it is worthwhile exploring all realistic possibilities that involve all kinds of detectors, and especially considering these questions when decisions are made about locating new detectors.

In this paper we survey what seem to us to be worthwhile possibilities for joint observations between bars and interferometers in the near future and further away, for the three main classes of gravitational wave signals: short bursts (including binary coalescences), pulsars and other continuous sources, and a stochastic background of gravitational waves from the Big Bang. In the latter case we give simple ways to estimate the sensitivity achievable by any pair of detectors, and we summarise the results of detailed computations performed elsewhere [6] for certain promising pairs.

1.1 Detectors

We will consider a number of detectors. In many cases, they are taken simply as examples of a class of similar detectors. Among the bars we consider

- NAUTILUS, a milliKelvin bar at Frascati [7]. When we refer to this bar, our remarks can usually refer equally well to other examples, such as AURIGA [8]. These bars will operate at good sensitivity in the very near future, and can be improved even more over the next 5 years.
- TIGA, the icosahedral design for an omni-directional detector [9]. Again, our remarks would apply to any of the new class of spherical solid-mass detectors. Such detectors are not likely to operate until after the next 5 years.

Interferometers include

- First generation interferometers: GEO600 [10], LIGO I [11], and VIRGO [12]. These differ in important details, but all should reach a sensitivity for bursts of near 10^{-21} by about the year 2000. They are of course not the first generation of interferometers to have been built, since prototypes have operated at Glasgow [13], Garching [14], Caltech [15], and elsewhere for many years, and have even conducted joint observations [2]. But they are the first generation to be capable of reaching astrophysically interesting sensitivities.
- Second generation interferometers. Sometimes called "advanced detectors", or in the American context LIGO II, they can be designed on paper but are not yet funded [5]. GEO600 and TAMA300 [16] are expected to play important roles in developing some of the techniques needed to make them possible, but are probably too short to reach second-generation sensitivity themselves.
- Narrow-banded detectors. By using resonant optical techniques, such as signal recycling [17], it will be possible to improve the sensitivity of interferometers in selected bandwidths, at the expense of their broad-band sensitivity. This could be desirable when working with bars. Narrow-banding is anticipated for GEO600.

Among these instruments are a number of natural bar-interferometer pairs, geographically near one another. The LIGO detector in Louisiana is near LSU, where a TIGA bar might be built. The NAUTILUS and AURIGA bars are near to both GEO600 and especially VIRGO. We will look carefully at what these pairs can do together.

1.2 Types of observing

Gravitational wave signals may be divided into three classes: bursts, which are short enough that the motion of the detector is not a consideration; continuous waves, which can be detected with maximum sensitivity only by correcting for the motion of the detector as the Earth turns and orbits the Sun; and a stochastic background, random gravitational waves left over from the early Universe.

Bursts seem to offer only limited opportunities for cooperation among bars and interferometers in the near future. Unstructured bursts, such as one might expect from a supernova explosion, might be much stronger in one detector's band than in the other's. If the two detectors have rather different burst sensitivity, then the weaker one also does not add much to the confidence of the detection. Coalescing binaries are more promising, because their signal rises in frequency in a predictable way from the natural bandwidth of interferometers to the natural operating frequencies of bars. It has been observed in many places that bars might be used to discover important information about the last stages of a coalescence that had been identified by interferometers [5]. This matter will be discussed elsewhere at this meeting (the talk by Coccia), so we will not consider it further in this paper.

Pulsars and other continuous sources offer better prospects for joint observing. We will see that bars and interferometers can have similar sensitivities in their common bandwidths, and joint observing can add confidence to a detection. There are many interesting kinds of sources. Searches should target not only known pulsars but also X-ray binaries and Be-giant stars [18]. All-sky, all-frequency searches for unknown sources should also be undertaken, to a sensitivity limit set by available computer resources,

A **stochastic background** can only be found by cross-correlation of two detectors, and here bars and interferometers work surprisingly well together [3,4]. We will calculate just how well for a number of cases.

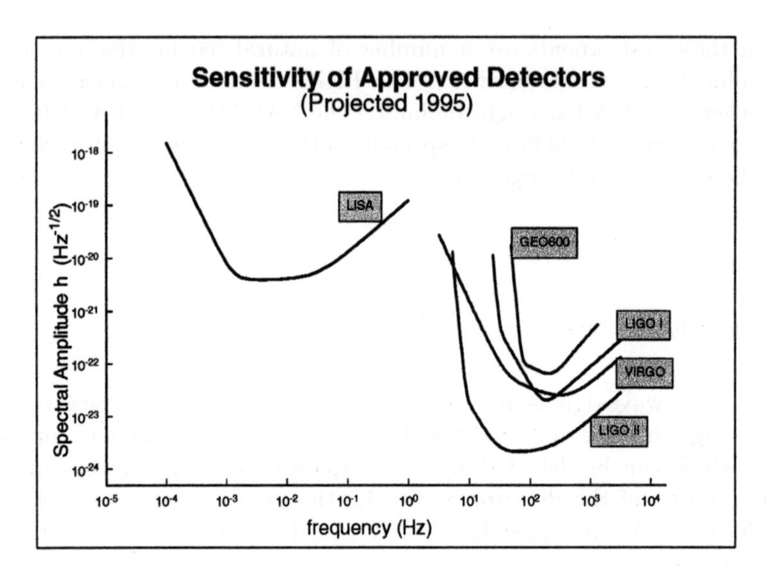

Figure 1: The expected sensitivity of several approved interferometers. For us the most interesting part is the sensitivity of the ground-based instruments above 100 Hz.

2 Approximate Characterisation of the Sensitivity of Different Detectors

For the rough estimates of working sensitivity in what follows, we need approximate characterisations of various detectors. We will use these only to generate our rough estimates. The more accurate calculations of Compton [6] are done with accurate sensitivity curves.

Bar detectors can be roughly approximated as having a square spectral noise density $S_h(f)$, with a constant value $S_h(f_b)$ in a bandwidth B_b about a central frequency f_b. (Many bars are instrumented to output data at two or more frequencies. We ignore this.) For our two example bars we have

Example	f_b	B_b	$S_h(f_b)$
NAUTILUS	900 Hz	2 Hz	$8 \times 10^{-45}\,\mathrm{Hz}^{-1}$
TIGA	1 kHz	1 Hz	$1.6 \times 10^{-47}\,\mathrm{Hz}^{-1}$ per mode

Both of these bars may operate with much larger bandwidths at these sensitivities, perhaps up to 100 Hz. This particularly affects correlation searches.

Interferometers have a more complicated spectrum. Figure 1 shows the spectral noise density of several detectors. For observing above about 200 Hz, which is mainly the region of interest for joint observing with bars, the interferometers may be characterised by a "knee" frequency f_k, where they have their best sensitivity, and by a rising shot-noise curve above this in which S_h is proportional to $(f/f_k)^2$. In fact, underlying this shot noise is a thermal noise from the vibrations of the test masses on which the mirrors are mounted. The interferometer's sensitivity is usually optimised so that this noise roughly equals the shot noise at f_k, and then it falls as $(f/f_k)^{-1}$. This is important, because when a detector is narrow-banded, the thermal noise sets the minimum on the achievable S_h at any frequency. Examples of the thermal noise expected in some detectors are shown as the dotted lines in Figure 2.

From Figure 1 we find the following simple characterisations of our example interferometers:

Example	f_k	$S_h(f_k)$
GEO600	200 Hz	$3 \times 10^{-45}\,\mathrm{Hz}^{-1}$
LIGO I	200 Hz	$4 \times 10^{-46}\,\mathrm{Hz}^{-1}$
VIRGO	200 Hz	$1 \times 10^{-46}\,\mathrm{Hz}^{-1}$
Advanced	100 Hz	$8 \times 10^{-48}\,\mathrm{Hz}^{-1}$

$S_{\mathrm{shot}} = S(f_k)(f/f_k)^2$, $S_{\mathrm{th}} = S(f_k)(f/f_k)^{-1}$

3 Looking for Neutron Stars

Provided that one correctly removes Doppler and amplitude modulation produced by the motion of the detector, as well as other effects such as any measurable proper motion of the source or its change of frequency during the observation, then the sensitivity limit on a continuous source during a time T_{obs} is

$$h_{PSR}^{1\sigma} = \left(\frac{S_h}{T_{obs}}\right)^{1/2} . \tag{1}$$

Targets of joint searches between bars and interferometers must be at a frequency near 1 kHz. Let us suppose that T_{obs} is about 1 yr. Then the sensitivity limits that our previous approximations for detector sensitivity give are

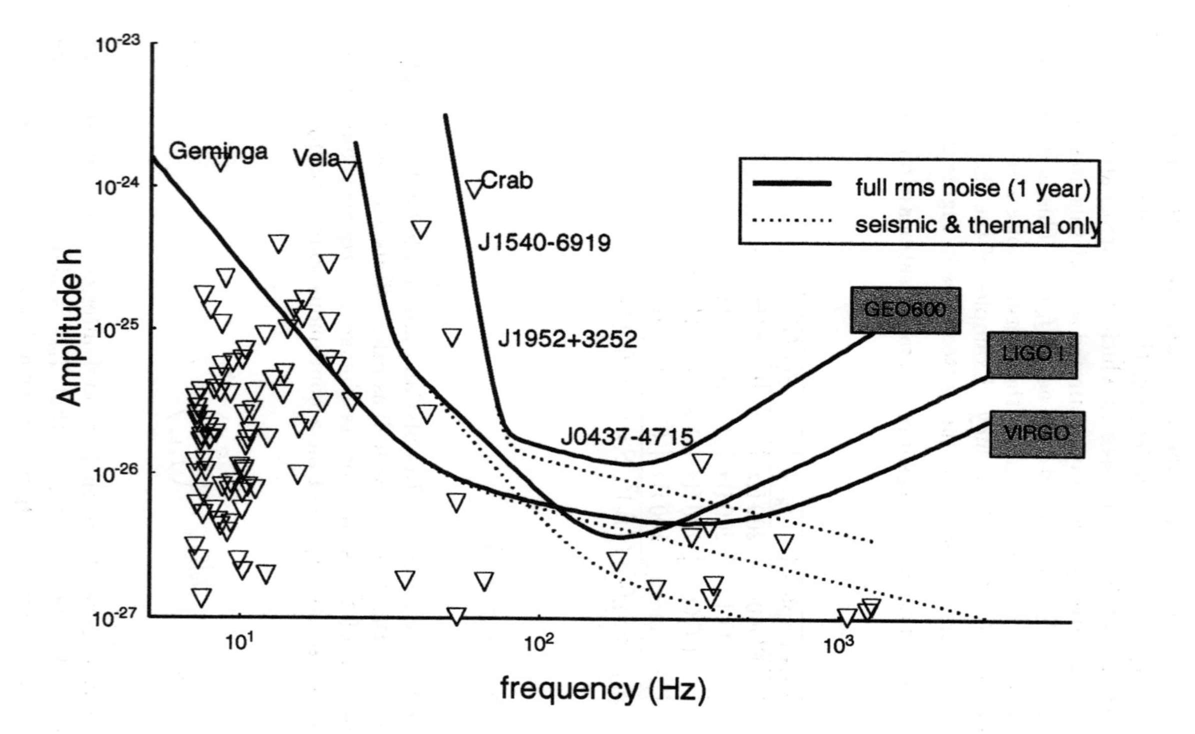

Figure 2: Upper limits on gravitational wave amplitudes from known pulsars, set by assuming all their spindown can be accounted for by the emission of gravitational wave energy. The points represent all pulsars in the November 1995 version of the database of Taylor *et al* [19], with gravitational wave frequencies above 7 Hz and amplitudes above 10^{-27}. Also shown are the expected sensitivities of three first-generation interferometers in a one-year observation, and the thermal noise limits on narrow-banding (dotted lines).

Detector	$h_{PSR}^{1\sigma}$ at 1 kHz for 1 yr
NAUTILUS	2×10^{-26}
TIGA	7×10^{-28}
GEO600 broad[a]	5×10^{-26}
GEO600 narrow[b]	5×10^{-27}
LIGO I broad	2×10^{-26}
(LIGO I narrow	2×10^{-27})
VIRGO broad	1×10^{-26}
(VIRGO narrow	1×10^{-27})
Advanced broad	3×10^{-27}
Advanced narrow	3×10^{-28}

We have placed the sensitivities for LIGO I and VIRGO operating in narrow-band modes in brackets because present plans for these detectors do not appear to envision implementing narrow-banding.

By comparing this table with Figure 2, we see that there are several pulsars whose spindown limits are above these sensitivities, and even a few that are in the right frequency range. The advanced bar detector TIGA has nearly as good a sensitivity as an advanced interferometer near 1 kHz. And NAUTILUS has respectable sensitivity, comparable with that of first-generation interferometers at this frequency. The NAUTILUS group does indeed plan to search for pulsars (see the talk by Astone at this meeting).

3.1 Narrow-banding interferometers as a strategy for pulsar searches?

A simple argument suggests that interferometers might do better at searching for unknown pulsars by narrow-banding. Narrowing the interferometer to a bandwidth B reduces the number of pulsars available to the instrument, to a number proportional to B if the distribution of pulsars in frequency is uniform (a strong and probably wrong assumption). But it also reduces the spectral noise density in this bandwidth by a factor proportional to B. This increases the range of the detector by $B^{-1/2}$. If pulsars are distributed in a plane, then the increase in range increases their numbers by B^{-1}, and the net effect of narrow-banding is neutral. But if pulsars are distributed more spherically, their number will be proportional to $B^{-3/2}$, and narrow-banding improves the number of detections in proportion to $B^{-1/2}$.

However, this argument is naive, and if we take into account other factors then we can reach the opposite conclusion. One difficulty is that narrow-banding must be done on the rising part of the shot-noise curve, since it is limited below by thermal noise. In order to make B smaller, one must go higher in f, and therefore in the base level of S_h. When this factor is taken into account, the number of pulsars available (if they are distributed spherically) is proportional to $B^{1/4}$, so it is a disadvantage to narrow-band. This assumes a uniform distribution in frequency, which is also almost certainly bad: as we push up in f, the density of pulsars is likely to decrease.

Finally, we mention a factor that should also be considered, but which has not yet been adequately studied. The range of a search for unknown pulsars is likely to be limited by computer power rather than observing opportunities: the computer power required to search a data set taken over a length of time T is proportional to [20] T^4. The range of the search, which is proportional to h^{-1}, goes as $T^{-1/2}$. But it is also proportional to bandwidth, since the computing requirements depend directly on the number of data points. So narrowing the bandwidth by a factor B allows the length of the data set to be increased by a factor of $B^{-1/4}$. The range increases, and the number of detections goes up by $B^{-3/8}$ if pulsars are distributed spherically. This wins over the factor of $B^{1/4}$ in the previous paragraph, so that the net effect is slightly favourable to narrow-banding, but not by much. This slight advantage might be overwhelmed by an unfavourable frequency distribution of pulsars.

4 Stochastic Searches Using Bars and Interferometers

According to Flanagan [21] and Compton [6], the 90% confidence limit on the energy density per unit logarithmic frequency that any pair of detectors located at the same site can set at any frequency is

$$\Omega_{gw}^{90\%} = \frac{8\pi}{G\rho_c} f^3 \left[\frac{S_1 S_2}{2TB}\right]^{1/2}, \tag{2}$$

where $\rho_c = 3H_0^2/8\pi G$ is the closure mass density S_1 and S_2 are spectral densities at the two detectors, T is the observing time, and B is the bandwidth. Note the f^3 factor, which reduces the effectiveness of high-frequency searches for backgrounds with constant Ω_{gw} (the so-called Harrison-Zel'dovich spectrum).

Putting in typical numbers, and using $H_0 = 75\,\mathrm{km\,s^{-1}\,Mpc^{-1}}$, we find

$$\begin{aligned} \Omega_{gw}^{90\%} &= 4\times 10^{-4}\left(\frac{f}{1\,\mathrm{kHz}}\right)^3 \left(\frac{S_1}{10^{-46}\,\mathrm{Hz}^{-1}}\right)^{1/2} \left(\frac{S_2}{10^{-46}\,\mathrm{Hz}^{-1}}\right)^{1/2} \\ &\times \left(\frac{T}{1\,\mathrm{yr}}\right)^{-1/2} \left(\frac{B}{1\,\mathrm{Hz}}\right)^{-1/2} . \end{aligned} \tag{3}$$

If the two detectors are not on the same site, then one divides by $|\gamma|$ as defined by Flanagan [21]. We consider this below.

We specialise this formula to the kinds of detectors whose sensitivity we estimated above:

- Two broadband interferometers:

$$\Omega_{gw}^{90\%} = 2 \times 10^{-7} \left(\frac{S_{k,1}}{10^{-46}\,\mathrm{Hz}^{-1}} \right)^{1/2} \left(\frac{S_{k,2}}{10^{-46}\,\mathrm{Hz}^{-1}} \right)^{1/2}, \tag{4}$$

with $f_k = 200\,\mathrm{Hz}$ and $T = 1\,\mathrm{yr}$.

- One bar and one broadband interferometer:

$$\Omega_{gw}^{90\%} = 2 \times 10^{-3} \left(\frac{S_b}{10^{-46}\,\mathrm{Hz}^{-1}} \right)^{1/2} \left(\frac{S_k}{10^{-46}\,\mathrm{Hz}^{-1}} \right)^{1/2}, \tag{5}$$

with $f_b = 1\,\mathrm{kHz}$, $B = 1\,\mathrm{Hz}$, $f_k = 200\,\mathrm{Hz}$ and $T = 1\,\mathrm{yr}$.

- Two bars:

$$\Omega_{gw}^{90\%} = 4 \times 10^{-4} \left(\frac{S_{b,1}}{10^{-46}\,\mathrm{Hz}^{-1}} \right)^{1/2} \left(\frac{S_{b,2}}{10^{-46}\,\mathrm{Hz}^{-1}} \right)^{1/2}, \tag{6}$$

with $f_b = 1\,\mathrm{kHz}$, $B = 1\,\mathrm{Hz}$, and $T = 1\,\mathrm{yr}$.

- One bar and one narrow-banded interferometer:

$$\Omega_{gw}^{90\%} = 2 \times 10^{-4} \left(\frac{S_b}{10^{-46}\,\mathrm{Hz}^{-1}} \right)^{1/2} \left(\frac{S_k}{10^{-46}\,\mathrm{Hz}^{-1}} \right)^{1/2}, \tag{7}$$

with $f_b = 1\,\mathrm{kHz}$, $B = 1\,\mathrm{Hz}$, $f_k = 200\,\mathrm{Hz}$ and $T = 1\,\mathrm{yr}$; note that S_{int} is limited by thermal noise.

- Single-detector noise limit (set just by its internal noise):

$$\Omega_{gw}^{90\%} = 2 \left(\frac{f}{1\,\mathrm{kHz}} \right)^{3} \left(\frac{S_h}{10^{-46}\,\mathrm{Hz}^{-1}} \right)^{1/2}. \tag{8}$$

These numbers can be translated into tables of values of sensitivity in terms of Ω_{gw} and in terms of spectral noise density of gravitational radiation,

$$S_{gw} = \left[\frac{S_1 S_2}{2TB} \right]^{1/2}.$$

These are given below.

Ideal (Same-Site) Limits on $\Omega_{gw}^{90\%}$

	N'LUS	TIGA	GEO600	LIGO I	VIRGO
N'LUS	3×10^{-2}	*	*	*	*
TIGA	1.4×10^{-3}	6×10^{-5}	*	*	*
GEO600	1×10^{-1}	4×10^{-3}	$[6 \times 10^{-6}]$	*	*
LIGO I	4×10^{-2}	2×10^{-3}	$[2 \times 10^{-6}]$	$[8 \times 10^{-7}]$	*
VIRGO	2×10^{-2}	1×10^{-3}	$[1 \times 10^{-6}]$	$[4 \times 10^{-7}]$	$[2 \times 10^{-7}]$
GEO nb	1×10^{-2}	4×10^{-4}	N/A	N/A	N/A

In this table, a * denotes entries that can be obtained from the symmetry of the table, "N'LUS" denotes the NAUTILUS detector, "nb" means "narrow-band", and square brackets [...] denote experiments that are not possible to perform on the same site. (Note, however, that at LIGO's Hanford site it will be possible to do a same-site experiment between a full-length and a half-length interferometer.) The TIGA results are per mode, so modes can be combined to make some improvements.

Some of these combinations were studied by Compton [6] in detail, allowing for more realistic instrumental sensitivities and, most importantly, for their geometrical separation and orientation. She found the following results for the limits on energy density:

Realistic Limits on $\Omega_{gw}^{90\%}$

Detector Pair	Realistic $\Omega_{gw}^{90\%}$, allowing for geometry
LIGO I — LIGO I	5×10^{-6}
GEO600 — N'LUS	4×10^{-2}, same site
GEO600 — N'LUS	8×10^{-1}, present locations
GEO600 (nb) — N'LUS	6×10^{-3}, same site
TIGA-TIGA	8×10^{-6}, same site, 5 modes

This table of estimates above can be converted to read spectral density limits:

Ideal (Same-Site) Limits on $|h_{25}| = \left[S_{gw}^{90\%}\right]^{1/2} / 10^{-25}\,\mathrm{Hz}^{-1/2}$

	N'LUS	TIGA	GEO600	LIGO I	VIRGO
N'LUS	9	*	*	*	*
TIGA	2	0.5	*	*	*
GEO600	7	2	[2]	*	*
LIGO I	4	1	[1]	[0.6]	*
VIRGO	3	1	[0.7]	[0.4]	[0.3]
GEO nb	2	0.5	N/A	N/A	N/A

There are a number of interesting conclusions that one can draw from these tables. For example, there is a substantial improvement in a possible GEO-NAUTILUS experiment if a NAUTILUS-type bar were built on the GEO site, and if GEO were run in narrow-band mode with it. A TIGA bar on a interferometer site, such as at the LIGO site in Louisiana, would be a little better than the GEO-NAUTILUS same-site combination, but among first-generation interferometers the best combination with a bar is a narrow-band GEO600 with a TIGA on the same site. These cannot beat, however, a same-site LIGO I experiment, using a half-length interferometer, which could do more than 2 orders of magnitude better than GEO-TIGA on Ω_{gw}, although it would be almost the same as GEO-TIGA in terms of spectral density. (The same-site experiments must be careful that they are not affected by common-mode environmental noise in the two detectors.)

When we look at the table of spectral density limits, the bars seem much better relative to the interferometers than they were in the energy-density comparison. This is because of the factor of f^3 in the energy density, which favours low-frequency observing. The best spectral limits will be set by a pair of TIGA's, but a narrow-band GEO would not be a bad companion for a TIGA.

Acknowledgements

We have been greatly helped in this work by our colleagues at Cardiff, particularly David Nicholson, John Watkins, Gareth Jones, and Chris Dickson. We also acknowledge important conversations with Pia Astone, Alberto Lobo, and James Hough.

References

1. Amaldi, E, Bonfazi, P, Pallottino, G V, Pizzella, G, "Coincidence Analysis among three resonant cryogenic detectors", in Michelson, P.F., Hu, En-Ke, Pizzella, G., eds., *Experimental Gravitational Physics*, (World Scientific, Singapore, 1991), p. 394–396.
2. D. Nicholson, C.A. Dickson, W.J. Watkins, B.F. Schutz, J. Shuttleworth, G.S. Jones, D.I. Robertson, N.L. Mackenzie, K.A. Strain, B.J. Meers, G.P. Newton, H. Ward, C.A. Cantley, N.A. Robertson, J. Hough, K. Danzmann, T.M. Niebauer, A. Rüdiger, R. Schilling, L. Schnupp, W. Winkler, "Results of the First Coincident Observations by Two Laser-Interferometric Gravitational Wave Detectors", *Phys. Lett. A*, accepted for publication (1996).
3. Astone, P., Lobo, J.A., and Schutz, B.F., "Coincidence Experiments Between Interferometric and Resonant Bar Detectors of Gravitational Waves", *Class. Quant. Grav.*, **11**, 2093–2112 (1994).

4. B.F. Schutz, "Sources of Gravitational Radiation for Detectors of the 21st Century", *in* Coccia, E., Pizzella, G., Ronga, F., eds., *Gravitational Wave Experiments*, (World Scientific, Singapore, 1995).
5. Thorne, K. S., *in Proceedings of Snowmass 1994 Summer Study on Particle and Nuclear Astrophysics and Cosmology*, Kolb, E. W. and Peccei, R., eds., (World Scientific, Singapore, 1995).
6. Compton, K.A., *Cross-Correlation Techniques for the Detection of Gravitational Radiation* (PhD Thesis, University of Wales Cardiff, 1996).
7. Coccia, E., "The NAUTILUS Experiment", in Coccia, E., Pizzella, G., Ronga, F., eds., *Gravitational Wave Experiments*, (World Scientific, Singapore, 1995), p. 161–175.
8. Cerdonio, M., "Status of the AURIGA Gravitational Wave Antenna and Perspectives for the Gravitational Waves Search with Ultracryogenic Resonant Detectors", in Coccia, E., Pizzella, G., Ronga, F., eds., *Gravitational Wave Experiments*, (World Scientific, Singapore, 1995), p. 176–194.
9. Coccia, E, Fafone, V., Frossati, G., "On the Design of Ultralow Temperature Spherical Gravitational Wave Detectors", in Coccia, E., Pizzella, G., Ronga, F., eds., *Gravitational Wave Experiments*, (World Scientific, Singapore, 1995), 463–478.
10. Danzmann, K., "GEO 600 – A 600-m Laser Interferometric Gravitational Wave Antenna", in Coccia, E., Pizzella, G., Ronga, F., eds., *Gravitational Wave Experiments*, (World Scientific, Singapore, 1995), p. 100–111.
11. Raab, F.J., "The LIGO Project: Progress and Prospects", in Coccia, E., Pizzella, G., Ronga, F., eds., *Gravitational Wave Experiments*, (World Scientific, Singapore, 1995), 70–85.
12. Giazotto, A., "The VIRGO Experiment: Status of the Art", in Coccia, E., Pizzella, G., Ronga, F., eds., *Gravitational Wave Experiments*, (World Scientific, Singapore, 1995), p. 86–99.
13. Ward, H, Hough, J, Kerr, G A, Mackenzie, N L, Mangan, J B, Meers, B J, Newton, G P, Robertson, D I, and Robertson, N A, "The Glasgow gravitational wave detector - present progress and future plans", in Michelson, P.F., Hu, En-Ke, Pizzella, G., eds., *Experimental Gravitational Physics*, (World Scientific, Singapore, 1991), p. 322–327.
14. Maischberger, K, Ruediger, A, Schilling, R, Schnupp, L, Winkler, W, and Leuchs, G, "Status of the Garching 30 meter prototype for a large gravitational wave detector", in Michelson, P.F., Hu, En-Ke, Pizzella, G., eds., *Experimental Gravitational Physics*, (World Scientific, Singapore, 1991), p. 316–321.

15. Abramovici, A., Althouse, W.E., Drever, R.W.P., Gursel, Y., Kawamura, S., Raab, F.J., Shoemaker, D., Sievers, L., Spero, R.E., Thorne, K.S., Vogt, R.E., Weiss, R., Whitcomb, S.E., and Zucker, M.E., "LIGO: The Laser Interferometer Gravitational-Wave Observatory", *Science*, **256**, p. 325–333 (1992).
16. Tsubono, K, "300-m Laser Interferometer Gravitational Wave Detector (TAMA300) in Japan", in Coccia, E., Pizzella, G., Ronga, F., eds., *Gravitational Wave Experiments*, (World Scientific, Singapore, 1995), p. 112–114.
17. Meers, B.J., "Recycling in laser-interferometric gravitational-wave detectors", *Phys. Rev. D*, **38**, p. 2317–2326 (1988).
18. Schutz, B.F., "The Detection of Gravitational Waves", in Marck, J.-A., Lasota, J.-P., eds., *Astrophysical Sources of Gravitational Radiation*, (Springer, Paris, 1996).
19. This is the updated electronic version of the database published in Taylor, J.H., Manchester, R.N., Lyne, A.G., *Astrophys. J. Suppl.* **88**, 529 (1993).
20. Schutz, B.F., "Data Processing Analysis and Storage for Interferometric Antennas", in Blair, D.G., ed., *The Detection of Gravitational Waves*, (Cambridge University Press, Cambridge England, 1991), p. 406–452.
21. Flanagan, E.E., *Phys. Rev.*, **D48**, 2389 (1993).

ARRAY OF DETECTORS

S. FRASCA
*Università di Roma "La Sapienza", Dipartimento di Fisica,
Rome, Italy*

Gravitational detectors that are now operating or in project are not enough sensitive in the frequency range above 1 or 2 kHz, that is predicted by theoreticians as very interesting. In this paper a project of a local array of stumpy cylinder resonant detectors is presented; it solves the problem and has many other interesting features.

1 Gravitational detectors and frequency

Since the early works on the astrophysical sources of gravitational waves, theoreticians have pointed out that the frequency range above $2 \simeq 3$ kHz up to 10 kHz is the most likely for the most effective impulsive sources. In fact the most effective quasi-normal mode of a black hole of mass M_{bh}, the lower quadrupolar, is at[1]

$$\nu_{\ell=2} \simeq 0.37 \cdot \frac{c^3}{2\pi G M_{bh}} \simeq 12 \frac{M_\odot}{M_{bh}} \; kHz, \tag{1}$$

and the expected values for the mass of more likely newborn black holes is $2 \simeq 3$ solar masses. Also in the case of neutron stars there are very effective processes in this frequency range. As examples, there is the non-axisymmetric instability in a rotating star [2], that has an efficiency of about 0.001 and a frequency of about 4.3 kHz, the final phase of the binary coalescence [3,4,5,6] (efficiency above 0.01 and frequency about $2.5 \simeq 6$ kHz) and the quasi-normal modes of the neutron stars [7] (frequency range over 1 kHz up to 10 kHz).

On the other hand, a general problem in designing gravitational antennas is that, in order to achieve high sensitivities, the size of the detectors must be very large. But, normally, bigger is the size of the device, lower is the frequency of best operation. In fact the maximum sensitivity of the proposed laser interferometers is under 1 kHz and the frequency of the proposed big resonant spheres is at about 900 Hz.

Namely, in the case of the resonant detectors, the energy flux sensitivity is proportional to the cross-section of the detector and this is proportional to the mass of the detector; on the other hand the frequency of the detector, for a given shape, is inversely proportional to its linear dimension and so to the cube root of the mass. So, for a given shape and a given material, the cross-section is inversely proportional to the cube of the frequency.

A solution to this dilemma is to build local arrays of small resonant antennas [8,9]; it is easy to see that if we sum the outputs of N equal antennas, placed in the same site, because the noises are incoherent and then are summed quadratically and the signals are coherent and then are summed linearly, the (quadratic) signal-to-nose ratio is N times larger than that of a single element antenna. Then an array of small antennas can be "matched" to the desired frequency and to the desired sensitivity, choosing the size and the number of elements.

In this paper I will discuss what is a good choice for the shape of the base element of the array, how the outputs can be combined to optimally detect the gravitational wave pulses and measure their characteristics and finally I will show some considerable advantages in this type of detector due to the small size and the high number of the element antennas.

2 The stumpy cylinder

The sensitivity of a resonant antenna is expressed normally by two numbers: the cross-section and the effective temperature. The last depends on the effectiveness of the transducer, the Q of the antenna, the thermodinamical temperature, the noise of the preamplifier and the filtering technique; it will not be discussed in this paper and when we do comparisons between different antennas, we assume that the effective temperatures are the same.

The cross-section is the ratio between the energy captured by the antenna and the incident spectral energy flux. It, for a given antenna, varies with the direction of the source and the polarization; we will consider the mean value $\overline{\Sigma}$ on all directions and polarizations. Because for a given geometry of the antenna and a given mode, the cross section is proportional to the mass M and the square of the sound velocity v of the material (the type of sound velocity depends on the mode), it was defined the *reduced cross-section* $\overline{\Sigma}_0$ as the cross-section divided by $M \cdot v^2$, and it depends only on the shape and the mode and not on the size and the material.

The shape is an important feature of a resonant detector. It should ensure

- high cross-section
- high Q factor
- practical feasibility.

For these reasons until now almost all resonant antennas have been cylinders. In the case of the most recent and sensitive antennas, Explorer (at CERN), Allegro (at LSU), Nautilus (in Frascati) and Auriga (in Legnaro) the cylinder has a length of 3 m and a diameter of 60 cm. Then we will call a cylinder of this size a "standard bar". In this discussion we will use the "shape factor" k_{shape}, depending on the shape and not the size of the antenna, that gives the enhancement in volume respect to the standard bar shape, for the same linear dimension (and then the same frequency).

The spherical shape was proposed by Wagoner and Paik [10] and the truncated icosahedron variant by Johnson and Merkowitz [11] for two reasons:

- it has a shape factor $k_{sphere} = 16.6$, so, in the case of a diameter of 3 m, which gives about the same operative frequency of the standard bar, it will have a mass 16.6 times bigger than a standard bar

- we can use the five degenerate modes for the detection (labelled with the index m from -2 to 2) and this gives two advantages: the cross-section integrated on all directions is 5 times larger and the data taken from five or six transducers in various places on the surface can be combined to obtain not only the amplitude of the incoming wave, but also the direction and the polarization.

In other words in the case of a sphere, as it is clear from its symmetry, there is the maximum mass for given linear dimension and no preferred direction of detection.

A problem is that the sphere is not too practical.

Let us consider a "stumpy" cylinder, i.e. a cylinder with the length equal to the diameter (this is normally the metal shape from which a sphere is cut). It has 1.5 times the mass of the inscribed sphere ($k_{stumpy} = 25$) and has five quadrupolar modes good for the detection of gravitational waves, that we label with the index m_k ($-2 \leq k \leq 2$), describing the type of symmetry.

m_0 is the "classical" longitudinal mode used in standard bars. To detect optimally this mode we need a transducer like a capacitive "mushroom" of Nautilus, on one of the two faces of the stumpy cylinder.

m_{-2} and m_2 are a couple of degenerate modes called "discoidal modes". They were studied in the past [12,13] and each of them has about the same reduced cross-section of the longitudinal mode and of each of the five spherical modes[10]. To detect them optimally we need a couple of transducers on the curved surface, at 45° each other.

m_{-1} and m_1 are a couple of degenerate modes and, because of their shape, can be called "pantograph" or "diamond" modes. To detect these modes the two detectors can be put on one edge of the cylinder, 90° each other.

From preliminary computations we found that the reduced cross-section of these modes is practically the same of the others (longitudinal, discoidal and spherical) and, with a finite element analysis program, we found that these modes are about at the same frequency. The results, for an alluminum stumpy cylinder with a length of 60 cm, at liquid helium temperature, gave the frequencies 4.2 kHz for m_0, 1 % less for the discoidal modes and 6 % less for the pantograph modes.

To optimize the cross-section of a resonant detector, an important role is played by the choice of the material. If the operative frequency of the detector is fixed, it is easy to show that, with good approximation, the expression that must be maximized is $Y^{2.5}/\rho^{1.5}$, where Y is the Young modulus and ρ is the density.

A list of materials is in table 1. The big detectors like the standard bars or the big spheres have been built or projected in alluminum because it is the most practical material for these big devices. A $4 \simeq 5$ kHz stumpy cylinder can be made easily and with a reasonable cost in molybdenum, with a gain in cross-section of more than 6. For higher frequencies (and then smaller antennas) the sapphire can be considered.

In general the cross-section for an array of N antennas, with shape factor k and with n observed modes (valid also for the case of a single antenna) is roughly

$$\Sigma \propto \frac{Y^{2.5}}{\rho^{1.5}} \cdot \frac{Nnk}{\nu^3} \tag{2}$$

Figure 1 shows the projected sensitivity for Virgo, the 3 meter alluminum sphere at 900 Hz and an array of 100 molybdenum stumpy cylinders at 4.5 kHz and with the same $T_{eff} = 10^{-6}K$ of the sphere. Note that in this computation, also if the value of the strain noise density is only a factor 2 lower, the bandwidth is five times larger than that of the sphere.

3 Detection with a local array

Let us discuss the data analysis procedures of a local array. Here we suppose that all the modes of all the antennas have the same detection band. If it is not so, the procedure becomes less simple and the sensitivity decreases (reasonably of about a factor 2 in energy).

The output of each of the $N \cdot n$ transducers is modelled by b_m in the following equation

$$\ddot{b}_m + \beta_m \dot{b}_m + \omega_m^2 b_m = D_m^{ij} \ddot{h}_{ij} = g_m \tag{3}$$

where β_m and ω_m are the dissipation coefficient and the angular frequency of the mode, h_{ij} is the metric tensor perturbation - i.e. the gravitational wave - and D_m^{ij} is a tensor describing the antenna mode and it is related to its quadrupole moment. From the observations $\mathbf{b}$ the estimated excitations $\hat{\mathbf{g}}$ can be obtained, by means of classical optimum filters.

The $\hat{g}_m$ can be combined linearly in many ways, in order to reach the sensitivity given by eq. 2 and to estimate the elements of the tensor of the wave. For this last purpose, it is convenient to define an "esa-vector" $\mathbf{w}$ in the band of the antenna

$$w_1 = \ddot{h}_{11} \quad w_2 = \ddot{h}_{22} \quad w_3 = \ddot{h}_{33} \quad w_4 = \ddot{h}_{12} \quad w_5 = \ddot{h}_{13} \quad w_6 = \ddot{h}_{23} \tag{4}$$

and a matrix $\mathcal{A}$ (having $j = 1, ..., N$ rows and 6 columns)

$$\mathcal{A}_{j1} = D_{11}^{(j)} \;\; \mathcal{A}_{j2} = D_{22}^{(j)} \;\; \mathcal{A}_{j3} = D_{33}^{(j)} \;\; \mathcal{A}_{j4} = 2D_{12}^{(j)} \;\; \mathcal{A}_{j5} = 2D_{13}^{(j)} \;\; \mathcal{A}_{j6} = 2D_{23}^{(j)} \tag{5}$$

such that

$$\mathcal{A}\mathbf{w} = \hat{\mathbf{g}} + \mathbf{e} \tag{6}$$

where $\mathbf{e}$ is the error in the estimate of $\mathbf{g}$. If $\mathbf{e}$ is gaussian, this equation admits an optimal solution if the rank of A is six:

$$\hat{\mathbf{w}} = (\mathcal{A}^T\mathcal{A})^{-1}\mathcal{A}^T\hat{\mathbf{g}} \tag{7}$$

and the error is given by the covariance matrix

$$V_w = \sigma^2(\mathcal{A}^T\mathcal{A})^{-1} \tag{8}$$

where σ^2 is the variance of the errors on the estimates of the driving excitation $\mathbf{g}$. This least mean square procedure can be used to reject very effectively signals that have not the pattern structure of a gravitational wave (recall that there are only four degrees of freedom and the observatons are $n \cdot N$). These signals are due normally to local disturbances and limit severely the sensitivity of the antennas. For the events that pass this selection, we can then deduce the two coordinates of the position of the source and the usual TT wave components h_+ and $h_\times$, the four degrees of freedom of the problem.

4 Practical features and conclusions

An array of stumpy cylinders has about the same cross-section of a sphere of the same mass and same material, but

- can be "matched" to the desired frequency
- local disturbances can be strongly rejected and the parameters of the pulse wave can be estimated
- because of the smaller size of the cylinders, materials with better performances can be chosen (like molybdenum or sapphire, depending on the frequency)

- because of the modular structure, high reliability can be achieved; this is a very important feature because very high sensitive devices like gravitational antennas are very delicate, as we learnt with all the antennas developed until now

- the development phase is easier because more prototypes can be used at the same time and the cryogenic times are much more shorter (for a today "standard bar" the time to warm-up, open, close and cool-down a cryostat is about six months; for a stumpy at 4500 Hz it can be of the order of two weeks).

Arrays of stumpy cylinders can be good candidates to cover high frequencies in the world wide network composed by all the gravitational detectors.

Acknowledgments

Some of the ideas in this paper were developed together with M.A.Papa and M.Bassan. I would also acknowledge the many discussions with the members of the Rome group.

References

1. Just two of many examples are M.Davis et al. *Phys. Rev. Lett.*, **27**, pag. 1466, 1971 and R.F.Stark, T.Piran, *Phys. Rev. Lett.*, **55**, pag. 891, 1985.
2. J.Houser et al. *Phys. Rev. Lett.*, **72**, pag. 1314, 1994.
3. K. Oohara and T. Nakamura, *Progr. Theor. Phys.*, **82**, pag. 535, 1989.
4. K. Oohara and T. Nakamura, *Progr. Theor. Phys.*, **82**, pag. 1066, 1989.
5. K. Oohara and T. Nakamura, *Progr. Theor. Phys.*, **83**, pag. 906, 1990.
6. T. Nakamura, M.Shibata, K. Oohara, *Progr. Theor. Phys.*, **89**, pag. 809, 1993.
7. K.D.Kokkotas, in these proceedings.
8. S.Frasca, M.A.Papa in Proceedings of Cornelius Lanczos Centenary Conference, (1993) publ. by SIAM.
9. S.Frasca, M.A.Papa, two papers in Proceedings of "First E. Amaldi Conference on Gravitational Wave Experiments" (1994), publ. by World Scientific.
10. R.V.Wagoner, H.J.Paik, in *Proc. of Int. Symp. on Experimental Gravitation*, pag. 257, Accademia Nazionale dei Lincei, Roma 1977.
11. W.W. Johnson and S.M. Merkowitz, Phys. Rev. Lett., **71**,4107, (1993).
12. H.J.Paik, *Phys. Rev. D*, **15**, pag. 409, 1977.
13. K.Tsubono et al. *Phys. Lett.* **67A**, 2, 1978

Table 1

Materials for GW resonant antennas

	Density (g/cm^3)	**Young mod.** (GPa)	**Cross sect. for fixed freq.** (normal. to Al)	**Length** (normal. to Al)	**Mass** (normal. to Al)	**Q** (million)
Al (5056)	2.67	70.6	1.000	1	1	60
B	2.35	441	118.100	2.664	16.641	
Be	1.85	318	74.656	2.550	11.484	0.2
Co	8.9	211	2.537	0.947	2.830	
Cr	7.1	279	7.159	1.219	4.818	
Fe	7.8	211	3.093	1.011	3.023	
Ir	22.4	528	6.295	0.944	7.061	
Mo	10.2	325	6.089	1.098	5.053	40
Nb	8.57	105	0.469	0.681	1.012	50
Os	22.5	559	7.211	0.969	7.675	
Ru	12.2	432	9.483	1.157	7.081	
Ti	4.54	116	1.561	0.983	1.615	
W	19.3	411	4.207	0.897	5.224	
Al_2 O_3	3.98	438	52.676	2.040	12.657	1000
Cu Be	8.2	130	0.855	0.774	1.426	
Si C	3.2	445	76.020	2.293	14.455	0.4

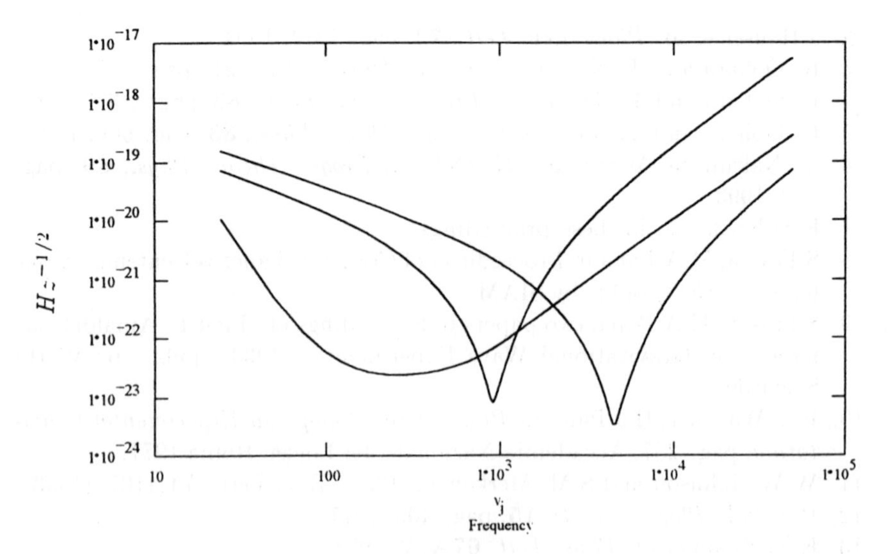

Figure 1

A COSMOLOGICAL BACKGROUND OF GRAVITATIONAL WAVES DETECTABLE BY CROSS CORRELATION OF BARS AND INTERFEROMETERS

DAVID BLAIR, LI JU
Department of Physics, The University of Western Australia
Nedlands, WA 6907, Australia

Abstract

Simple arguments demonstrate that the rate of supernovae within a red shift horizon $z \sim 2$ is at least of the order of 10^{10} per year or 1000 per second. This rate could be enhanced by more than an order of magnitude if the supernova rate in the early universe is enhanced as predicted by star formation models, metallically observations, and the recent observations of an abundance of faint blue galaxies at high red shift. The gravitational waves from supernovae in the early universe create a continuous stochastic background. The amplitude of this background depends on the efficiency of gravitational wave production in supernovae, which in turn depends on the fraction of collapses which create neutron stars and black holes, the dynamics of the collapse, and the post collapse evolution of the system. It is shown that the stochastic supernova background is detectable by cross correlation of nearby detectors if the efficiency of gravitational wave production exceeds 10^{-5}. The expected spectrum is in the frequency band well suited both laser interferometer and resonant mass detectors and cross correlation between advanced bars and interferometers provides an appropriate means of detection.

Introduction

Supernovae, in which a star collapses to form a black hole or a neutron star, have long been considered likely sources of gravitational waves. Numerical modelling of gravitational collapses have lead to rather low estimates of the conversion efficiency to gravitational waves [1]. However models of the post collapse evolution of neutron stars leads to higher conversion efficiency estimates. For example Lai and Shapiro [2] consider the post-collapse evolution of a neutron star in which non-axisymmetrical instabilities lead to efficient radiation of angular momentum. They predict gravitational wave amplitudes $\sim 10^{-21}$ at $\sim$ 30 Mpc distance ($f \sim$ few hundred hertz), corresponding to gravitational efficiency $\varepsilon \sim 10^{-3}$. Houser *et. al*[3], also obtain a $\varepsilon \sim 10^{-3}$ for a model which predict peak emission gravitational wave amplitudes $\sim 10^{-22}$ at $\sim$ 20 Mpc ($f \sim$ few thousand hertz).

Here we present a first order analysis[4,5] of the combined effect of all the supernovae in the universe up to a red shift distance $z \sim 2$. We show that there can be little doubt that a continuous background of gravitational waves is created by these events. The amplitude of this background could be within the range of detectability of proposed advanced detectors if $\varepsilon > 10^{-5}$. It almost certainly will mask predicted cosmological backgrounds from the big bang in the frequency range of terrestrial detectors. It provides a powerful probe of early epochs of star formation. For example, observation of the spectrum and duty cycle can allow investigation of star formation rates, supernova rates, branching ratios between black hole and neutron star formation processes, the mass distribution of black hole births, angular momentum distributions and even black hole growth mechanisms in regions of the universe completely inaccessible to individual stellar observations with electromagnetic astronomy.

The Supernova Duty Cycle

We first estimate the rate of supernovae in the universe as a function of distance, based on observations of supernovae in external galaxies. We assume that for distances greater than 10 Mpc the universe is sufficiently isotropic and homogeneous that the rate of supernovae beyond this distance scales proportional to the enclosed volume. Roughly 50 supernovae are discovered each year in external galaxies. Some of these discoveries are serendipitous, and they certainly do not represent a complete survey. Most of these supernovae are between 5 and 50 Mpc. As reported by Giazotto [6], the number of detected supernovae within ~10 Mpc is ~ 5 per year. The detection efficiency is probably much less than 50%, so a conservative estimate for the rate of supernovae within 10 Mpc is 10 per year. We shall use 10 Mpc as the fiducial scaling distance, denoted r_o, for our analysis. It corresponds to a mean interval T_o between events of 0.1 years, or 3×10^6 seconds, and a mean gravitational wave amplitude h_o.

The gravitational wave burst from a gravitational collapse is expected to have duration $\tau \sim 1$ ms. The *duty cycle* of gravitational waves from supernovae within a distance r, denoted $D(r)$ is given by

$$D(r) = \tau / T.$$

Assuming flat space-time,

$$D(r) = \tau / T_o (r/r_o)^3 .$$

Since the mean amplitude h of a gravitational wave burst from a supernova at distance r scales as $1/r$, it follows that $r/r_o = h_o/h$, and we may express D as a function of the mean observable gravitational wave amplitude h,

$$D(h) = \tau / T_o (h_o/h)^3 .$$

The duty cycle is a significant parameter here because it is only in the case where D approaches unity that the gravitational waves from short bursts can be considered to form a stochastic background. In addition, the ability to dig out such a background from the noise by cross correlation analysis also depends critically on the value of D. Cross correlation analysis should be able to extract the curve $D(h)$ for values of $D >$ 0.01 and this, as discussed above, will allow much astrophysical information to be obtained.

To complete an estimate of $D(h)$ we need to estimate the value of h_o. It is not possible to be definitive regarding ho because no modelling has encompassed the full collapse and post-collapse evolution. Compare the results of Stark and Piran [1] who modelled a rotating gravitational collapse, with those of Lai and Shapiro [2] who modelled the post-collapse evolution of a rapidly rotating neutron star. It appears that the strongest gravitational radiation may occur in the post-collapse period when non-axisymmetric deformation causes essentially all the star's angular momentum to be radiated as gravitational waves. One can make several observations regarding this proposition. First, the existence of millisecond pulsars, which have very low spin down rates tells us that not all neutron stars can experience such deformations. Second, the low angular momentum of most pulsars is very conveniently explained by the Lai and Shapiro mechanism. Third, the above apparently contradictory statements can be reconciled. Very hot newly formed pulsars may be prone to strong non-axisymmetric deformations, driven by convection and rotation, while cooler stars spun up during binary interaction may remain axis-symmetric. Finally, it is interesting to note that a value of $\varepsilon \sim 10^{-3}$ follows naturally if neutron stars are born with a spin frequency of 400 Hz, and lose most of their angular momentum by gravitational emission. The strain amplitude is then consistent with Lai and Shapiro's estimate of $h_o \sim 10^{-21}$. We shall adopt these values for our analysis but finally will consider the consequence of lower values of h_o.

Figure 1 shows the form of $D(h)$. The duty cycle increases as h^{-3}. Using the above numerical values, D reaches 0.3 for $h \sim 1.5 \times 10^{-24}$. Various factors can modify this. If the supernova rate is enhanced in the early universe, as predicted from metallically studies [7], stellar evolution models [8] and the observation of faint blue galaxies[9], D will easily exceed unity. Typical estimated enhancements of 10 - 100 fold mean that the mean supernova frequency will be at least $10/\tau$ for $h \sim 10^{-24}$, corresponding to a supernova rate of 10 kHz. In this regime the excess supernova rate does not increase the duty cycle (since a duty cycle above unity is meaningless). Instead it increases the rms strain amplitude leading to an *enhancement* as shown in figure 1.

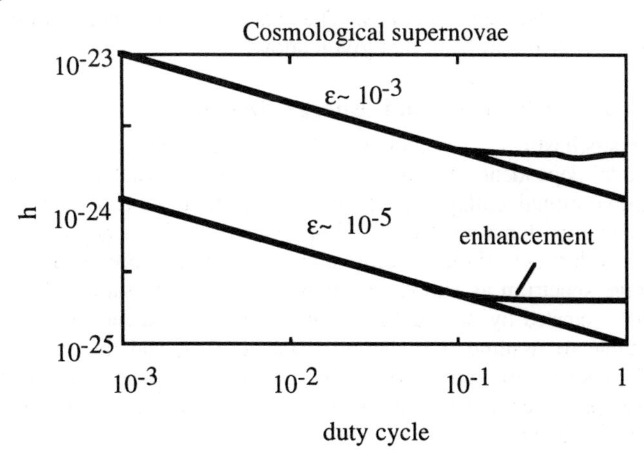

Figure 1: The dependence of duty cycle on strain amplitude for supernovae at cosmological distances. Two curves are shown, one for gravitational wave conversion efficiency 10^{-3}, the other for two orders of magnitude lower efficiency.

The above analysis does not consider cosmological red shift effects nor more realistic cosmological models. Since the dominant contribution to the background is from supernovae at $z \sim 2$, these effects will be substantial.

As emphasised by Schutz[10] stochastic backgrounds can be detected by cross correlation of signals from detectors spaced within less than one reduced wavelength. Fortuitously several bar-interferometer pairs exist or are planned which will be ideal for cross correlation. Cross correlation leads to an increase of the signal to noise ratio determined by the geometric mean of the detector sensitivities, and increasing as the 1/4 power of the integration time:

$$S/N = [S^2/(S_1 \cdot S_2) \cdot B \cdot t]^{1/4},$$

where S is the spectral strain amplitude of the stochastic background, S_1 and S_2 are the spectral strain sensitivities of the two detectors, B is the overlapping bandwidth of the detectors, and t is the integration time. When applied to a bar-interferometer pair Schutz showed that S/N is independent of the bar's bandwidth. For *signal* = *noise* the detectable stochastic background spectral density can be expressed in terms of the burst sensitivity of the instruments, h_b, for a given burst duration t:

$$h = [h_{b1} \cdot h_{b2}]^{1/2} \cdot \tau^{3/4} / t^{1/4}.$$

Assuming $h_{b1} = h_{b2} = 10^{-21}$, and t = 10^8s, it follows that a background with a spectral strain amplitude of $10^{-25}/\sqrt{\text{Hz}}$ can be marginally detected in 3 years observation. Advanced detectors are proposed to be able to achieve about one order of magnitude improvement over these figures.

We point out that the above analysis assumes Gaussian noise. The signals considered here are only likely to be a good approximation to Gaussian noise in the low amplitude high duty cycle regime. For higher amplitudes the background will be *popcorn* noise containing widely spaced bursts. The analysis of cross correlation for such signals is likely to be different from that given above.

Stochastic Background Spectrum and Energy Density

The spectrum of the stochastic background is difficult to estimate since predicted spectra are highly model dependent. We have considered two alternatives. (a) we assume a standard gravitational collapse spectrum similar to the Stark and Piran result[1], for which the energy density scales as $f^{2.5}$ up to a cut off frequency (~ 5 kHz for about 3 solar mass collapses). If the structure above cut off is ignored and $\varepsilon = 10^{-3}$, this leads to an average spectrum as shown in figure 2. (b) We assume that the dominant background is created by post collapse evolution as discussed above. In this case it is reasonable to assume only neutron star sources, and the stochastic spectrum is identical in shape to that predicted by Lai and Shapiro[2]. While the precise shape of these spectra cannot be trusted, but the average magnitude can be compared with planned detectors.

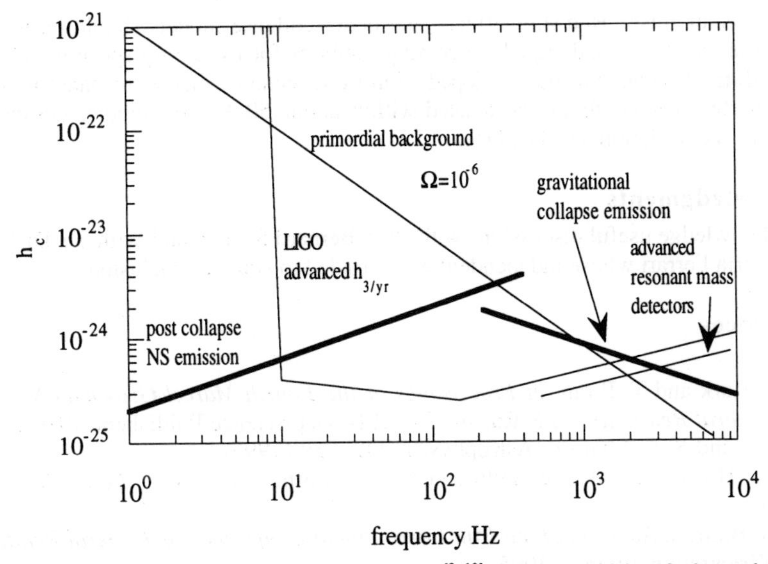

Fig. 2. Comparison of stochastic background signals [2, 13] and the sensitivity of advanced laser interferometer[11,13] and resonant mass detectors[12].

It is useful to consider the energy density of this background, by comparison with predicted cosmological backgrounds from the big bang. A background with 10^{-6} of closure density per decade has an amplitude of 10^{-24} at 100 Hz, falling to 10^{-25} at 1 kHz[13]. The spectrum predicted here corresponds to a comparable density. The gravitational wave energy density from supernovae Ω_{sn} can be independently estimated from the relation: $\Omega_{sn} = \Omega_m \cdot f_s \cdot f_{sn} \cdot \varepsilon$, where Ω_m is the mass density of baryonic matter in the universe, f_s is the fraction of this which has taken part in star formation, f_{sn} is the mass fraction of stars which undergo supernovae and ε is the mean gravity wave conversion efficiency in supernovae. $\Omega_{sn} = 10^{-6}$ can be achieved if $\Omega_m \cdot f_s = 10^{-2}$, $f_{sn} = 10^{-1}$ and $\varepsilon = 10^{-3}$.

Conclusion

The cosmological background of gravitational waves from supernovae is certain to exist. Its magnitude could be in the range of 10^{-24} to 10^{-25}, detectable by advanced instruments. This implies that a relatively large fraction of the matter in the universe must pass through supernovae, and could imply that more baryonic matter than conventionally believed is locked up in collapsed objects. This topic is worthy of further consideration. The analysis presented here is a first approximation. However, it leads to one strong and interesting conclusion. Pairs of detectors capable of detecting identical burst events at 30 Mpc are capable of detecting the stochastic background of such events at cosmological distances as long as the duty cycle for this background approaches unity. Further study should examine the cross correlation of non-Gaussian noise. Red shift effects, and better collapse models should be used to model the spectrum, and signal processing methods for extracting spectral and duty cycle information need to be developed. Future detectors, whether resonant mass of laser interferometers, should be located within about 50 km of another detector to allow cross correlation to take place.

Acknowledgments

We acknowledge useful discussions with Prof. Bernard Schutz, and with. Dr Di Fazio and Valeria Ferrari whose independent work has led to similar conclusions.

References

1 R. F. Stark and T. Piran, *in Proceedings of the Fourth Marcel Grossman Meeting on General Relativity,* ed R Ruffini, 327 (Elsevier Science Publishers, 1986).
2. D. Lai and S. L. Shapiro, Astrophys. J. 442, 259 (1995).
3. J. L. Houser, J. M. Centrella and S. C. Smith, *Phys. Rev. Lett.* **72**, 1314 (1994).
4. D. G. Blair, in *Abstract of the 15th International Conference on General Relativity and Gravitation*, Firenze, 1995.
5. L. Ju and D. G. Blair, "The detection of Gravitational Waves", *Int. J. Mod. Phys.* in press (1996).
6. A. Giazotto, this proceedings.
7. J. P. Ostriker and L. L. Cowie, *Astrophys. J.* **243**, L127 (1981).
8. Di Fazio and Ferrari, this proceedings.
9. M. Longhair, *Proceedings of the International Conference on Gravitation and Cosmology*, Pune 1995.
10. B. F. Schutz, in *Proceedings of the First Edoarado Amaldi Conference on Gravitational Wave Experiments*, eds. E. Coccia, G. Pizzella and G. Tonga (World Scientific 1995).
11. A. Abramovici, W. E. Althouse, R. W. P. Drever, Y. Gü rsel, S. Kawamura, F. J. Raab, D. Shoemaker, L. Sievers, R. E. Spero, K. S. Thorne, R. E. Vogt, R. Weiss, S. E. Whitcomb and M. E. Zuker, *Science*, **256**, 325 (1992).
12 . W. W. Johnson and S. M. Merkowitz, *Phys. Rev. Lett.* **70**, 2376 (1993).
13. K. S. Thorne, in *300 years of Gravitation*, eds. S. Hawking and W. Israel, (Cambridge University Press 1987).

Experimental Perspectives

ASTROPHYSICS WITH A SPHERICAL GRAVITATIONAL WAVE DETECTOR

E. COCCIA

Dipartimento di Fisica, Universitá di Roma "Tor Vergata"
and INFN, via della Ricerca Scientifica 1, I-00133 Roma
e-mail: COCCIA@ROMA2.INFN.IT

Spherical gravitational wave detectors offer a wealth of so far unexplored possibilities to detect gravitational radiation. They allow isotropic sky coverage and determination of the source direction and wave polarizations. The cross section is high at two frequencies, allowing to determine the features of coalescing binaries. These detectors appear powerful instruments for measuring a stochastic background of gravitational radiation.

1 Introduction

Thirty-five years after the beginning of the experimental search for cosmic gravitational waves (GW), several resonant-mass detectors (cryogenic cylindrical bars) are today monitoring the strongest potential sources in our Galaxy and in the local group [1]. The sensitivity of such detectors is $h \simeq 6 \times 10^{-19}$ for millisecond GW bursts, or, in spectral units, 10^{-21} $\mathrm{Hz}^{-1/2}$ over a bandwidth of a few Hz around 1 kHz.

Even though the direct and unambiguous detection of GWs remains an important goal in today's experimental physics, the results from 20 years' observation of PSR 1913 + 16 [2] is moving the scientific interest of GW hunters towards the possibility of studying the physical and astrophysical features of the radiation. An observatory constituted by the present cylindrical bars has serious limitations on reaching this goal. Projects for omnidirectional resonant-mass GW observatories of enhanced sensitivity, able to reconstruct the polarization and the direction of propagation of the wave, besides its intrinsic amplitude, emerged in the last few years in the resonant-mass community [3,4,5,6]. We review here the possibilities offered by large resonant spheres to detect gravitational radiation. We discuss the signal-to-noise ratio of such detectors to various GW signals.

2 Cross section and eigenfrequencies of spherical detectors

In a cylindrical bar, only the first longitudinal mode of vibration interacts strongly with the GW and consequently only one wave parameter can be measured: the amplitude of a combination of the two polarization states.

Each quadrupole mode of a spherical mass is five-fold degenerate (its angular dependence can be described in terms of the five spherical harmonics $Y_{lm}(\theta,\varphi)$ with l=2 and m=-2...+2) and presents an isotropic cross section. The cross section of the lowest order ($n = 1$) mode is the highest and is larger than that of a cylindrical antenna made of the same material and with the same resonant frequency by a factor of about $0.8(R_s/R_b)^2$ [5,6], where R_s and R_b are the radius of the sphere and of the bar, respectively. This means a factor 20 respect to the present bars.

Moreover, a spherical detector allows one to determine the GW amplitudes of the two polarization states and the two angles of the source direction. Let us focus on the lowest order quadrupole mode. The method first outlined by Forward[7] and specified by Wagoner and Paik[8], consists of measuring the sphere vibrations in at least five independent locations on the sphere surface so as to determine the vibration amplitude of each of the five degenerate modes. The Fourier components of the GW amplitudes at the lowest quadrupole frequency and the two angles defining the source direction can be obtained as proper combinations of these five outputs [4,5,9,10].

The signal deconvolution is based on the assumption that in the wave reference frame (the frame in which the z axis is aligned along the wave propagation direction) only the modes with l=2 and m=+2 and -2 are excited by the GW, as the helicity of a GW is 2 in general relativity. One can take advantage of this to deconvolve the Euler angles between the laboratory frame and the wave frame, thus finding the wave propagation direction and the GW amplitudes in the wave frame.

The GW energy absorbed by a resonant detector can be expressed in term of the integrated cross section, which for the quadrupole modes of a spherical detector can be written as [6]

$$\sigma_n = F_n \frac{G}{c^3} M_s v_s^2 \tag{1}$$

where n is the order of the quadrupole mode, M_s is the sphere mass, v_s is the speed of sound and F_n is a dimensionless coefficient which is characteristic of each quadrupole mode. Numerically has been found F_1=2.98, F_2=1.14, F_3=0.107. It is remarkable that the second-order ($n = 2$) quadrupole mode cross section is only a factor 2.61 lower than that of the first order ($n = 1$) quadrupole mode. This means that this detector can be used to advantage at two frequencies.

The sphere quadrupole eigenfrequencies can be found to be [8,11,12]

$$\omega_n = \frac{c_n}{R_s} v_s \tag{2}$$

The dimensionless coefficients c_n are numerically found to be c_1=1.62 and c_2=3.12. A Poisson ratio value of 1/3, common to most materials including the ones we are interested in, has been assumed in the reported numerical results.

Values of the first and second quadrupole resonant frequencies for spheres of different diameter and materials are reported in table 1. The reported materials are Al 5056 (the high mechanical quality factor material used for the present bars), and a copper aluminium alloy, indicated as CuAl, under study at Kamerlingh Onnes Laboratory [13]. Speed of sound are 5400 m/s for Al 5056 at low temperature, and 4700 m/s for CuAl. The masses range from 38 tons (3 m dia Al 5056) up to 250 tons (4 m dia CuAl). The corresponding integrated cross section values of the lowest order quadrupole mode are 9 10^{-22} m^2 Hz, and 4.5 10^{-21} m^2 Hz. The cross section of the typical 3 m long bars is 4.5 10^{-23} m^2 Hz for the best orientation and most favourable polarization. In the following we will often refer to the sensitivity of the lowest order mode of a 3 m dia, CuAl spherical detector (which mass is about 100 ton), as that of a typical large spherical detector.

3 Sensitivity of spherical detectors

The mechanical oscillation induced in the antenna by interaction with the GW is transformed into electrical signals by a set of transducers and then amplified by electrical amplifiers. Unavoidably, Brownian motion noise associated with dissipation in the antenna and the transducer, and electronic noise from the amplifiers, limit the sensitivity of the detector.

The total noise at the output of each resonant mode can be seen as due to an input noise generator having spectral density of strain $S_h(f)$, acting on a noiseless oscillator. $S_h(f)$ represents the input GW spectrum that would produce a signal equal to the noise spectrum actually observed at the output of the detector instrumentation. In a resonant-mass detector, this function is a resonant curve [14] and can be characterized by its value at resonance $S_h(f_n)$ and by its half height width. $S_h(f_n)$ can be written as:

$$S_h(f_n) = \frac{G}{c^3} \frac{4kT}{\sigma_n Q_n f_n} \tag{3}$$

here T is the thermodynamic temperature of the detector (plus a back-action contribution from the amplifiers, which is very small for the present readout systems), and Q_n is the quality factor of the mode. The predicted sensitivity of the lowest mode of a typical large spherical detector, cooled at 20 mK, and made of a high Q material ($\simeq 10^7$), is in the range of $[S_h(f_1)]^{1/2} \simeq 10^{-24}$ $\mathrm{Hz}^{-1/2}$.

The half height width of $S_h(f)$ gives the bandwidth of the resonant mode[15]:

$$\Delta f_n = \frac{f_n}{Q_n}\Gamma_n^{-1/2} \tag{4}$$

Γ_n is the ratio of the wideband noise in the n-th resonance bandwidth to the narrowband noise. In practice $\Gamma_n \ll 1$ and the bandwidth is much larger than the pure resonance linewidth f_n/Q_n. In the limit $\Gamma_n \to 0$, the bandwidth becomes infinite. The bandwidth of the present resonant bars is of the order of a few Hz. If a quantum limited readout system were available, values of the order of 200 Hz could be reached.

The last two equations characterize the sensitivity of the quadrupole modes of spherical resonant-mass detector. The optimum performance of a detector is obtained by filtering the output with a filter matched to the signal. The energy signal-to-noise ratio (SNR) of the filter output is given by the well-known formula

$$SNR = \int_{-\infty}^{+\infty} \frac{|H(f)|^2}{S_h(f)} df \tag{5}$$

where $H(f)$ is the Fourier transform of $h(t)$. Let us consider the SNR of a spherical detector for various GW signals.

3.1 Burst

We model the burst signal as a featureless waveform, rising quickly to an amplitude h_0 and lasting for a time τ_g much shorter then the detector integration time $\Delta t = \Delta f_n^{-1}$. Its Fourier transform will be considered constant within the detector bandwidth: $H(f) \simeq H(f_n) = H_0$. From (5) we get [14]

$$SNR = \frac{2\pi \Delta f_n H_0^2}{S_h(f_n)} \tag{6}$$

for $SNR = 1$, and using the equation $H_0^{min} = h_0^{min}\tau_g$, we find

$$h_0^{min} = \tau_g^{-1}[\frac{S_h(f_n)}{2\pi\Delta f_n}]^{1/2} \tag{7}$$

The requirement of a large bandwidth makes a strong coupling of the transducer to the antenna desirable. The range $h_0^{min} \simeq 10^{-22}$, could be reached by the lowest order mode of a typical large spherical detector with bandwidth of the order of 10 Hz. This sensitivity would allow the detection of the GW collapses occurring in the Virgo cluster at the level of an energy release of $10^{-4}M_\odot$.

3.2 Monochromatic signal

We consider a sinusoidal wave of amplitude h_0 and frequency f_s constant for the entire observation time t_m. The Fourier transform amplitude at f_n is $\frac{1}{2}h_0 t_m$ with a bandwidth given by t_m^{-1}. For $SNR = 1$ we obtain the minimum detectable value of h_0, which at $f_s = f_n$ is

$$h_0^{min} = [\frac{2S_h(f_n)}{t_m}]^{1/2} \tag{8}$$

The minimization of the factor T/Q maximizes the SNR of the detector over a narrow bandwidth centered on the resonant frequency f_n.

For instance, the nearby pulsar [16] PSR J0437-4715 at a distance of 150 pc should emit at 347 Hz a GW amplitude (optimistically) of 2×10^{-26}. This would give SNR /simeq 400 on a spherical detector having $M = 100$ tons (we may think to a hollow sphere, to get such a low resonant frequency), $Q = 10^7$, $T = 20$ mK, after integrating the signal for 1 month.

3.3 Chirp

We consider here the interaction of the spherical detector with the waveform emitted by a binary system, consisting of either neutron stars or black holes, in the inspiral phase. The system, in the Newtonian regime, has a clean analytic behaviour, and emits a waveform of increasing amplitude and frequency that can sweep up to the kHz range of frequency.

From the resonant-mass detector viewpoint, the chirp signal can be treated as a transient GW, depositing energy in a time-scale short with respect to the detector damping time [17]. We can then use (6) to evaluate the SNR, where the Fourier transform $H(f_n)$ at the resonant frequency f_n can be explicitly written as

$$H(f_n) = \{[\int h(t)cos(2\pi f_n t)dt]^2 + [\int h(t)sin(2\pi f_n t)dt]^2\}^{1/2} \tag{9}$$

with $h(t)$ indicating $h_+(t)$ or $h_\times(t)$. Substituting in (9) the well-known chirp waveforms for optimally oriented orbit of zero eccentricity [18], the SNR for chirp detection is [19]:

$$SNR = \frac{2^{1/3}5}{48} G^{2/3} \frac{\sigma_n Q_n 2\pi \Delta f_n}{4kT} \frac{1}{r^2} M_c^{5/3} (2\pi f_n)^{-4/3} \tag{10}$$

M_c is the chirp mass defined as $M_c = (m_1 m_2)^{3/5}(m_1 + m_2)^{-1/5}$, m_1 and m_2 are the masses of the two compact objects (neutron stars or black holes) and r is the distance to the source. The chirp mass is the only parameter that determines the overall frequency acceleration of the chirp signal. It follows that the measurement of the time delay $\tau_2 - \tau_1$ between excitations of the first and second quadrupole modes on a spherical detector will determine the frequency acceleration of the chirp signal and allow the chirp mass to be measured automatically. An explicit formula has been found [19]:

$$M_c = 2^{8/5}(\frac{5}{256})^{3/5}\frac{c^3}{G}(\frac{\omega_2^{-8/3} - \omega_1^{-8/3}}{\tau_2 - \tau_1})^{3/5} \tag{11}$$

where ω_1 and ω_2 are the angular frequencies of the first and second quadrupole modes. In Fig. 1 we report the distance at which binaries of the indicated masses will give $SNR = 1$ at the detector second quadrupole mode. For example, a system of two neutron stars or black holes having $m_1 = m_2 = 1.4$ $M_\odot$ at 100 Mpc distance will give SNR=24 in the $n = 1$ mode and $SNR = 4$ in the $n = 2$ mode of a CuAl spherical detector of 4 m diameter, 250 tons of weigth, having quantum limited sensitivity.

In table 2 we show the time delays between the excitation of the two modes. Time delays are of the order of some hundredths of seconds. Since the resonant frequencies are usually known with a relative precision better than 10^{-6}, the main relative error in the estimation of the chirp mass is due to timing. It has been shown that a timing accuracy better than 1 millisecond is attainable in a resonant mass detector, even at low SNR [20].

Other consequences of the multimode and multifrequency nature of a spherical detector are the possibilities to determine the orbit orientation by the measurement of the relative proportion of the two polarization amplitudes, and the source distance r by the measurement of the intrinsic GW amplitudes [19].

These capabilities of a spherical detector can provide a fundamental astrophysical measurement. In fact, if the binary source position is determined precisely enough to locate the binary in a particular galaxy, then the comparison of the galaxy redshift with the GW distance determines the Hubble constant [21].

3.4 Stochastic background

In this case $h(t)$ is a random function and we assume that its power spectrum, indicated by $S_{gw}(f)$, is flat and its energy density per unit logarithmic frequency is a fraction $\Omega_{gw}(f)$ of the closure density ρ_c of the universe:

$$\frac{d\rho_{gw}}{dlnf} = \Omega_{gw}\rho_c \tag{12}$$

Table 1
Values of the first and second quadrupole resonance frequencies

	ϕ (m)	ν_1 (Hz)	ν_2 (Hz)
Al5056	3	950	1826
	3.5	814	1566
	4	712	1370
CuAl	3	725	1395
	3.5	622	1196
	4	544	1046

Table 2
Time delays between the excitation of the two lowest quadrupole modes for binaries having $m_1 = m_2 = m$

	$m / M_\odot$	$\Delta\tau$ (ms)		
		$\phi = 4$ m	$\phi = 3.5$ m	$\phi = 3$ m
Al5056	0.61	37.3	26.2	17.3
	0.96	17.7	12.4	8.2
	1.12	13.6	9.5	–
	1.28	10.9	–	–
CuAl	0.61	76.6	53.6	35.6
	0.96	36.2	25.3	16.9
	1.40	19.3	13.5	–
	1.67	14.3	–	–

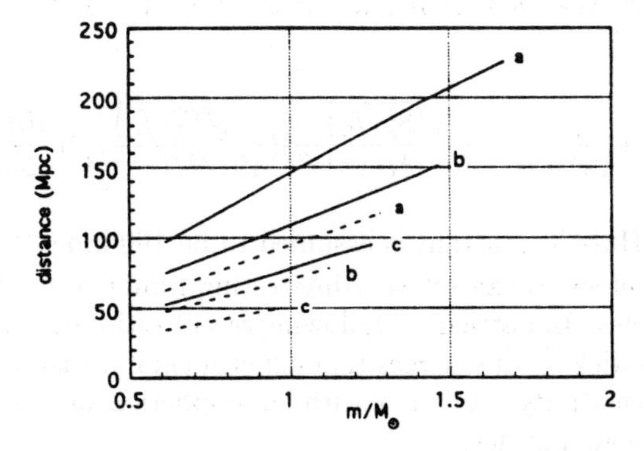

Fig. 1. Distance at which binaries of the indicated mass ($m = m_1 = m_2$) give SNR = 1 at the resonance frequency ω_2. When observed at the frequency ω_1 the same sources gives a SNR about 6. Dotted lines refer to Al5056 spheres, continuous lines to CuAl spheres. The considered diameters are: 4 m (a), 3.5 m (b) and 3 m (c).

$S_{gw}(f)$ is given by

$$S_{gw}(f) = \frac{2G}{\pi} f^{-3} \Omega_{gw}(f) \rho_c \tag{13}$$

The measured noise spectrum $S_h(f)$ of a single resonant-mass detector automatically gives an upper limit to $S_{gw}(f)$ (and hence to $\Omega_{gw}(f)$).

Two different detectors with overlapping bandwidth Δf will respond to the background in a correlated way. The SNR of a GW background in a cross- correlation experiment between two spherical detectors located near one another and having a power spectral density of noise $S_h^1(f)$ and $S_h^2(f)$ is [22]

$$SNR = \left(\frac{S_{gw}^2}{S_h^1 S_h^2} \Delta f t_m\right)^{1/4} \tag{14}$$

where t_m is the total measuring time.

Detectors that are separated by some distance are not as well correlated, because GWs coming from within a certain cone about the line joining the detectors will reach one detector well before the other. The fall-off in the correlation with separation is a function of the ratio of the wavelength to the separation, and has been studied for pairs of bars and pairs of interferometers [23,24].

Supposing two identical large spherical detectors are co-located for optimum correlation, the background will reach a SNR = 1 if Ω_{gw} is

$$\Omega_{gw} \simeq 2 \cdot 10^{-7} \left(\frac{f_n}{10^3 Hz}\right)^3 \left(\frac{\sqrt{S_h^1(f_n)}}{10^{-24} Hz^{-1/2}}\right) \left(\frac{\sqrt{S_h^2(f_n)}}{10^{-24} Hz^{-1/2}}\right) \left(\frac{10 Hz}{\Delta f_n}\right)^{1/2} \left(\frac{10^7 s}{t_m}\right)^{1/2} \tag{15}$$

where the Hubble constant is assumed to be 100 km s^{-1}.

Spherical detectors can set very interesting limits on the GW background at kHz frequencies. In particular, following recent estimations based on cosmological string models [25], it emerges that experimental measurements performed at the level of sensitivity attainable with large spherical detectors would be true tests of Plank-scale physics.

Eqs. (14) and (15) hold for whichever cross-correlation experiment between two GW detectors adjacent and aligned for optimum correlation. An interesting consequence is that it may be worthwhile in the near future to move an advanced resonant mass detector very near to a large interferometer, like VIRGO, to perform stochastic searches near 1kHz [26].

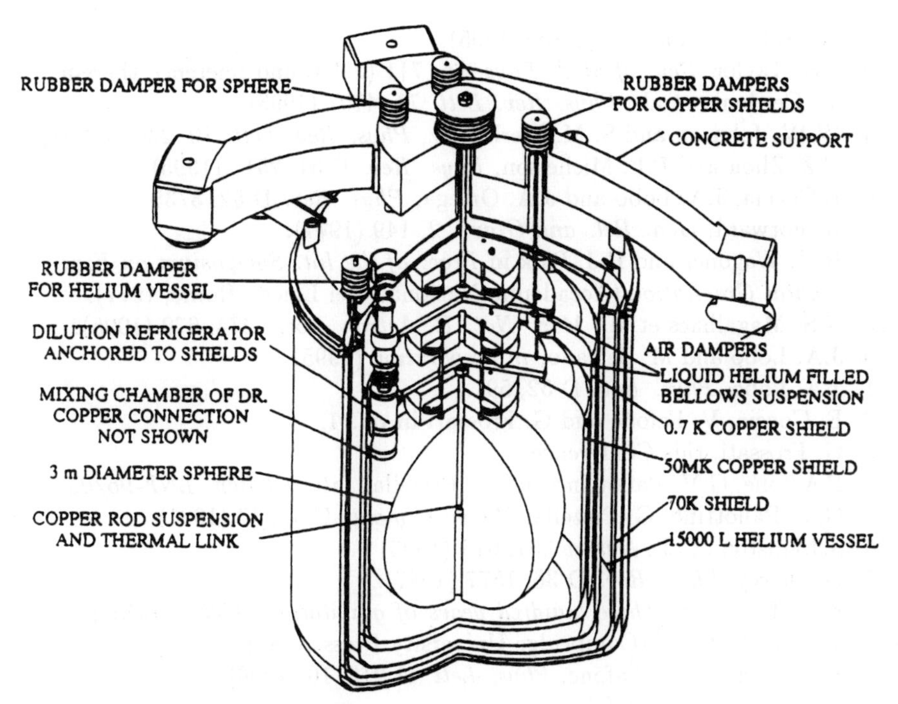

Fig. 2 Schematic layout of an ultracryogenic, 3-m-diam spherical detector (drawing by G. Frossati).

4 Conclusions

We reported on the sensitivity of large spherical detectors to various GW signals.

From the viewpoint of the project feasibility, it seems possible to cool a 100 ton mass to 10 mK [27] (see fig. 2). Fabrication methods of large spherical masses of high-Q materials are under investigation [12]. Proposals in the USA, Netherlands, Italy, and Brazil are under discussion.

References

1. E. Coccia, G.Pizzella, F.Ronga (eds.), *Gravitational Wave Experiments*, Proceedings of the First Edoardo Amaldi Conference, Frascati 1994

(World Scientific, Singapore, 1995).
2. J.H. Taylor, *Rev. of Mod. Phys.* 66, 711 (1994) and reference therein.
3. M. Cerdonio et al., *Phys. Rev. Lett.* 71, 4107 (1993).
4. W.W. Johnson and S. M. Merkowitz, *Phys. Rev. Lett.* 70, 2367 (1993).
5. C.Z. Zhou and P.F. Michelson, *Phys. Rev.* D 51, 2517 (1995).
6. E.Coccia, J.A. Lobo and J.A. Ortega, *Phys. Rev.* D 52, 3735
7. R. Forward, *Gen. Rel. and Grav.* 2, 149 (1971).
8. R.V. Wagoner and H.J. Paik in *Proc. of the Int. Symposium on Experimental Gravitation* (Accademia Nazionale dei Lincei, Rome, 1977).
9. N.S. Magalhaes et al., *Mon. Not. R. Astron. Soc.* 274, 670 (1995).
10. J.A. Lobo and M. Serrano, *in preparation* (1995).
11. J.A. Lobo, *Phys. Rev.* D 52, 591.
12. E. Coccia, V. Fafone and G. Frossati in ref. 1.
13. G. Frossati, this Conference.
14. P.Astone, G.V. Pallottino and G. Pizzella, *Internal Rep. LNF-96/001.*
15. G.V. Pallottino, G. Pizzella, *Nuovo Cimento* C 4, 237 (1981).
16. S.Johnston et al., *Nature* 361, 613 (1993)
17. D. Dewey, *Phys. Rev.* D 36, 1577 (1987).
18. K.S. Thorne in *Three hundred years of gravitation*, S.W. Hawking and W. Israel editors (Cambridge University Press, 1987).
19. E. Coccia and V. Fafone, *Phys. Lett. A* 213, 16 (1996).
20. S. Vitale et al., in ref 1.
21. B. Schutz, *Nature* 323, 310 (1986).
22. J.S.Bendat and A.G.Piersol, *Measurement and analysis of random data*, (John Wiley & Sons, New York, 1966).
23. P.F.Michelson, *Mon. Not. R. Astr. Soc.* 227, 933 (1987).
24. E. E. Flanagan, *Phys. Review* D48, 2389 (1995).
25. R. Brustein, m. Gasperini, M. Giovannini and G. Veneziano, *Phys. Lett. B* 36, 45 (1995).
26. P. Astone, J.A. Lobo and B.F. Schutz, *Class. Quantum Grav.* 11, 2093 (1994).
27. G. Frossati and E. Coccia, *Cryogenics, ICEC supplement* 34, 9 (1994).

THE DEVELOPMENT OF TECHNOLOGY FOR HIGH PERFORMANCE LASER INTERFEROMETER GRAVITATIONAL WAVE DETECTORS

C.N.ZHAO, J.WINTERFLOOD, M.TANIWAKI, M.NOTCUTT, J.LIU, L.JU, D.G.BLAIR

Department of Physics, The University of Western Australia, Nedlands, WA 6907, Australia

In this paper we report progress in the development of technology for high performance laser interferometer gravitational wave detectors. This includes: (a) the development of ULF pre-isolators which are expected to greatly simplify laser interferometer control systems, (b) the demonstration of successful locking of an interferometer using multistage cantilever spring vibration isolators, and (c) the design of sapphire test masses capable of achieving a 16 fold reduction in test mass thermal noise.

1. Introduction

Laser interferometer gravitational wave detectors require high performance vibration isolation to reduce seismic noise. In previous work we have shown that vibration isolation with corner frequency down to about 10 Hz is relatively easy to achieve using mass-cantilever spring isolators [1, 2], and also that it is possible to achieve substantially improved interferometer performance if a single ultra low frequency (ULF) pre-isolator stage is used [3]. A local control and locking system is essential to damp the low frequency normal modes of the vibration isolator and keep the interferometer locked at dark fringe.

High performance suspension systems are also crucial for reducing thermal noise. The thermal and mechanical properties of sapphire make it an interesting material for use as test masses. The high thermal conductivity of sapphire means that thermal lensing [4] can be minimised, while the combination of its very high Young's modulus and low acustic loss will ensure that the internal resonant modes have high frequency and low thermal noise. The recent progress on polishing and coating of sapphire mirrors [5] seems also to indicate that the material is suitable.

In this paper we report progress in all of the areas discussed above. A new two dimensional ULF isolator structure is described. The first successful locking of a fully suspended interferometer using multistage cantilever spring isolators is reported. New measurements on sapphire are also reported along with modelling results which demonstrate that an interferometer with arms of only 400 m can achieve sensitivity comparable to that predicted for VIRGO and LIGO.

2. Ultra Low Frequency Pre-isolators

It has been shown that relatively simple ultra low frequency pendulum-like structures can be cascaded in front of multistage isolators to greatly improve the seismic isolation performance in the frequency range from tens of millihertz through tens of hertz as shown in figure 1. The internal modes of these ULF structures bypass their isolation above a few tens of hertz. However the multistage low-frequency isolator provides such enormous isolation at these frequencies that this is no disadvantage. The advantage provided by pre-isolation is a large reduction in the seismic motion driving the isolator normal modes and also an extension to the low frequency end of the detection band. The normal mode motion is so reduced that damping of these resonances is not required. Indeed if viscous damping of any sort is applied with respect to a seismic reference, the normal mode motion may be worsened since even the peaks can be reduced below the seismic level.

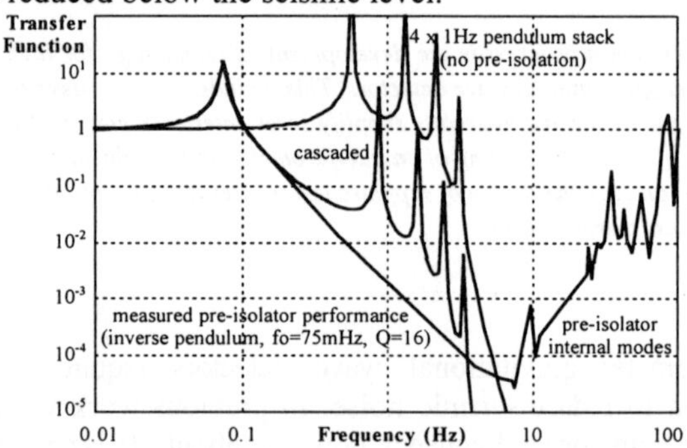

Fig.1 Typical transfer functions of multistage isolator, pre-isolator, and the combination.

We have investigated several approaches. Inverted pendulum [6] or counter-sprung structures are mechanically the simplest. However in order to get low resonant frequencies, there has to be a large cancellation of gravitational force with elastic spring constant. For instance to tune a 1 metre inverted pendulum to 30mHz resonant frequency requires a 99.6% cancellation of forces. At this level time dependent and non-linear effects begin to dominate [7] and the temperature dependence of the elastic spring constant becomes prominent. This system is also inconveniently sensitive to mass loading.

A more suitable approach for horizontal isolation is to use a linkage structure carefully aligned with gravity which simulates the motion of a very long radius pendulum [8, 9], some examples of which are shown in figure 2. Ideally all joints should be perfectly flexible and the structure perfectly rigid. This would make the stability and resonant frequency independent of temperature and mass load. In practice the stiffness of the flexures and the limited rigidity of the mechanical structure are both significant at the 30mHz level. However these non-ideal elastic spring constants have comparable magnitudes to the required residual restoring force so the balance is very stable with temperature and time and is only slightly sensitive to the suspended mass.

The folded pendulum or Watt's linkage (Fig.2a) is one of many linkages which simulate a long radius pendulum in one dimension. Using this structure we have demonstrated 16 mHz corner frequency and isolation exceeding 100dB at 10 Hz [8]. A few long-radius linkages may be generalised to cylindrical symmetry to mimic a very long *conical* pendulum achieving xy isolation in a single stage. One such is the

Scott-Russel linkage shown in figure 2b. It consists of a normal pendulum of length r joined near the mid-point of a beam of length $a+b$. The normal pendulum is under tension and supports the entire weight of the structure and suspended mass. The top section of the beam supports the suspended mass under compression (and bending). The lower end of the beam is merely constrained to move in a vertical line directly under the main support. This is shown as a sliding joint but in practice taut wire guidance functions acceptably as the vertical motion is extremely small.

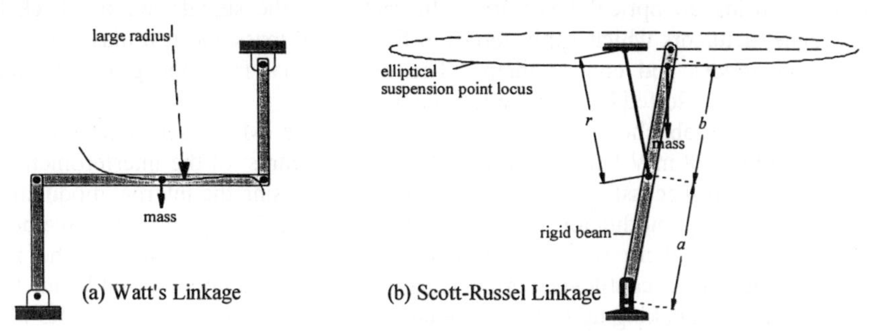

Fig.2 Linkages simulating the large radius of a very long pendulum.

If length a is made equal to r then the upper section of the beam b may be considered as an inverted pendulum which is constrained to follow the same angle of deflection as the normal pendulum a. If b is also equal to r then the effect of the normal and inverted pendulums cancel and the suspension point follows a straight horizontal line. If a slightly lower suspension point is chosen, then the suspension point follows an elliptical path as shown which may in principle be set to any large radius for small displacements. This linkage may be given cylindrical symmetry to produce spherical motion. There are some spatial conflicts between supports and suspended masses but these are readily overcome with a little ingenuity. We have tested a small model with very encouraging results [10]. A full sized pre-isolator for our existing isolators is under construction.

3. The 8 m Interferometer and Its Control System`

Multistage low frequency isolators based on the cantilever spring-mass isolation elements have been developed at UWA and described elsewhere[1]. Four isolator stages of about 90 kg each support an 11 kg compound pendulum test mass suspended with a thin membrane [11]. The isolator system has four vertical modes (1.3, 3.2, 4.5, and 6.5 Hz), four horizontal modes (0.47, 1.8, 1.9 and 2.5 Hz), four rocking modes (below 1.2 Hz) and four low frequency torsional modes (below 0.4 Hz). The high frequency performance of the isolator has been tested using a sapphire transducer [12]. The vertical and horizontal response of the isolator at high frequencies (above the transducer mechanical resonance of about 60 Hz) reaches the noise floor of the sapphire transducer of 3×10^{-15}m/$\sqrt{\text{Hz}}$ [1].

Since demonstrating the performance of the individual isolators, we have concentrated on demonstrating that this isolator structure along with membrane suspended compound pendulum test masses can be used in a practical interferometer. To suppress the normal mode amplitudes sufficiently to be able to achieve locking of the interferometer output to a dark fringe we have used a computer controlled PID servo system [13]. Shadow sensors are used to sense the normal mode motion of the translation of the control mass and the tilt of the test mass. Rotational motion is monitored using an optical lever arm. In each case the signals are fed back to coil/magnet actuators which apply forces to the control mass and test mass to damp the normal modes and control them. The servo signal from the global locking interferometer is also fed back to the test mass actuators.

We have established an 8 m suspended simple Michelson interferometer illuminated by a 58 mW Nd:YAG laser. The fringe contrast of the interferometer is 0.98. It has been successfully locked to the dark fringe using the internal modulation technique [14] with a modulator in one arm. We are particularly pleased to have been able to show that there is no apparent problem in terms of control of the low frequency multistage cantilever spring isolators, nor the compound pendulum test mass suspension. Locking up is easy. We can see some excess noise around 1.5 kHz which we are currently investigating and noise introduced by the control system below 1 kHz.

4. Sapphire Test Masses

Sapphire with its high Q-factor, high sound velocity, high thermal conductivity and low absorption is a very attractive material for use as beamsplitters and test masses. For a cylindrical sapphire test mass with diameter of d = 200 mm, thickness of H =200 mm (~ 25 kg) and a beam size of 2 cm, the internal resonances are higher than 22 kHz [17]. The high internal resonant frequencies in the sapphire test mass are very important in reducing their thermal noise contribution. Integrating the thermal noise over the first 200 normal modes, taking into account that the Q-factor of sapphire is about 43 times greater than that of silica [15,16], and comparing the results with that of a silica test mass of the same dimensions, the thermal noise amplitude of this sapphire test mass will be a factor 16 times better than that of a silica test mass with the same dimensions [17].

Optical absorption of test masses and mirrors causes thermal lensing which limits the maximum optical power in interferometers. Sapphire, with thermal conductivity about 17 times higher than silica, produces less thermal lensing. The distortion is determined by the ratio du/dt/K and this is 30 times less for sapphire than for silica. Recent measurements [18] have shown that the optical absorption coefficient of sapphire can be as low as 3.5 ppm/cm at 1 μm wavelength. This is superior to most samples of fused silica, although a silica sample with 1 ppm/cm loss has recently been demonstrated [19].

The intrinsic birefringence of sapphire is a disadvantage. However analysis has shown [17] that in a well controlled interferometer with misalignment angles less than

100 μrad, the loss due to birefringence will be significantly less than other loss sources such as mirror losses and curvature mismatch. Test masses are normally controlled to ≤100 μrad in prototype instruments and even better in long baseline designs, so this aspect of birefringence seems to be tolerable, although accurate metrology is required to define the crystal axis of optical components.

On the other hand, inhomogeneity and localised stress can cause inhomogeneous birefringence. The birefringence of a small sapphire sample has been investigated[17]. The sample shows stable performance in the central region but is degraded at the corners where machining has created high stress. For a large sapphire test mass, it is expected that stress associated with machining will be greatly reduced. The inhomogeneous birefringence of the small sapphire sample is estimated to be less than 0.04°/cm. It is of great importance that further more accurate measurements on large sapphire samples be made to obtain definite results.

5. Test Mass Suspension

To fully benefit from the low thermal noise acoustic properties of sapphire it is essential to develop greatly improved suspension systems. For this reason there is great interest in the development of "monolithic" suspension systems. The bonding of a niobium flexure to sapphire seems quite promising. We have caculated that niobium flexure bonded to sapphire test masses can allow pendulum Q factors as high as 10^{10} and vertical mode Q-factors $\geq 3\times10^{8}$. This requires strength low acoustic loss bonding of niobium and sapphire. A preliminary bonding test using Incusil-ABA sheet between niobium and sapphire showed a bonding strength >150 MPa. The effect of the bonded flexure on the internal Q-factor of the sapphire will be investigated.

6. Conclusion

We have shown that pre-isolators can greatly improve isolator performance and that two-dimensional isolator structures are well suited to use in interferometers. Progress in the study of sapphire test masses leads to the expectation that thermal noise can be reduced by about one order of magnitude compared with previous estimates. The locking of the 8 m suspended laser interferometer has shown the practicality of multistage cantilever spring isolators. The results on the prototype interferometer system at UWA combined with the above modelling lead to a prediction [11] that the mid-baseline laser interferometer AIGO 400 using multistage cantilever spring isolators and sapphire test masses should be capable of approaching the sensitivity level of stage one long baseline (3~4 km) instruments over a bandwidth of a few hundred Hertz.

References

1. L. Ju and D. G. Blair, *Rev. Sci. Instrum.* **65**, 3482 (1994)

2. *VIRGO Report, PJT94 008*, "Seismic Isolation: The use of Blade Springs".
3. D. G. Blair, L. Ju and H. Peng, *Class. Quantum Grav.*, **10**, 2407-2418 (1993)
4. W. Winkler, K. Danzmann, A. Rüdiger and R. Schilling, *.Phys. Rev. A*, **44**, 7022 (1991)
5. D. G. Blair, M. Notcutt, C. T. Taylor, E. K. Wong, C. Walsh, A. Leistner, J. Seckold and J.-M. Mackowski, "Development of low sapphire mirrors", submitted to *Appl. Opt.*(1995)
6. M. Pinoli, D. G. Blair and L. Ju, *Meas. Sci. Technol.*, **64**, 995 (1993)
7. P. R. Saulson, R. T. stebbins, F. D. Dumont and S. E. Mock, *Rev. Sci. Instrum.* **65**, 182 (1994)
8. J. Liu, J. Winterflood, and D.G. Blair, *Rev. Sci. Instrum.* **66** 3216 (1995).
9. N. Kanda, M.A. Barton and K. Kuroda, *Rev. Sci. Instrum.* **65** 3780 (1994)
10. J. Winterflood, D.G. Blair, "A long-period conical pendulum for vibration isolation", *Phys. Lett. A* (accepted 1996).
11. D. G. Blair, L. Ju and M. Notcutt, *Rev. Sci. Instrum.* **64**, 1899 (1993)
12. H. Peng, D. G. Blair, E. Ivanov, *J. Phys. D: Appl. Phys.* **27**, 1150 (1994)
13. J. Winterflood, D.G. Blair, R. Schilling and M. Notcutt, *Rev. Sci Instrum.* **66,** 2763 (1995)
14. A. K. Strain, *PhD Thesis*, University of Glasgow (1991)
15. V. B. Braginsky, V. P. Mitrofanov and O. I. Panov, *Systems with Small Dissipation*, University of Chicago Press, 1985.
16. V. B. Braginsky, V. P. Mitrofanov and O. A. Okhrimenko, *JEPT Lett.*, 55, 432 (1992)
17. L. Ju, M. Notcutt, D. Blair, F. Bondu and C. N. Zhao, "Sapphire beamsplitters and test masses for advanced laser interferometer gravitational wave detectors", *Phys. Lett.* (accepted 1996)
18. D. G. Blair, F. Cleva and C. N. Man, "optical absorption measurements in monocrystalline saphire at 1 micron", submitted to *Optical Materials* (1995)
19. C. N. Man, Private Comunication

IMPROVING THE SENSITIVITY OF THE UWA RESONANT-MASS DETECTOR

M. E. TOBAR, E. N. IVANOV, D. G. BLAIR
The University of Western Australia, Nedlands 6907, WA. Australia

The two-mode resonant-mass gravitational wave (GW) antenna at the University of Western Australia (UWA) is currently operating continuously with a noise temperature of a few mK. This low noise temperature is partly due to the successful operation of a readout based on 9.5 GHz superconducting re-entrant cavity. The amplitude and phase noise of the pump oscillator limit the sensitivity of the detector. Newly developed cryogenic microwave oscillators are now available with ultra-low noise levels of -165 dBc/Hz. With the advent of new noise suppression schemes this can be further reduced to -185 dBc/Hz. This type of pump oscillator can reduce the detector noise temperature to 20 μK. Further improvements to the superconducting re-entrant cavity system are limited. We show by using an ultra-low loss Sapphire Dielectric Transducer the mode temperature can be reduced to about 2 μK.

1 Introduction

The UWA GW detector consists of a niobium bending flap of 0.45 kg effective mass tuned near the fundamental resonant frequency of a 1.5 tonne resonant-bar of 710 Hz. The acoustic Q of the bar is 2×10^8 while the bending flap is 10^7. The displacement of the bending flap is monitored with a 9.5 GHz re-entrant cavity transducer, which has been described in detail previously[1,2]. The transducer has an electrical Q of 10^5 and a displacement sensitivity of $df/dx = 3.4\times10^{14}$ Hz/m. This system has achieved a noise temperature of less than 2 mK, which represents a 3-fold increase in noise energy sensitivity over SQUID based systems operated to date. A schematic is shown below in figure 1.

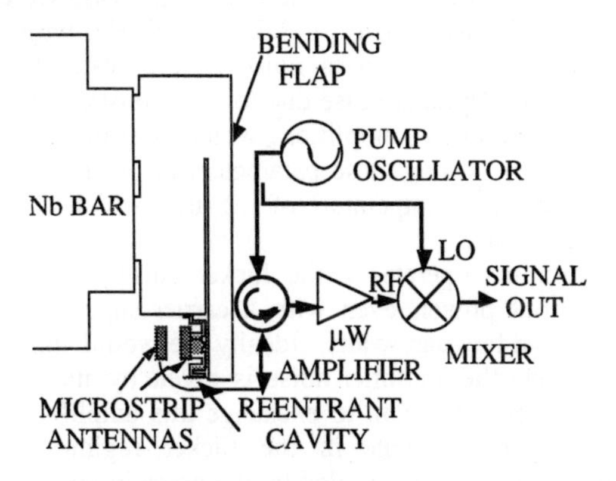

Figure 1: Schematic of the UWA GW detector with a superconducting re-entrant cavity readout.

We have implemented a lumped-element model of the resonant-mass antenna interacting with a capacity transducer to calculate the detector sensitivity. First we show that our model accurately predicts the current performance. Then we determine the expected improvements to the sensitivity by replacing the pump with a new low noise oscillator. Finally we show by replacing the superconducting re-entrant cavity transducer with a low loss sapphire dielectric transducer a noise temperature of less than 2 μK can be achieved in a liquid helium cooled detector.

2 Detector Sensitivity

The way we calculate the sensitivity of the detector is described in detail in[2]. To optimise the detector sensitivity and bandwidth one balances the narrow band noise due to a spectral density of force [N^2/Hz] exciting the resonant-mass system with the broad band electronic noise [m/√Hz] supplied by the readout. Important parameters that effect the sensitivity are the incident power (P_{inc}) to the re-entrant cavity, the cavity electrical Q factor (Q_e), the cavity displacement sensitivity (df/dx), the pump amplitude (S_{am}) and phase noise (S_{pm}), and the readout effective noise temperature (T_{ro}), which is dominated by the microwave amplifier shown in figure 1.

The force fluctuations that excite the bending flap are proportional to $P_{inc}^2 Q_e^2 (df/dx)^2 S_{am}$, while the power in the signal sidebands is proportional to $P_{inc} Q_e^2 (df/dx)^2$. By increasing P_{inc} and Q_e(df/dx) the broad band noise due to T_{ro} will be reduced at the expense of enhanced narrow band noise due to S_{am}. Also, S_{pm} adds to the broad band noise, however it is independent of P_{inc} because the ratio of the signal sidebands to the reflected noise power remains constant. This noise term can only be reduced by increasing df/dx or reducing S_{pm}.

The pump oscillator is based on a commercial synthesiser with an amplitude and phase noise spectral density of -140 and -120 dBc/Hz respectively. Typically it supplies a P_{inc} of 10 μW to the re-entrant cavity. In this case we calculate that the amplitude noise supplies a force density of 8.5×10^{-27} N^2/Hz, while the phase noise contributes 2×10^{-18} m/√Hz to the series noise of the readout. At present the sensitivity is limited by the Nyquist noise due to the acoustic losses in the bending flap and T_{ro} due to the microwave amplifier[2], which contribute 5×10^{-26} N^2/Hz and 3×10^{-17} m/√Hz respectively. Using these parameters in our system model leads us to calculate a detector effective temperature of 1.2 mK, which is very close to the measured effective temperature.

The microwave amplifier operates in the flicker noise regime where the noise level increases with input power, even with a carrier suppression system which limits the input power to a few nanowatts. Ideally we would like to operate in the small signal regime where the amplifier noise is given by its noise figure and is independent of input power. We plan to exchange this amplifier with a low noise figure device that does not operate in the flicker regime. Achieving this improvement will leave the detector limited by the noise in the pump oscillator and will reduce the noise temperature to about 0.2 mK.

3 Low Noise Cryogenic Microwave Oscillators

Recently we built the lowest noise microwave oscillator ever, which exhibited a phase noise of -165 dBc/Hz @ 1 kHz offset from a 9 GHz carrier[3]. The oscillator is based on a liquid nitrogen cooled sapphire dielectric resonator with a Q_e of 6×10^7, and a servo that suppresses the oscillator phase fluctuations[4].

We have recently developed a new type of phase noise suppression technique based on the Ivanov-Tobar-Woode (ITW) phase detector[6]. This detector is several orders of magnitude more sensitive than the conventional phase detector. At room temperature we have already shown that the ITW detector can suppress the noise in a free running oscillator by at least 50 dB[5], whereas the conventional detector only supplies at most 30 dB suppression[3,7]. Thus by incorporating a compact liquid nitrogen cooled ITW detector into the oscillator described in reference[3], a phase noise of order -185 dBc/Hz can be achieved.

These noise levels will enable an increase in P_{inc} improving the noise floor due to T_{ro} without degrading the spectral density of force caused by S_{am}. However the re-entrant cavity can not handle much more than a few hundred μW before Q_e starts to be severely degraded. Assuming P_{inc} is 300 μW the detector noise temperature can be improved to 20 μK with this type of pump oscillator.

4 Two-Mode Sapphire Transducer

Recently we have developed a high Q sapphire dielectric transducer for operation in a resonant-mass GW detector[8]. The transducer consists of a single piece of low-loss sapphire crystal which acts as both the acoustic oscillator and the dielectric transducer[9], with a displacement sensitivity of $df/dx = 2\times10^{12}$ Hz/m and electrical quality factor of $Q_e = 2\times10^8$. The acoustic Q is expected[10] to be on the order of 10^9. A schematic of the transducer is shown in figure 2.

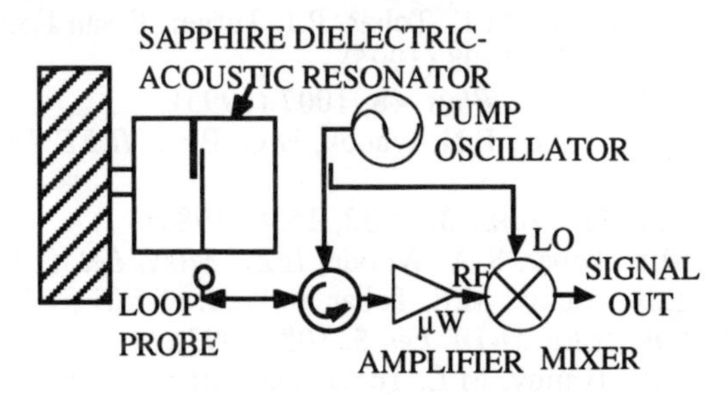

Figure 2: Schematic of the sapphire transducer.

The sapphire transducer can handle a much larger P_{inc} compared to the re-entrant cavity transducer. Generally the factor $P_{inc}Q_e^2(df/dx)^2$ can be up to 3 orders of magnitude greater for the sapphire transducer. For this system the noise contribution of T_{ro} can be rendered insignificant. However the increased operating power and reduced df/dx renders the transducer more susceptible to the purity of the pump. With an incident power of 5 mW we calculate that the noise temperature of the detector can be reduced to about 2 μK assuming both an amplitude and phase purity of -185 dBc/Hz. This is just above the Nyquist limit of 1.5 μK governed by the acoustic losses in the resonant-mass detector.

We plan to implement a two-mode transducer by attaching the sapphire to an intermediate mass before bonding it to the Nb bar. Thus we will operate a three-mode detector which has the potential to increase the bandwidth of the detector from 1 Hz to 90 Hz.

5 Conclusion

By implementing a newly developed sapphire transducer pumped by an ultra low-noise oscillator, a liquid helium cooled GW detector with a noise temperature of 2μK can be realised. This can be achieved in the short term with very little expense in comparison to the construction of a 3 km laser interferometer GW detector. The improved detector will be ideal for data correlation with the proposed AIGO interferometer GW detector.

Acknowledgement

This work was funded by the Australian Research Council.

References

1. D.G. Blair, E.N. Ivanov, M.E. Tobar, P.J. Turner, F van Kann, I.S. Heng,*Phys. Rev. Let.* **74**, 1908 (1995).
2. M.E. Tobar *et al.*, *Aust. J. Phys.* **48**, 1007 (1995).
3. R.A. Woode, M.E. Tobar, E.N. Ivanov, D.G. Blair, *IEEE Trans. on UFFC* (1996).
4. Z. Galani *et al.*, *IEEE Trans. MTT* **32**, 1556 (1984).
5. E.N. Ivanov, M.E. Tobar, R.A. Woode, *IEEE MWG Let.* (1996).
6. R.A. Woode, E.N. Ivanov, M.E. Tobar, in Proc. of *IEE EFTF* (1996).
7. M.E. Tobar *et al.*, *IEEE MGW Let.* **5**, 108 (1995).
8. I.A. Bilenko, E.N. Ivanov, M.E. Tobar, D.G. Blair, *Phys. Let. A* **211**, 139 (1996).
9. M.E. Tobar, E.N. Ivanov, D. Oi, B.D. Cuthbertson, D.G. Blair, *Applied Phys. B*, (1996).
10. V.B. Braginsky, V.P. Mitrofanov, *Systems with Small Dissipation*, (University of Chicago Press, 1985).

EXPERIMENT TO MEASURE THE PROPAGATION SPEED OF GRAVITATIONAL INTERACTION

W. D. WALKER and J. DUAL
Mechanics, ETH, Zurich, Switzerland

The purpose of this project is to develop a laboratory experiment which can measure the propagation speed of gravitational interaction (PSGI). The experiment entails vibrating a mass near another mass and monitoring the gravity-induced vibration. By changing the distance between the masses, a phase shift in the observed gravity-induced vibration will result due to finite PSGI. PSGI can then be determined from the measured change in the distance between the masses, the vibration frequency, and the measured phase shift. PSGI has never been measured, but it is assumed to be equal to the speed of light. During the 1940s - 1960s several physicists proposed that a laboratory experiment to measure PSGI might be possible. Since then, various technologies have been developed that could enable this experiment to be currently feasible. We have investigated several experimental possibilities. We present here a new bending beam gravitational-interacting system capable of generating nanometer gravity-induced vibrations. In addition, we present a phase measurement technique that appears capable of measuring nanodegree phase shifts.

The aim of this project is to experimentally measure the propagation speed of gravitational interaction (PSGI) which, as yet, has never been measured. During the 1940s - 1960s physicists considered the possibility of measuring PSGI by vibrating a transmitter (Tx) mass near a receiver (Rx) mass, and measuring the phase shift of the gravitational-induced vibration. Because of the limited technology at that time, no experiments were performed. Since then, several technologies have been developed that may enable this type of experiment to be performed in the near future. In 1947, J. Cook proposed this type of experiment in a Ph.D. thesis [1]. Later in 1958, Q. Kerns independently proposed the same experiment and developed a phase measurement technique capable of measuring 10^{-6} degree phase shifts [2]. In 1962, J. Cook published the last paper on this topic, summarizing all the work performed up to that time [3]. After 1962 this topic was never discussed in the literature and was apparently forgotten. In 1991 J. Dual independently proposed to experimentally measure PSGI as previously described. Since then we have researched several experimental possibilities. In this paper we present a new bending beam gravitationally-interacting system capable of generating nanometer gravity-induced vibrations. We also present a phase measurement technique that appears capable of measuring nanodegree phase shifts, which would enable PSGI to be measured to 1%.

Since 1962 several groups have constructed gravitational Tx/Rx systems for the purpose of calibrating their gravity wave detectors or to verify the $1/R^2$ gravitational force law, but not to measure PSGI. In 1967 J. Sinsky and J. Weber reported observing 10^{-16} m gravitationally-induced vibrations between two longitudinally-aligned rods, spaced 20 cm apart [4]. The Tx rod was longitudinally vibrated at 1.66 kHz. The resultant gravity-induced Rx vibration was monitored by piezoelectric crystals bonded to the center node. In 1980, H. Hirakawa, K. Tsubono, K. Oide reported observing 10^{-13} m gravitationally-induced vibrations between a rotating rod and a square plate with cuts on each side [5]. The Tx beam was rotated at 30 Hz and the 60 Hz gravity-induced vibration in the plate was monitored with a capacitive sensor mounted in one of the cuts. P. Astone in 1991 reported observing 10^{-14} m gravity-induced vibrations between a rotating beam and a cryogenic rod [6]. The Tx beam was rotated at 450 Hz and the 900 Hz gravity-induced vibration in the rod was monitored with a resonant capacitive transducer and a D.C. SQUID amplifier attached to the end of the rod. Note that in all of the above experiments both Tx and Rx were housed in separate vacuum chambers to reduce acoustic crosstalk.

Because of the small masses and velocities involved in a typical laboratory experiment, the gravitational effects predicted by General Relativity reduce to the retarded Newtonian approximation, where PSGI is the speed of light. This was assumed by all the early PSGI papers during the 1940s - 1960s. Many of the important experimental issues can be clearly observed by analyzing the gravitationally-induced vibration of a simple mass spring system (Rx), at resonance, due to the vibration of a nearby mass (Tx)

$$Y_{\max} = \frac{2GQ_{Rx}M_{Tx}X_{\max}}{\omega^2 d^3}\ ; \qquad \theta = \frac{\omega d}{c}\ .$$

This result was obtained by inserting a simple Newtonian gravitational force in a damped mass spring 2nd order differential equation, and solving it for small vibrations ($X_{max} << d$). It clearly indicates that the receiver's gravitationally-induced vibration amplitude (Y_{max}) is increased by maximizing the receiver mass's quality factor (Q_{Rx}), the vibrating mass (M_{Tx}), and the amplitude of the transmitting mass's vibration (X_{max}). In addition, minimizing the vibration frequency (ω) and the distance between the two masses (d) is especially effective. This result also indicates a tradeoff between Y_{max} and the resultant phase shift (θ). Large vibration frequencies generate large phase shifts but small induced vibration amplitudes, whereas low vibration frequencies generate large induced vibration amplitudes but, small phase shifts. It should be noted that phase shifts due to the measurement instruments can be minimized by placing the Rx mass at two different distances from the Tx mass, and subtracting the measured phase shifts. From the relative phase shift, the vibration frequency, and the relative change in distance, one can calculate PSGI (c).

Since 1991 we have investigated several gravitational system setups, including a longitudinal rod Tx/Rx system. Our current and most successful gravitational experimental system consists of two parallel beams spaced 2.5 cm apart. The Tx beam is a 1 m long, 1 cm square brass beam, electromagnetically vibrated near its first mode frequency of 43 Hz. 1 cm end displacements are easily obtained with this setup. The vibration amplitude is monitored with another electromagnet and amplitude-controlled with a feedback loop. The quartz temperature-stabilized oscillator driving the Tx beam is stable to 1E-6 Hz over long periods of time. The Rx beam is a 1 m long, 8 mm dia. quartz glass beam, with a measured Q factor of about

200,000 in vacuum. The resonance frequency of the Rx beam is controlled by a temperature control loop, which maintains the Rx beam temperature to 50 $\pm$ 0.003 deg C. To reduce the acoustic crosstalk and to increase the Q factor, the Rx beam is housed in a vacuum chamber capable of 1E-2 mbar. Mechanical isolation of the system is accomplished by pinning the Tx beam at its two nodal points, to an aluminum tray, and suspending the tray with springs to an overhead crossbeam which is attached to the floor of the experiment. In addition, the Rx beam is suspended by elastic rubber at its two nodal points from the inside of the vacuum chamber. The vacuum chamber, with the Rx beam inside, is placed on top of an optical isolation table. The vibration of the Rx beam is monitored by a laser interferometer through a quartz glass port hole in the side of the vacuum chamber. The interferometer, which is also mounted on the optical table, has a sensitivity of 2E-9 m/Sqrt Hz. The signal is then processed with a 30 sec time constant, 4 pole lock-in resulting in a vibration sensitivity of 1E-10 m. The Rx beam is electromagnetically isolated by grounding the vacuum chamber. Theoretical analysis of the gravitational-interaction between the beams indicates that the Rx gravitationally-induced vibration should be 4E-10 m. The theoretical analysis entailed inserting a simple Newtonian gravitational force in a Bernoulli-Euler damped bending vibrating beam 2nd order differential equation, and solving it for small vibrations ($X_{max} << d$). The resultant 6th order integral was solved numerically. Experimentally we observe a 1E-9 m Rx beam vibration, which is not affected by changing the acoustic, mechanical, and electromagnetic coupling. In addition, when the gap distance (d) between the two beams is increased from 2.5 cm to 5 cm, we observe a $1/d^{1.6}$ decrease in the vibration amplitude (within 10%) as expected from Newtonian gravitation theory, for the given setup. The phase of the observed Rx beam vibration is stabile to 4 deg over 5 hrs, due to slow temperature drifts of the high Q Rx beam.

In addition to the gravitational system, we present a phase detection technique that appears capable of measuring the expected 5E-7 degree phase shifts with 5E-9 degree accuracy. The technique is a digital implementation of Q. Kerns' phase detection system [2]. Both the oscillator sinusoidal signal and the gravitationally phase shifted sinusoidal signal are sampled by an analog-to-digital converter and fed into a computer which subtracts the two signals. The oscillator signal is then converted into a cosine and multiplied by the subtraction signal. Finally, the output signal is time averaged to eliminate the 2nd harmonic term produced by the multiplication, and to reduce the noise by bandwidth reduction. The technique has been simulated by specifying the two input signals mathematically in a computer loop along with the described algorithm and stepping timewise through the signals. This algorithm simulates the gravitational signals, the sampling of the A/D converter, and the computer signal processing. In addition, white noise was added to the input signals to study the effect of SNR. The results show that with A/D sampling rates of about

9 kHz, and with an input SNR greater than 1E7, phase shifts of 5E-7 degree can be measured with an accuracy of 5E-9 degree. Convergence times of less than 30 sec can be achieved if the input signals have the same amplitude within 1 ppm. The amplitude requirement can be eliminated by multiplying the oscillator sinusoid with the subtraction signal and time averaging the signal. This signal can then be used to control the amplitude of the phase-shifted signal via a PI controller. It should be noted that experiments with analog electronic implementations of the above system resulted in a maximum phase sensitivity of 1E-6 degrees due to thermal drift of the electronic components.

To measure PSGI to 1%, the following system specifications need to be achieved with the current system. Note that future research into many of these technologies may result in significant specification changes. The oscillator frequency must be measured to an accuracy of 4E-2 Hz. The gap distance between the beams must be measured to an accuracy of 1E-5 m. The gravitationally-induced phase shift must be measured to an accuracy of 5E-9 degree. The frequency stability of the system oscillator must be stable to less than 1E-14 Hz. The temperature stability of the Rx beam must be less than 2E-12 degrees over the measurement time. The acoustic crosstalk must be reduced to less than 2E-18 m Rx induced vibration. The mechanical crosstalk must be less than 2E-17 m Rx induced vibration. The electromagnetic crosstalk must be reduced to less than 2E-12 m Rx induced vibration. The gravitational SNR must be greater than 1E7. This, in turn, requires the vibration sensor to have a sensitivity of better than 1E-16 m/Sqrt Hz. The environmental mechanical noise must be less than 1E-16 m/Sqrt Hz. The Rx beam vibration due to Brownian motion must be less than 1E-16 m/Sqrt Hz.

In conclusion, we are presently unable to measure PSGI. Specifications for the various systems have been determined and several technologies have been developed which achieve these specifications, but some of the technologies need improvement. The gravitationally-induced vibration measured in this experiment is 4 orders of magnitude larger than previously accomplished above 1 mHz. The phase measurement technology being developed appears to be 7 orders of magnitude more sensitive than conventional lock-in phase measurement technology. From our research, it appears that measurement of PSGI may be possible to within 1%.

---

[1] J. Cook, On measuring the phase velocity of an oscillating gravitational field, Thesis, Penn. State (1947).

[2] Q. Kerns, Proposed laboratory measurement of the propagation velocity of gravitational interaction, UCRL-8438, U.C. Berkeley, Dec. (1958).

[3] J. Cook, On measuring the phase velocity of an oscillating gravitational field, J. of the Franklin Inst., Vol. 273, No. 6, June (1962).

[4] J. Sinsky and Weber, New source for dynamical gravitational fields, Physical Review Letters, Vol. 18, No. 19, May (1967).

[5] H. Hirakawa, K. Tsubono, K. Oide, Dynamical test of the law of gravitation, Nature, Vol. 283, Jan. (1980).

[6] P. Astone, Evaluation and preliminary measurement of the interaction of a dynamical gravitational near field with a cryogenic gravitational wave antenna, Z. Phys. C 50, pp. 21-29 (1991).

ADVANCES IN LINEAR TRANSDUCERS FOR RESONANT GRAVITATIONAL WAVE ANTENNAS

M.BASSAN, Y.MINENKOV, R.SIMONETTI

Dipartimento di Fisica- Universitá Tor Vergata and I.N.F.N.- Sezione Roma II
00133 Roma, Italy

We discuss the sensitivity of resonant gravitational wave detectors equipped with linear transducers and with SQUID amplifiers and we show that a small gap is a key element for improved devices. We then report on recent experimental advances in the fabrication of high performance resonant capacitive transducers and in particular on a prototype with a gap less than $10\mu m$

1 Introduction

Resonant gravitational wave detectors presently operating have noise temperatures of about $10mK$. Ultracryogenic antennas will soon be operational at noise levels of a fraction of a mK. The Standard Quantum Limit (SQL) for such antennas, i.e. the detection of one single quantum of energy innovation corresponds, at a frequency around $1kHz$, to about $10^{-7}K$. This means that there is still ample room for improvement of linear, d.c. biased transducer before reaching the sensitivity where Quantum Non Demolition strategies will be needed to further improve performances. In particular the resonant capacitive transducer, that has been employed on the antennas of the Roma group for many years, can be used, with minor changes, to lower the noise temperature of the Nautilus detector down to the μK range, i.e. within an order of magnitude from the SQL. We recall [1] some relations useful to evaluate the sensitivity of an antenna of mass m_x and decay time τ, in equilibrium at a temperature T, coupled to an amplifier (typically a SQUID superconducting device) of wide band noise ϕ_n via a transducer of mass m_y and coupling constant α_ϕ (i.e. providing α_ϕ Weber of output signal flux for one meter of vibrational amplitude). The analysis is greatly simplified if one recalls that d.c. SQUIDs (to which linear transducers are invariably coupled, because of their unrivaled low noise) seem to be exempt, to the extent of what is currently measurable, from input noise, i.e. they give no back action force on the mechanical system. In this case, as long as the transducer resonator m_y is sufficiently massive:

$$m_y \geq m_y^{opt} = \frac{2\alpha_\phi}{\omega_x^2 \phi_n}\sqrt{\frac{k_B T m_x}{\tau}} \tag{1}$$

($m_y^{opt} \leq 100g$ for present technology) we can express the noise temperature and the signal bandwidth of the antenna as:

$$k_B T_{\text{eff}} = \frac{4\phi_n \omega}{\alpha_\phi} \sqrt{\frac{m_y k_B T}{\tau}} \; ; \qquad \Delta\omega = 2 \frac{\alpha_\phi}{\phi_n \omega} \sqrt{\frac{k_B T_e}{m_y \tau}} \tag{2}$$

T_{eff} grows rapidly to unacceptable values as m_y is reduced below its optimum value. It is clear that both noise temperature and bandwidth improve linearly with the transduction coefficient α_ϕ, hence the importance of increasing its value as much as technology allows. When α_ϕ is worked out explicitely for the capacitive transducer [2] and the superconducting matching circuit [3] of the Explorer and Nautilus antennas, one finds that it grows roughly linearly with the active capacitance C_t. The main goal of our transducer project is to vastly increase this parameter.

2 New Transducer Design

The surface of the transducer resonating mass is limited by practical problems: so we chose to reduce the gap, much below the $50\mu m$ of existing devices. For a new design of a resonant capacitive transducers we followed these guidelines:
a) — A capacitor with the smallest gap, down to $10\mu m$ and lower.
b) — A mechanical resonator where resonant frequency and mass can be independently adjusted, so that the technology developed can be preserved as α_ϕ is improved and a larger mass required (see eq. 1).
c) — A mechanical design where the resonator and the outer rim end surfaces lie on the same plane, in order to avoid machining errors due to longitudinal offsets, and to permit visual inspection of the gap in the last assembly stage.
d) —a device with a minimum number of parts and in particular of bolts, in order to reduce risks related to differential contraction and Q degradation.

We have also redesigned the facing electrode, according to these specifications:
a) — remove all insulators (typically Teflon sheet) from the axial direction, i.e. from the path of the elastic wave.
b) —Separate the electrical insulation from the determination of the gap: this means doing without the insulating spacers in the gap plane that separate *and* insulate the two electrodes.
c) — Surround the charged electrode by a grounded Faraday cage, to avoid any electrostatic interaction with the surrounding (dust collection, induced charges on the thermal shield around the antenna, etc.).

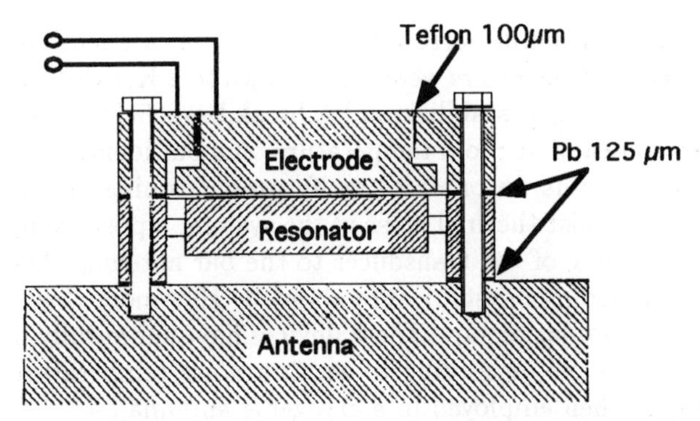

Figure 1: Schematics of the capacitive transducer and its mounting on the antenna

The mechanical oscillator was derived from the rosette resonator [4]. It consists of a central disc supported on six curved cantilevers: in a lumped parameter representation, the disc represents the loading mass while the flexion of the cantilevers provides the spring constant. It is then simple to independently set either the resonator mass (by adjusting the thickness of the disc) or the resonant frequency (by adjusting the cantilever thickness). This design (see fig. 1) yields, with respect to the classical mushroom geometry, a thicker resonator, so that precision machining of the electrode surface results more accurate and within the tight tolerances needed to achieve a small gap.

The facing plate has also been redesigned according to the above cited requirements: the electrode itself mounted at the center of a massive support ring, from which it is electrically insulated by means of 0.1 mm thick PTFE foil. The fastening by thermal contraction provides a very stable and reliable way of assembling the two parts with no bolts, and it well stands subsequent machining. Assembling of the transducer then requires to clamp together two metal (in our case Al) surfaces: we have found that, despite our efforts in machining two very flat surfaces, this joint is usually source of Q degradation. A satisfactory solution to this problem consists of inserting some Pb washers, 0.12 mm thick, between the two plates. The electrode can be easily covered by a ground plate (not shown in figure), that can also host some of the ancillary components of the read-out circuit, like the bias resistor or the decoupling capacitor.

3 Experimental Results

We have built two prototypes of the device described above, both with a resonating disc of 132 mm in diameter and about 700 g in mass. The transducers have been assembled to a gap of 25 and 9 μm respectively, thank to a diamond tool final machining and to extreme attention to the cleanliness of the surfaces.

We believe that gaps still narrower, down to $5\mu m$, are feasible with adequate equipment. Both transducers were tested down to 2 K, exhibiting a mechanical Q in excess of 10^6, gap stability and a breakdown field larger than $10^7 V/m$. They were also tested at room temperature on a bar identical, for material and geometry, to Nautilus, in order to verify proper tuning of the oscillator. We measured the Nyquist thermal noise of the system, a positive proof of the good mechanical coupling of the transducer to the bar antenna. The coupled Qs of the antenna-transducer system were about $3 \cdot 10^4$, more than twice the value reported in any similar set-up. These devices have succesfully passed thorough testing and will allow, according to eq. 2, a noise temperature of $70\mu K$ with present Squids when employed in a cryogenic antenna,.

4 Perspectives and Conclusions

A third prototype of this family of devices is in production now: we plan to face its resonator with one electrode on each side, in a push pull configuration. This will allow us to achieve transducer capacitance to 26 nF, to be compared with the 4 nF of present devices. With modest, foreseable improvements in SQUID technology, the capacitive transducer can reach within an order of magnitude of the Standard Quantum Limit. Moreover, this resonator, where the disc moves of parallel motion and is therefore immune from mechanical stresses, is suitable for future developments that require insertion of a different component at the resonator center: we cite as examples the inductive (with a Nb disc), bimodal (with a third, light mass oscillator) or optical transducer (with a mirror). The technology developed for the transducers reported here can be usefully applied to a variety of related devices.

Acknowledgments

The seed of this project was in the thesis of G.Zaccarian. We are grateful to the entire Roma g.w. group, for continued motivation and support, and to P.Rapagnani, for sharing his vast experience.

References

1. M.Bassan, G.Pizzella, submitted to *Measur.Sci. and Techn.*
2. P.Rapagnani, *Nuovo Cimento* C **5**, 385 (1982).
3. E.Amaldi *et al.*, *Nuovo Cimento* C **4**, 829 (1986).
4. M.Bassan, Y.Minenkov, G.Zaccarian in *Proceeding of the First E. Amaldi Conf. on G.W. Experiments*, (World Scientific, Singapore, 1995).

Astrophysics

GRAVITATIONAL WAVES FROM ACCRETING NEUTRON STARS

S. BONAZZOLA, E. GOURGOULHON
Département d'Astrophysique Relativiste et de Cosmologie,
UPR 176 C.N.R.S.,
Observatoire de Paris,
F-92195 Meudon Cedex, France

We show that accreting neutron stars in binary systems or in Landau-Thorne-Zytkow objects are good candidates for continous gravitational wave emission. Their gravitational radiation is strong enough to be detected by the next generation of detectors having a typical noise of 10^{-23} Hz$^{-1/2}$.

1 Introduction

A crude estimate of the gravitational luminosity of an object of mass M, mean radius R and internal velocities of order V can be derived from the quadrupole formula:

$$L \sim \frac{c^5}{G} s^2 \left(\frac{R_s}{R}\right)^2 \left(\frac{V}{c}\right)^6 , \qquad (1)$$

where $R_s := 2GM/c^2$ is the Schwarzschild radius associated with the mass M and s is some asymmetry factor: $s = 0$ for a spherically symmetric object and $s \sim 1$ for an object whose shape is far from that of a sphere. According to formula (1), the astrophysical objects for which $s \sim 1$, $R \sim R_s$ and $V \sim c$ may radiate a fantastic power in the form of gravitational waves: $L \sim c^5/G = 3.6 \times 10^{52}$ W, which amounts to 10^{26} times the luminosity of the Sun in the electromagnetic domain!

A neutron star (hereafter NS) has a radius quite close to its Schwarzschild radius: $R \sim 1.5 - 3\, R_s$ and its rotation velocity may reach $V \sim c/2$ at the equator, so that they are a priori valuable candidates for strong gravitational emission. The crucial parameter to be investigated is the asymmetry factor s. It is well known that a uniformly rotating body, perfectly symmetric with respect to its rotation axis does not emit any gravitational wave ($s = 0$). Thus in order to radiate gravitationally a NS must deviate from axisymmetry. Moreover, CW emission is possible only if the NS accreets angular momentum from an angular momentum reservoir.

Low Mass X-ray Binary systems (LMXB) and High Mass X-rays Binary systems (HMXB) are a good examples of a NS coupled with an angular momentum reservoir. These systems are formed by a NS and an ordinary companion.

If the two stars are close enough, the NS accretes matter (and angular momentum) from the companion and consequently can radiate CW if its axisymmetry is broken.

The fate of such a system depends on the mass of the companion. For the HMXB for which the companion is a massive star ($\geq 8M_{\odot}$) and consequentely the life time is quite short ($\approx 10^6$ yr) the companion evolves until when the nuclear fuel is exhausted and becomes a supernova. If the binary system is not disrupted by the explosion, the outcome is a binary pulsar. PSR B1913+16 is a good illustration of this scenario. If the companion is kicked away by the explosion, then the outcome is an isolated pulsar.

If the mass of the companion is lower than $1 M_{\odot}$ (LMXB), the NS is spun up by the accretion of matter and angular momentum and, provided the axisymmetry is broken, the NS radiates gravitational waves steadily if the accreted angular momentum is evacuated via gravitational radiation. The light companion is evaporated by the electromagnetic emission of the NS and the final outcome is an isolated millisecond pulsar.

In the intermediate case (mass of the companion between $1M_{\odot}$ and $8M_{\odot}$) the final state is a binary system formed by a white dwarf and a millisecond pulsar. The important point is that by measuring the period modulation of the pulsar in a binary system it turns out to be possible to measure the mass of the NS. The mass of the NS is a fundamental parameter as will be explained later.

Landau-Thorne-Zytkow objects (LTZO) constitute another example of a NS coupled with an angular momentum reservoir. These objects, introduced by Landau [1] to explain the stellar source of energy, have been discussed in details by Thorne and Zytkow [2]. They look as ordinary red supergiant stars, the main difference being the core which is a NS instead of being white-dwarf like. The origin of these objects (if they exist) is beleived to be a HMXB during the phase in which the NS is orbiting into the envelop of the companion. Another possible origine are *aborted supernovae*, i.e. supernovae for which the explosion is not strong enough to eject the envelop. The NS forming the core of these objects accrets matter from the envelop at the maximum rate, i.e. at the Eddington limit : $10^{-8}\, M_{\odot}\, \mathrm{yr}^{-1}$. If some amount of angular momentum is stored in the envelop, the NS is spun up by this accretion. The life of a LTZO is about 10^8 yr, until the mass of the NS reaches the maximum value $M_{\max}$ and the NS collpases into a black hole. It must be noticed that the mass of NS, during its life, varies between the values of the mass at which the NS is born to the value of $M_{\max}$ that depends on the equation of state (EOS).

2 Symmetry Breaking Mechanisms

As already said, gravitational waves are radiated by a rotating NS only if its axisymmetry is broken. Two distinct classes of symmetry breaking mechanisms exist: The axisymmetry can be broken spontaneously (via some kind of instability of the NS) or the axisymmetry can be broken via some external mechanism. Both cases are pertinent to what follows and therefore will be discussed in some detail.

A rotating NS can break spontaneously its axial symmetry if the ratio of the rotational kinetic energy T to the absolute value of the gravitational potential energy, $|W|$, exceeds some critical value. When the critical threshold $T/|W|$ is reached, two kinds of instabilities may drive the star into the non-axisymmetric state:

1. the *Chandrasekhar-Friedman-Schutz instability* (hereafter *CFS instability*) [3,4,5] driven by the gravitational radiation reaction.

2. the viscosity driven instability [6].

Let us recall some classical results from the theory of rotating Newtonian homogeneous bodies. It is well known that a self-gravitating incompressible fluid rotating rigidly at some moderate velocity takes the shape of an axisymmetric ellipsoid: the so-called *Maclaurin spheroid.* At the critical point $T/|W| = 0.1375$ in the Maclaurin sequence, two families of triaxial ellipsoids branch off: the *Jacobi ellipsoids* and the *Dedekind ellipsoids.* The former are triaxial ellipsoids rotating rigidly about their smallest axis with respect to an inertial frame, whereas the latter have a fixed triaxial figure in an inertial frame, with some internal fluid circulation at constant vorticity (see ref. [7] or [8] for a review of these classical results). The Maclaurin spheroids are dynamically unstable for $T/|W| \geq 0.2738$. Thus the Jacobi/Dedekind bifurcation point $T/|W| = 0.1375$ is dynamically stable. However, in presence of some dissipative mechanism such as viscosity or gravitational radiation (CFS instability) that breaks the circulation or angular momentum conservation, the bifurcation point becomes secularly unstable against the $l = 2, m = 2$ "bar" mode. Note also that a non-dissipative mechanism such as a magnetic field with a component parallel to the rotation axis breaks the circulation conservation [9] and may generate a spontaneous symmetry breaking. If one takes into account only the viscosity, the growth of the bar mode leads to the deformation of the Maclaurin spheroid along a sequence of figures close to some Riemann S ellipsoids[a] and whose final state is a Jacobi ellipsoid [11]. On the opposite, if

[a]The *Riemann S* family is formed by homogeneous bodies whose fluid motion can be

the gravitational radiation reaction is taken into account but not the viscosity, the Maclaurin spheroid evolves close to another Riemann S sequence towards a Dedekind ellipsoid [12].

The CFS instability is due to the coupling between the degrees of freedom of the star and gravitational waves: the star can loose angular momentum (and kinetical energy) via gravitational radiation. The formation of waves on the sea when the wind blows is due to an analougous mechanism: in the frame of reference of the wind, the water looses momentum because of its coupling with the atmosphere. Two conditions must be fulfilled to allow for the growth of the CFS instability: (i) the phase velocity of the pertubation must be less then the rotation velocity at the equator of the star, (ii) the viscosity must be less than a threshold value μ_{crit}.

The first condition is always met: it turns out that the phase velocity of the gravity waves (the so-called f modes) is $\propto\ l^{-1/2}$, where l is the "quantum" number of the wave in the harmonic functions expansion (all that in a complete analogy with the sea waves). On the contrary, the second condition is hardly fulfilled: the dumping effect due to the viscosity grows as l^2. Therefore, taking into the account the viscosity of nuclear matter, only the mode $l = 2$ can survive. Recent computations [13] show that this kind of instability can exist only during a short period in the life of the star: in fact if the NS is too hot (resp. too cool), the bulk viscosity (resp. the shear viscosity) inhibits the instability. Actually the interior of a NS is more complicated: it is superfluid and type 2 superconductor. Superfluid vortices are coupled with magnetic fluxoides via their own magnetic field. Vortices and fluxoides are strongly pinned in the solid crust of the star. All that results in an effective viscosity higher than the one computed in the absence of magnetic field. Moreover, any mechanism that tends to rigidify the rotation of star (for example the magnetic field) acts against the CFS instability. From the above it turns out that the CFS instability seems to be very unlikely.

The viscosity driven instability seems to be more promising: in fact, its rising time decreases when the viscosity increases. The physical mechanism of this instability is very simple: consider the rotational kinetical energy of the NS at fixed angular momentum L: $T = L^2/I$ where I is the moment of inertia with respect to the rotation axis. The kinetical energy T decreases if I increases. It turns out that for a large enough L, the total energy of the star (sum of the kinetical and gravitational energy) decreases when I increases. The natural

---

decomposed into a rigid rotation about a principal axis and a uniform circulation whose vorticity is parallel to the rotation vector. Maclaurin, Jacobi and Dedekind ellipsoids are all special cases of Riemann S ellipsoids (for more details, cf. Chap. 7 of ref. [7] or Sect. 5 of ref. [10]).

way to increase I is to let the configuration to be tri-axial. It is worth to note that the final stellar configuration is again an equilibrium configuration (in a rotating frame). The transition between Maclaurin and Jacobi configurations is a real Landau phase transition of the second order as was shown by Bertin and Radicati [14]. The reader can find more details in our lecture on the subject at Les Houches School [15].

The main problem is that this instability can work, as already said, only if the NS rotates fast enough. The maximun angular velocity of a rigidly rotating star is achieved when the velocity at the equator is equal to the Keplerian velocity. It turns out that if the EOS of the fluid forming the star is too soft, the Keplerian velocity is less than the critical velocity for which the axisymmetry breaks. For a polytropic EOS ($P \propto \rho^{\gamma}$) and in the Newtonian theory, γ must be greater than $\gamma_{\text{crit}} = 2.238$ [16,17,18].

3 Results for realistic equations of state

Recently, we have generalized the above results to the existing "realistic" EOS in a General Relativistic frame [17]. Table 1 shows the results: among the 12 EOS taken under consideration, five are stiff enough to allow for the transition toward a 3-D configuration. In table 1, the EOS are labeled by the following abbreviations: PandN refers to the pure neutron EOS of Pandharipande [19], BJI to model IH of Bethe & Johnson [20], FP to the EOS of Friedman & Pandharipande [21], HKP to the $n_0 = 0.17$ fm^{-3} model of Haensel et al. [22], DiazII to model II of Diaz Alonso [23], Glend1, Glend2 and Glend3 to respectively the case 1, 2, and 3 of Glendenning EOS [24], WFF1, WFF2 and WFF3 to respectively the AV_{14} + UVII, UV_{14} + UVII and UV_{14} + TNI models of Wiringa et al. [25], and WGW to the $\Lambda^{00}_{\text{Bonn}}$ + HV model of Weber et al. [26].

From the above results it appears that only NSs whose mass is larger than $1.64\,M_{\odot}$ meet the conditions of spontaneous symmetry breaking via the viscosity-driven instability. The above minimum mass is quite below the maximum mass of a fast rotating NS for a stiff EOS ($3.2\ M_{\odot}$ [27]). Note that the critical period at which the instability happens ($P = 1.04$ ms) is not far from the lowest observed one (1.56 ms). The question that naturally arises is: do these heavy NSs exist in nature ? Only observations can give the answer; in fact, the numerical modelling of a supernova core and its collapse [28] cannot yet provide us with a reliable answer. The maximum critical rotation period (1.2 ms) at which the the instability appears is compatible with the rotation period of the fastest known pulsar (1.56 ms); moreover the age of these pulsar spans between 10^7 and 10^9 yr.

The real problem is the minimum mass, $1.64 M_{\odot}$, for the triaxial instability

Table 1: Neutron star properties according to various EOS: $M_{\rm max}^{\rm stat}$ is the maximum mass for static configurations, $M_{\rm max}^{\rm rot}$ is the maximum mass for rotating stationary configurations, $P_{\rm K}$ is the corresponding Keplerian period, $P_{\rm break}$ is the rotation period below which the symmetry breaking occurs, $H_{\rm c,break}$ is the central log-enthalpy at the bifurcation point and $M_{\rm break}$ is the corresponding gravitational mass. The EOS are ordered by decreasing values of $M_{\rm max}^{\rm stat}$.

EOS	$M_{\rm max}^{\rm stat}$ $[M_\odot]$	$M_{\rm max}^{\rm rot}$ $[M_\odot]$	$P_{\rm K}$ [ms]	$P_{\rm break}$ [ms]	$H_{\rm c,break}$	$M_{\rm break}$ $[M_\odot]$
HKP	2.827	3.432	0.737	1.193	0.168	1.886
WFF2	2.187	2.586	0.505	0.764	0.292	1.925
WFF1	2.123	2.528	0.476	0.728	0.270	1.741
WGW	1.967	2.358	0.676	1.042	0.170	1.645
Glend3	1.964	2.308	0.710		stable	
FP	1.960	2.314	0.508	0.630	0.412	2.028
DiazII	1.928	2.256	0.673		stable	
BJI	1.850	2.146	0.589		stable	
WFF3	1.836	2.172	0.550	0.712	0.327	1.919
Glend1	1.803	2.125	0.726		stable	
Glend2	1.777	2.087	0.758		stable	
PandN	1.657	1.928	0.489		stable	

to develop. This is not in very good agreement with the measured masses (all in binary systems)[29], except for PSR J1012+5307 which appears to be a heavy NS: $1.5\,M_\odot < M < 3.2\,M_\odot$ [30]. Four NS masses (all in binary radio pulsars) are known with a precision better than 10% and they turn out to be around $1.4\ M_\odot$ [29]. Among the X-ray binary NSs, two of them seem to have a higher mass: 4U 1700-37 and Vela X-1 ($1.8 \pm 0.5\ M_\odot$ and $1.8 \pm 0.3\ M_\odot$ respectively). These objects show that NSs in binary systems may have a mass larger than $1.64\ M_\odot$.

A natural question that may arise is: why do X-ray binary NSs, which are believed to be the progenitors of binary radio pulsars, have a mass larger than the latter ones ? We have not yet any reliable answer to this question. A first (pessimistic) answer is that the measurements of X-ray NS masses are bad (compare the error bars of the masses of the binary radio pulsars with the ones of the X-ray binaries in Fig. 3 of ref.[29]), and consequently not reliable. Actually it should be noticed that the error bars of the X-ray pulsars do not have the same statistical meaning as the error bars of the binary radio pulsars[31]: they give only the extremum limits of NS masses in the X-ray binary. Consequently $1.4\ M_\odot$ is not incompatible with these masses.

A related question arises naturally: why are the observed masses of millisecond radio pulsars almost identical ? Following the standard model, a millisecond radio pulsar is a recycled NS, spun up by the accretion of mass and angular momentum from a companion. The observed mass and angular velocity are those of the end of the accretion process. Consequently the accreted mass depends on the history of the system and on the nature of the companion. By supposing "per absurdo" that all NSs are born with the same mass, it is difficult to understand why the accreted mass is the *same* for all NSs. A possible answer is that this could result from some observational selection effect. For example, suppose that accreted matter quenches the magnetic field, it is then easy to imagine that the final external magnetic field depends on the mass of the accreted plasma. If the accreted mass is large enough, the magnetic field can be lower than the critical value for which the pulsar mechanism works. On the contrary, if the accreted mass is quite small, the magnetic field is large and the life time of the radio pulsar phase is shorter and consequently more difficult to observe.

4 Detectability

If the nuclear matter EOS is stiff enough and accreting NS in binary systems have a mass large enough for the symmetry breaking to take place, accreting NS are efficient gravitational wave emitters. It is very easy to compute the amplitude of the emitted gravitational waves. By equating the rate of the accreted angular momentum to the rate of radiated angular momentum one obtains[32]

$$h = 1.3 \times 10^{-27} \left(\frac{1\ \mathrm{kHz}}{\nu}\right) \left(\frac{F_X}{10^{-8}\ \mathrm{erg\,cm^{-2}\,s^{-1}}}\right)^{1/2} \tag{2}$$

where h is the strain of the gravitational wave, ν the rotation frequency of the source, F_X the X-ray flux received on Earth. Note that the distance of the source does not appear in the above formula. From (2), the signal-to-noise ratio S/N can be easily computed in terms of the observation time T and the sensitivity of the detector B. For the brighest X-ray source, Sco X-1 ($F_X = 2 \times 10^{-7}\ \mathrm{erg\,cm^{-2}\,s^{-1}}$), we obtain

$$\left(\frac{S}{N}\right)_{\mathrm{Sco\ X-1}} = \left(\frac{0.17}{B/(10^{-23}\ \mathrm{Hz}^{-1/2})}\right) \left(\frac{1\ \mathrm{kHz}}{\nu}\right) \left(\frac{T}{1\ \mathrm{day}}\right)^{1/2} \tag{3}$$

From the above formula, we see that one month of observation is sufficient to obtain $S/N = 1$ with a detector of the $10^{-23}\ \mathrm{Hz}^{-1/2}$ class. This is however a misleading result: in fact, because the frequency of the CW emission is not known, a signal-to-noise of about 7 is required in order to have a detection with a confidence level equivalent to the ordinary 3σ criterium. This means that 2.5 years of observation time with *one* detector are necessary to detect the gravitational radiation emitted by rotating NS. With 3 detectors (e.g. 2 LIGO + VIRGO) the situation appears more favorable: 10 months of observation time instead of 30. Moreover a less naive strategy can be used to couple the 3 detectors; we do not discuss this possibility here.

LTZO objects are also good candidates. The radiation mechanism is analogous to that of the accreting binary sources: the NS forming the core is spun up by the accreted angular momentum from the envelop. The main avantage of these sources is that the mass range of the inner NS spans from the initial mass of the NS ($\sim 1.4\ M_\odot$) up to the critical mass ($\geq 2\ M_\odot$). The drawback is that we do not if these objects exist.

Finally, note that a deformation (ellipticity) as small as $\approx 10^{-8}$ is sufficent to radiate the accreted angular momentum at the Edington mass accretion rate ($10^{-8}\ M_\odot\,\mathrm{yr}^{-1}$). A question naturally arises: do there exist any other mechanism able to deform the NS by a such a amount ? No alined magnetic field can do the job. The accreted matter is funelled by the magnetic field onto the crust of the NS, and spreads out on the surface, but magnetic field acts as a magnetic brake for this process. The efficiency of this magnetic braking depends on the conductivity of the plasma and on the strength of the magnetic field. A rough estimation of the typical spreading time τ of the accreted plasma on the surface of the NS gives $\tau \gg 1$ yr. This means that 3-D asymetries can be larger than 10^{-8} for the accreting rate of $10^{-9} - 10^{-8}$ $M_{\odot}$ yr^{-1}. The above encouraging result is correct only if the effecive conductivity of the plasma is equal to the microscopic one. Indeed plasma instabilities can reduce the effective conductivity by orders of magnitude. The most dangerous of them is the instability generating the reconnection of the magnetic field lines. Fortunately, no X or O point exists in the magnetic field configuration. Therefore this kind of instability seems to be excluded. More investigation is needed to clarify this important question (the Authors thank Prof. E. Spiegel and Dr. A. Mangeney for illuminating discussions on this point).

5 CONCLUSION

Accreting NS in binary systems or in LTZOs can be good gravitational CW emitters. Their positions on the sky are known, therefore data can be easily reduced to the solar system barycentric frame and the Doppler shift induced by the motion of the Earth can be properly taken into account. The amplitude of the predicted gravitational waves is large enough to be detected with the 10^{-23} Hz$^{-1/2}$ class of detectors. Positive detection will give to us important informations on the equation of state of the nuclear matter in NS. The proof of existence of the LTZOs will be a major discovery leading to important informations on the stellar evolution during the common envelop phase.

References

1. Landau L., *Nature* **141**, 333 (1938).
2. Thorne K.S., Zytkow A.N., *Astrophys. J.* **212**, 832 (1977).
3. Chandrasekhar S., Phys. Rev. Lett. **24**, 611 (1970).
4. Friedman J.L., Schutz B.F., *Astrophys. J.* **222**, 281 (1978).
5. Friedman J.L., *Commun. Math. Phys.* **62**, 247 (1978).
6. Roberts P.H., Stewartson K. *Astrophys. J.* **137**, 777 (1963).
7. Chandrasekhar S., *Ellipsoidal figures of equilibrium* (Yale University Press, New Haven, 1969).
8. Tassoul J.-L., *Theory of rotating stars* (Princeton University Press, Princeton, 1978).
9. Christodoulou D.M., Kazanas D. Shlosman I., Tholine J.E., *Astrophys. J.* **446**, 510 (1995)
10. Lai D., Rasio F.A., Shapiro S.L., Astrophys. J. Suppl. **88** , 205 (1993)

11. Press W.H., Teukolsky S.A., *Astrophys. J.* **181**, 513 (1973).
12. Miller B.D., *Astrophys. J.* **187**, 609 (1974).
13. Lindblom L., *Astrophys. J.* **438**, 265 (1995).
14. Bertin G., Radicati L.A., *Astrophys. J.* **206**, 815 (1976).
15. Bonazzola S., Gourgoulhon E., in *Astrophysical Sources of Gravitational Radiation (Les Houches 1995)*, eds. J.-A. Marck, J.-P. Lasota, to be published (preprint: astro-ph/9605187).
16. James R.A., *Astrophys. J.* **140**, 552 (1964).
17. Bonazzola S., Frieben J., Gourgoulhon E., *Astrophys. J.* **460**, 379 (1996).
18. Skinner D., Linblom L., *Astrophys. J.*, in press.
19. Pandharipande V.R., *Nucl. Phys. A* **174**, 641 (1971).
20. Bethe H.A., Johnson M.B., *Nucl. Phys. A* **230**, 1 (1974).
21. Friedman B., Pandharipande V.R., *Nucl. Phys. A* **361**, 502 (1981).
22. Haensel P., Kutschera M., Prószyński M., *Astron. Astrophys.* **102**, 299 (1981)
23. Diaz Alonso J., *Phys. Rev.* D **31**, 1315 (1985).
24. Glendenning N.K., *Astrophys. J.* **293**, 470 (1985)
25. Wiringa R.B., Fiks V., Fabrocini A., *Phys. Rev. C* **38**, 1010 (1988)
26. Weber F., Glendenning N.K., Weigel M.K., *Astrophys. J.* **373**, 579 (1991).
27. Salgado M., Bonazzola S., Gourgoulhon E., Haensel P., *Astron. Astrophys.* **291**, 155 (1994).
28. Müller E., in *Astrophysical Sources of Gravitational Radiation (Les Houches 1995)*, eds. J.-A. Marck, J.-P. Lasota, to be published.
29. Thorsett S.E., Arzoumanian Z., McKinnon M.M., Taylor J.H., *Astrophys. J.* **405**, L29 (1993).
30. van Kerkwijk M.H., Bergeron P., Kulkarni S.R., submitted to *Astrophys. J.*(preprint: astro-ph/9606045)
31. Lindblom L., private communication.
32. Wagoner R.W., *Astrophys. J.* **278**, 345 (1984).

PULSATING RELATIVISTIC STARS: WHAT CAN WE LEARN FROM FUTURE OBSERVATIONS?

K. D. KOKKOTAS
Max-Planck-Society, Research-Unit Theory of Gravitation, Jena D-07743, Germany
Department of Physics, Aristotle University of Thessaloniki, Greece.

N. ANDERSSON
Department of Physics, Washington University, St. Louis, MO 63130, USA
Department of Physics and Astronomy, UWCC, Cardiff CF2 3YB, U.K.

We discuss new results for pulsating stars in general relativity. These results extend previous research in two important directions. First we show that the so-called gravitational-wave modes of a neutron star can be excited when a gravitational wave impinges on the star. This result is interesting because the gravitational-wave modes are not present in a Newtonian description — they represent a purely relativistic effect. As an extension of this result, we discuss how observations of the stellar pulsation modes can lead to estimates of both the radius and the mass of a neutron star. This discussion provides the first step towards a rigorous investigation of the extraction of neutron star parameters from a gravitational-wave signal.

1 Introduction

Pulsating stars have been studied within Einstein's general theory of relativity for thirty years. Hence it is somewhat surprising that recent results have altered our understanding of these problems considerably. Recent work has shown that a compact star exhibits two distinct sets of pulsation modes. One is slowly damped and corresponds to the well-known fluid modes. The modes in the second set are rapidly damped. These are pure "spacetime" modes that *have no analogue in Newtonian theory*. The new modes have been termed w-modes because they are closely associated with gravitational waves. The existence of such modes was first illustrated for a simple model problem [1], but they have also been found for realistic stellar models [2,3,4].

We now know that the w-modes are directly associated with the curvature of spacetime. But it is not clear yet whether these modes are of any relevance to astrophysics. In order to demonstrate their relevance one must show that the modes can be excited in a realistic astrophysical situation, such as gravitational collapse to form a neutron star. That calculation is, however, still well beyond our capability. Hence, we discuss a much simpler problem (of somewhat dubious relevance to astrophysics): We show that the w-modes are excited when a gravitational wave scatter off the star.

The equations that describe perturbed stars (fluid + spacetime) in general relativity split into two classes [5,6]. The *polar* perturbations correspond to compressions of the star, whereas the *axial* ones induce differential rotation. In general, the polar problem corresponds to two coupled wave equations: One equation represents the fluid motion and the other describes the gravitational waves. On the other hand, axial perturbations are described by a single, homogeneous wave equation [5].

The axial problem is thus considerably simpler than the polar one. Nevertheless, it attained little attention until recently. This is mainly because an axial perturbation cannot induce pulsations in the stellar fluid. But if the star can be made ultra-compact ($R < 3M$) the peak of the exterior curvature potential barrier (that is familiar from black-hole perturbation theory) will be unveiled. Then gravitational waves that impinge on the star can be trapped. That such trapped modes exist has been demonstrated [7,8], but they are unlikely to be of astrophysical relevance: A sufficiently compact star will probably never form.

More importantly, one can argue that the axial problem should be relevant also for less compact stars. For black holes, where the gravitational waves are the only active agent, the axial problem leads to a pulsation spectrum identical to the polar one. Since the character of the w-modes depends mainly on the spacetime curvature there is no reason why such modes should not exist in the axial case. A recent mode-survey for uniform density stars showed that the axial and the polar w-mode spectra are surprisingly similar [9,10]. This has important implications for physical interpretations of the w-modes. Since the axial modes do not couple to the fluid, one cannot invoke the fluid in an explanation of them. They depend only on the curvature of spacetime (as can be illustrated by simple model problems [11]). The w-modes arise because gravitational waves can be temporarily trapped in the "bowl of spacetime curvature" provided by the mass of the star (for a detailed exposition on the above ideas we refer to a recent review [12]).

2 Will the w-modes ever be excited?

An important question is whether the w-modes can be excited and play a role in astrophysics. To test this we have studied the scattering of axial gravitational wave-packets by a compact star. This problem is almost identical to the black-hole one that was studied by Vishveshwara in 1970 [13]. The result of a typical simulation is shown in Figure 1. The exponential ringdown at late times (from $t \approx 100M$) is perfectly described by the first axial w-mode. By analyzing the power spectrum for the simulation in Figure 1 one can show that (1) the

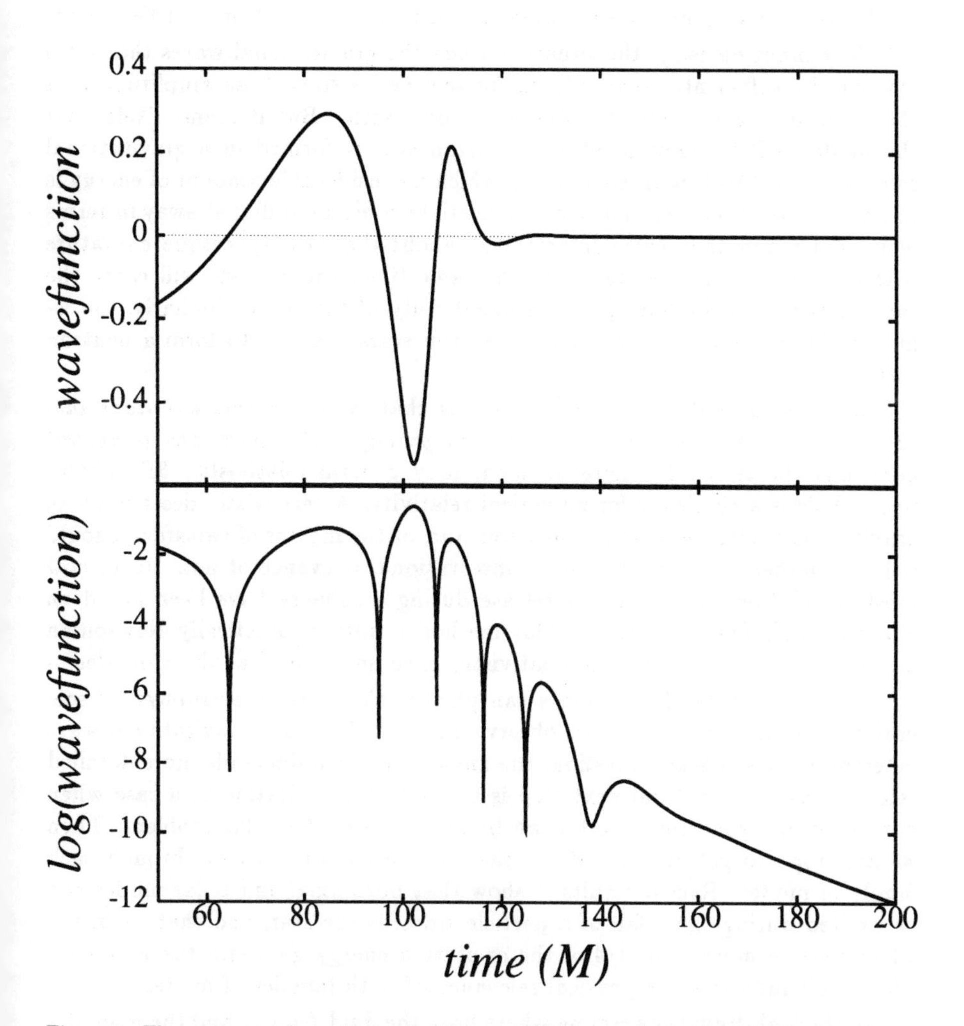

Figure 1: The response of a uniform density star ($M/R = 0.2$) to a Gaussian pulse of axial gravitational waves. The top panel shows the actual form of the axial perturbation as seen by a distant observer. The lower panel shows the same function on a logarithmic scale.

first three axial w-modes are excited and (2) the stellar modes are excited to roughly the same level as the quasinormal modes of an equal mass black-hole would be. These results are in agreement with more recent ones of Ferrari [14].

We cannot envisage the situation when the gravitational waves that hit a star will be sufficiently strong to excite the modes to such an amplitude that the resulting waves could be observable on Earth. But it seems likely that the modes will be excited when a neutron star is formed in a gravitational collapse: And this is a process during which a considerable amount of energy is released. The initial deformation of spacetime could be radiated away in terms of w-modes and they carry information about the local spacetime curvature (the compactness of the star). At the same time, the f-mode will carry the characteristics of the fluid (the average density of the star). Similarly, the w-modes should be excited when two neutron stars coalesce to form a neutron star.

What is absolutely clear at present is that *we are asking questions that can never be answered within Newtonian gravity.* The assertions presented here must be tested by more detailed, fully general relativistic, 3D simulations. This is a challenge for numerical relativity. A relativistic description of gravitational collapse to form a neutron star, or the merger of two stars, should tell us whether the w-modes are of observational relevance or not. Up to now most calculations of the energy release during a collapse have been based on quadrupole formula estimates. This method results in practically Newtonian formulae and it <u>cannot</u> handle relativistic phenomena such as the w-modes.

We believe that the w-modes can play a role in many astrophysical scenarios, but will we be able to observe them with future gravitational-wave detectors? This is a key question, the answer of which demands more detailed calculations. The natural next step is to study the excitation in a case when both fluid and spacetime modes can be excited, *i.e.* the polar problem. Then we can hope to get an estimate of the relative energy emitted through each family of modes. Recent results [14] show that both axial and polar modes can be excited during the infall of a particle towards the star, and that — in the latter case — more than 3/4 of the excitation energy goes into the w-modes. This result indicates the physical relevance of both families of modes.

In the high-frequency regime where both the fluid f-mode and the w-modes reside (1-3 kHz and 8-12 kHz, respectively) one would not expect too much from the new generation of laser-interferometric gravitational-wave detectors. Pulsating neutron stars though are ideal sources for the resonant detectors that are currently operating. Moreover, the w-modes provide an interesting source for the proposed array of smaller bar detectors which should be sensitive in the few kHz regime [15].

Finally, we want to mention that in the distance of 0.5 Mpc (within the Local galaxy group) the expected event rate of supernova collapse is approximately one per year [16]. Thus, there is a good chance that we will see a number of quite non-spherical collapses from which we can detect the stellar pulsation modes.

3 What can we learn from observations of f and w-modes?

Suppose that we detect a gravitational-wave signal from a compact star, what information can we hope to extract from it? Let us assume that we detect a signal and manage to infer both the fundamental polar w-mode and the fluid f-mode from it. Then spectral studies for thirty three stellar models and seven different realistic equations of state (we use the same EOS and notation as Lindblom and Detweiler [17]) suggest the following empirical relations (the corresponding data is shown in Figures 2a,b and d)

- the oscillation frequency of the f-mode scales with the average density of the star as $\sqrt{M/R^3}$ and for the data in Figure 2a we get the following relation:

$$\omega_f(kHz) \approx 0.247 + 2.2\left(\frac{\bar{M}}{\bar{R}^3}\right)^{1/2} \tag{1}$$

- the damping of the f-mode scales as $R(R/M)^3$ and we find

$$\frac{1}{\tau_f(sec)} \approx \left(\frac{\bar{M}^3}{\bar{R}^4}\right)\left[31.63 - 31.28\left(\frac{\bar{M}}{\bar{R}}\right) + 7.58\left(\frac{\bar{M}}{\bar{R}}\right)^2\right] \tag{2}$$

- the frequency of the w-mode scales with the radius of the star [10]. But in this case it is not easy to get a useful approximate relation. On the other hand, the data in Figure 2c can be useful for identifying the EOS once the compactness of the star is known.

- the damping rate of the w-mode depends linearly on the compactness ratio M/R of the star [2] and we get the approximate relation:

$$\frac{1}{\tau_w(msec)} \approx 103.7 - 63.3\left(\frac{\bar{M}}{\bar{R}}\right) \tag{3}$$

In all these relations we have used $\bar{M} = M/1.4M_\odot$ and $\bar{R} = R/10$ km.

In principle, one should be able to infer the average density of the star from an observation of the f-mode, cf. eqn (1). Similarly, an observation of

Table 1: Using the relations (1), (2), and (3) we estimate the masses and the radii for polytropic stars. The errors of these estimates are shown. The notation ω_f-τ_w (etcetera) means that we use the frequency of the f-mode, eqn (1), and the damping of the w-mode, eqn (3), to approximate the stellar parameters.

N	R (Km)	ω_f - τ_w error %	ω_f - τ_f error %	$M/M_\odot$	ω_f - τ_w error %	ω_f - τ_f error %
0.8	10.026	.96	.95	1.084	6.30	.76
0.8	9.493	1.02	1.08	1.357	1.46	7.46
1.0	8.863	5.98	7.72	1.266	8.72	13.69
1.0	7.415	.99	.31	1.351	.43	1.63
1.2	12.771	4.34	5.10	1.237	4.26	1.80
1.2	8.968	5.18	3.46	1.459	1.54	7.16

the w-mode leads to an estimate of the stars compactness through eqn (3). But — more importantly — the combination of both results, *i.e.* eqns (1) and (3), gives an estimate of both the mass and the radius of the star. This result can be verified as accurate by combining equations (1) and (2). As a final step one can use this information together with the data in Figure 2c to put constraints on the equation of state.

This idea seems promising and is simple enough, but will it be useful in practice? Let us give an example for stars with polytropic equations of state: We have constructed a set of independent polytropic stellar models ($p = K\rho^{1+1/N}$) with varying polytropic index ($N = 0.8; 1; 1.2$). We have determined the f-mode and the slowest damped polar w-mode for each of these models. This data represents the "observed" gravitational-wave signal. Then we used the empirical relations (1), (2) and (3) to estimate the mass and radius of each star. Finally, the resulting values were compared to the true ones. The results of this comparison — shown in 1— clearly demonstrate the robustness of the empirical relations: The error is usually smaller than 10%.

The presented results provide the first step towards a full discussion of this problem. Much future work is required before any real conclusion about the usefulness of the proposed scheme can be drawn. We must incorporate the estimated effect of statistical and measurement errors in the analysis. It will, for example, be much more difficult to infer the w-mode damping rate from a data set than to find the f-mode pulsation frequency. It is also important to obtain fits similar to (1), (2) and (3) for a larger number of realistic EOS. Nevertheless, it seems likely that the general principle discussed here will prove useful. Moreover, if the star is rotating the mode-spectrum changes and the

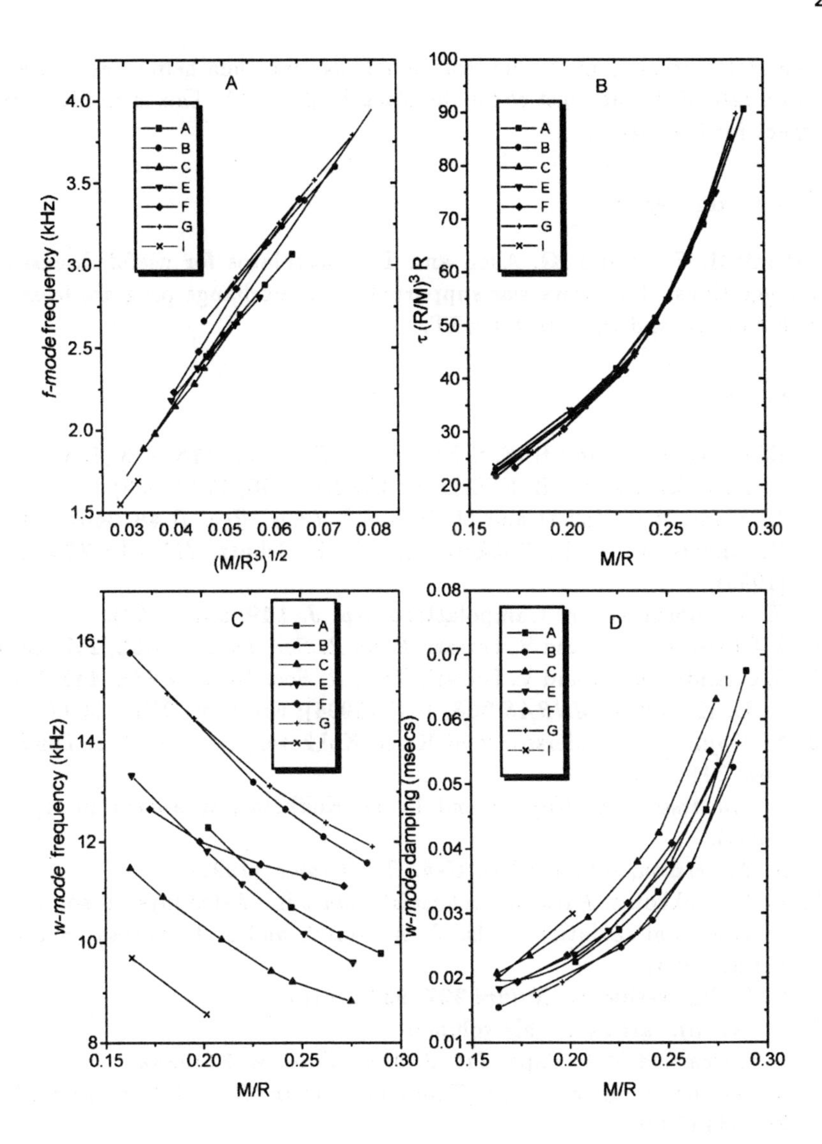

Figure 2: Data for 33 stellar models and seven EOS[17] are plotted. We show (A) the pulsation frequency of the f-mode as a function of the average density, (B) the e-folding time for the f-mode as a function of the stellar compactness, (C) the frequency of the w-mode as function of the stellar compactness, (D) the e-folding time for the slowest damped w-mode as a function of the stellar compactness (dimensionless).

degree of deviation from the non-rotational case can then help us estimate the rotation rate of the star just as for the black-hole case[18]. This is an interesting prospect for the future.

Acknowledgments

We thank B. F. Schutz, G. Allen and T. Apostolatos for useful discussions and suggestions. This work was supported by an exchange program from the British Council and the Greek GSRT.

References

1. K. D. Kokkotas and B. F. Schutz, *Gen. Rel. Grav.* **18**, 913 (1986).
2. K. D. Kokkotas and B. F. Schutz, *MNRAS* **255**, 119 (1992).
3. M. Leins, H-P. Nollert and M. H. Soffel, *Phys. Rev. D* **48**, 3467 (1993).
4. N. Andersson, K. D. Kokkotas and B. F. Schutz, *MNRAS* **274**, 1039 (1995).
5. K. S. Thorne and A. Campolattaro, *Ap. J.* **149**, 591 (1967).
6. S. Chandrasekhar and V. Ferrari, *Proc. R. Soc. Lond. A* **432**, 247 (1991).
7. S. Chandrasekhar and V. Ferrari, *Proc. R. Soc. Lond. A* **434**, 449 (1991).
8. K. D. Kokkotas, *MNRAS* **268**, 1015 (1994); Erratum: **277**, 1599 (1995).
9. Y. Kojima, N. Andersson and K. D. Kokkotas, *Proc. R. Soc. Lond. A* **451**, 341 (1995).
10. N. Andersson, Y. Kojima and K. D. Kokkotas, to appear in *Ap. J.* (1996).
11. N. Andersson, submitted to *Gen. Rel. Gravit.* (1996).
12. K. D. Kokkotas, *Pulsating relativistic stars* in "Astrophysical sources of gravitational radiation" Eds: J. A. Marck and J. P. Lasota (Springer Verlag 1996).
13. C. V. Vishveshwara, *Nature* **227**, 936 (1970).
14. V. Ferrari, article in this volume.
15. S. Frasca and M.A. Papa, *Int. J. Mod. Phys.* **4**, 1 (1995).
16. S. van der Bergh and G.A. Tammann, *Annu. Rev. Astron. Astrophys.* **29**, 363 (1991).
17. L. Lindblom and S.L. Detweiler, *Ap. J. Suppl.* **53**, 73 (1983).
18. L.S. Finn *Phys. Rev. D* **46**, 5236 (1992).

Data Analysis

SOLUTION TO THE INVERSE PROBLEM FOR GRAVITATIONAL WAVE BURSTS

MASSIMO TINTO
Jet Propulsion Laboratory, MS 161-260, Pasadena, CA 91109, USA

A review of the work on networks of three and four laser interferometers, searching in coincidence for gravitational wave bursts, is presented.[1,2] These data analysis methods allow us to locate in the sky the source of a gravitational wave burst, and to optimally reconstruct the wave's two amplitudes without any *a priori* assumption on their time dependence. In fact these techniques work for gravitational-wave bursts of any kind. Results from numerical simulations, covering different network configurations, are presented.

1 Introduction

Coincidence experiments with networks of gravitational wave detectors provide the only reliable way for discriminating gravitational wave events from spurious fluctuations induced by noise. In order to measure the gravitational waveform more than one detector is also necessary. This is because the response of a gravitational wave detector depends on the location of the source in the sky (given by two angles (θ, ϕ)), and the wave's two amplitudes $h_+(t)$, $h_\times(t)$ corresponding to the two independent polarization states referenced to a given coordinate system. In order to estimate these four unknowns, i.e. to solve the inverse problem in gravitational wave detection, a sufficiently large network of detectors widely separated on the Earth and running in coincidence is necessary.

2 Solution to the Inverse Problem

2.1 Three-Detector Networks

The key to our analysis is the characterization of any given detector by its response function $R_\Lambda(t)$ to a gravitational wave burst. This is a linear combination of the wave's two amplitudes and can be written as follows

$$R_\Lambda(t) = F_+(\theta, \phi, \alpha, \beta, \gamma)\ h_+(t)\ +\ F_\times(\theta, \phi, \alpha, \beta, \gamma)\ h_\times(t)\ +\ \Lambda(t)\ , \qquad (1)$$

where the functions F_+ and $F_\times$ are linear combinations of spin-weighted spherical harmonics in the angles (θ, ϕ), which locate the source in the sky, and in the parameters (α, β, γ) representing the geographic orientation, latitude and longitude of the detector respectively.[3,4] Here $\Lambda(t)$ is the random process associated with the noise in the detector.

In order to understand how the inverse problem can be solved with three wide-band detectors, let us assume for the moment noiseless responses. These can be written as

$$
\begin{aligned}
R_1(t) &= F_{1+}(\theta,\phi)\,h_+(t) + F_{1\times}(\theta,\phi)\,h_\times(t) &(2)\\
R_2(t+\tau_{12}) &= F_{2+}(\theta,\phi)\,h_+(t) + F_{2\times}(\theta,\phi)\,h_\times(t) &(3)\\
R_3(t+\tau_{13}) &= F_{3+}(\theta,\phi)\,h_+(t) + F_{3\times}(\theta,\phi)\,h_\times(t)\,, &(4)
\end{aligned}
$$

where we denote by τ_{12} and τ_{13} the relative time delays between detectors 1 and 2, and detectors 1 and 3 respectively. We choose detector 1 as the reference detector without loss of generality. The responses of the detectors $R_1(t)$, $R_2(t)$, $R_3(t)$ are also assumed to last only for a finite interval of time.

Let us assume for the moment that we know the exact time delays τ_{12}, τ_{13}. We shall drop this assumption later, when we derive our method for solving the inverse problem. From Eqs. (2-4) we notice that the waveforms h_+, $h_\times$ are the same functions in the three detectors at times related by the relative time delays τ_{12}, τ_{13}, while the responses are different because they are different linear combinations of the wave's amplitudes. The two independent relative time delays τ_{12}, τ_{13} provide two source directions that are mirror images of each other relative to the plane defined by the position of the detectors. Let us denote by (θ_A, ϕ_A), (θ_B, ϕ_B) these two directions. For each source direction we can obtain $h_{J+}(t)$, $h_{J\times}(t)$, $J = A, B$ as linear combinations of two of the three responses. For instance by inverting Eqs. (3, 4) we get the following expressions for $h_{J+}(t)$, $h_{J\times}(t)$

$$
h_{J+}(t) = \frac{F_{J3\times}\,R_2(t+\tau_{12}) - F_{J2\times}\,R_3(t+\tau_{13})}{F_{J2+}\,F_{J3\times} - F_{J2\times}\,F_{J3+}}\,, \qquad (5)
$$

$$
h_{J\times}(t) = \frac{F_{J2+}\,R_3(t+\tau_{13}) - F_{J3+}\,R_2(t+\tau_{12})}{F_{J2+}\,F_{J3\times} - F_{J2\times}\,F_{J3+}}\,. \qquad (6)
$$

Since we know the explicit form of the output from detector 1 (Eq. 2), by substituting Eqs. (5, 6) into Eq. (2) we derive two expressions for the output from detector 1 in terms of the outputs from detector 2 and 3. The reconstructed response of detector 1 and the actual response measured by detector 1 will agree exactly only at the correct source location. In other words we can discriminate the real location of the source from the spurious one.

From the above considerations we conclude that there exists a linear combination of the three responses which, in the noiseless case and at the correct source location, is identically equal to zero. In Ref. [1] it has been shown that such a linear combination can be written in the following way

$$I(t,\theta,\phi) = K_1(\theta,\phi)R_1(t) \; + \; K_2(\theta,\phi)R_2(t+\tau_{12}) \; + \; K_3(\theta,\phi)R_3(t+\tau_{13}) \; , \quad (7)$$

where the three functions $K_1(\theta,\phi)$, $K_2(\theta,\phi)$, $K_3(\theta,\phi)$ are equal to

$$K_1(\theta,\phi) = F_{2+}\, F_{3\times} - F_{2\times}\, F_{3+} \; , \quad (8)$$

$$K_2(\theta,\phi) = F_{3+}\, F_{1\times} - F_{3\times}\, F_{1+} \; , \quad (9)$$

$$K_3(\theta,\phi) = F_{1+}\, F_{2\times} - F_{1\times}\, F_{2+} \; . \quad (10)$$

If we now drop our initial assumption of knowing the correct time delays, we can still regard the function $I(t,\theta,\phi)$ as a two-parameter family of templates in which the two parameters are the angles (θ,ϕ), and the function we are trying to fit is the function identically equal to zero. Since we know that at the correct source location the template $I(t,\theta,\phi)$ will fit exactly the zero value, the search for the correct source location can be performed by minimizing the following expression

$$L(\theta,\phi) = \int_{-\infty}^{+\infty} I^2(t,\theta,\phi)\; dt \; . \quad (11)$$

This integral is identically equal to zero at the correct source location. Note that this method works for any gravitational wave burst, independently of the time dependence of the wave's two amplitudes $h_+(t)$, $h_\times(t)$.

In the presence of noise we have shown [1] that the function provided in Eq. (11) can be transformed into a Least Squares Function $L_\Lambda(\theta,\phi)$ as follows

$$L_\Lambda(\theta,\phi) = \frac{1}{t_1 - t_0} \int_{t_0}^{t_1} \frac{{I_\Lambda}^2(t,\theta,\phi)}{{K_1}^2(\theta,\phi)\sigma_1^2 + {K_2}^2(\theta,\phi)\sigma_2^2 + {K_3}^2(\theta,\phi)\sigma_3^2}\; dt \; , \quad (12)$$

where $I_\Lambda(t,\theta,\phi)$ is defined as follows

$$I_\Lambda(t,\theta,\phi) = K_1 R_{1\Lambda}(t) + K_2 R_{2\Lambda}(t+\tau_{12}) + K_3 R_{3\Lambda}(t+\tau_{13}) \; , \quad (13)$$

and the functions $R_{i\Lambda}(t)$; $i = 1,2,3$ are the noisy detector responses to a gravitational wave burst. The normalizing factor ${K_1}^2(\theta,\phi)\sigma_1^2 + {K_2}^2(\theta,\phi)\sigma_2^2 + {K_3}^2(\theta,\phi)\sigma_3^2$ does not contain any cross-terms since we have assumed the random processes of the noises in the detectors to be uncorrelated, and $t_1 - t_0$ is an optimally determined interval of integration containing the signal in the three responses.

2.2 Four-Detector Networks

A network with four detectors has significant advantages over a network of only three detectors. In the three detector network, the resolution of the source location determination is worse when the source location lies near the plane defined by the location of the three detectors.[1] The solution of the inverse problem is not possible whenever one of the detectors does not produce a response above the noise threshold due to the unfavorable orientation of that detector with respect to the source location. A network of four detectors does not have a *preferred* plane in general. Such networks have a better sky coverage since the region of the sky which produces two simultaneous, below-threshold responses is smaller. A four detector network supplies four independent detector responses in general. From what we described in the previous section we know that three of these responses can be used to uniquely identify the source location. The remaining independent detector response can then be used to improve the accuracy of the method.

Let us consider again the problem for noise-free detectors. Their responses can be written as follows

$$
\begin{aligned}
R_1(t) &= F_{1+}(\theta,\phi)\ h_+(t)\ +\ F_{1\times}(\theta,\phi)\ h_\times(t) && (14)\\
R_2(t+\tau_{12}) &= F_{2+}(\theta,\phi)\ h_+(t)\ +\ F_{2\times}(\theta,\phi)\ h_\times(t) && (15)\\
R_3(t+\tau_{13}) &= F_{3+}(\theta,\phi)\ h_+(t)\ +\ F_{3\times}(\theta,\phi)\ h_\times(t) && (16)\\
R_4(t+\tau_{14}) &= F_{4+}(\theta,\phi)\ h_+(t)\ +\ F_{4\times}(\theta,\phi)\ h_\times(t)\ , && (17)
\end{aligned}
$$

where τ_{14} is the relative time delay between detectors 1 and 4.

Let us assume again to know the time delays τ_{12}, τ_{13}, τ_{14}. In this case the source location is uniquely determined. We can solve for $h_+(t)$, and $h_\times(t)$ with any pair of the equations (14 - 17). By proceeding in this way, we obtain 6 expressions for $h_+(t)$, and 6 expressions for $h_\times(t)$. By taking all possible differences of the reconstructions for $h_+(t)$, we obtain 15 equations, involving the detector responses, that vanish. Similarly we can derive another 15 equations by taking all possible differences of the reconstructed $h_\times(t)$. These 30 expressions are all equal to zero only at the correct source location. It turns out [2] that of these 30 expressions only 2 are algebraically independent, each involving only three of the detector responses. They are given by the following expressions

$$
\begin{aligned}
N(t,\theta,\phi) &= a(\theta,\phi)R_1(t)+b(\theta,\phi)R_2(t+\tau_{12})+c(\theta,\phi)R_3(t+\tau_{13}) && (18)\\
O(t,\theta,\phi) &= d(\theta,\phi)R_1(t)+e(\theta,\phi)R_2(t+\tau_{12})+c(\theta,\phi)R_4(t+\tau_{13})\ , && (19)
\end{aligned}
$$

where the functions $a(\theta,\phi)$, $b(\theta,\phi)$, $c(\theta,\phi)$, $d(\theta,\phi)$, $e(\theta,\phi)$ are equal to

$$a(\theta,\phi) = F_{2+}\,F_{3\times} - F_{2\times}\,F_{3+}\,, \tag{20}$$

$$b(\theta,\phi) = F_{3+}\,F_{1\times} - F_{3\times}\,F_{1+}\,, \tag{21}$$

$$c(\theta,\phi) = F_{1+}\,F_{2\times} - F_{1\times}\,F_{2+}\,, \tag{22}$$

$$d(\theta,\phi) = F_{2+}\,F_{4\times} - F_{2\times}\,F_{4+}\,, \tag{23}$$

$$e(\theta,\phi) = F_{4+}\,F_{1\times} - F_{4\times}\,F_{1+}\,. \tag{24}$$

The least-squares function we will minimize for reconstructing the source location should include only the *noisy* functions $N_\Lambda(t,\theta,\phi)$, $O_\Lambda(t,\theta,\phi)$. We derived an optimal least-squares function for a network of four detectors, and the details of its derivation will be given in a forthcoming paper.[2]

3 Results

The methods for three and four detector networks discussed above have been implemented numerically [1,2] using simulated data as the detector responses for detectors with a bandwidth of 2 kHz.

We showed that, for broad-band signals centered around 1 kHz with a signal-to-noise ratio of 10 in each detector, a three-detector network could locate the source within a solid angle of 1×10^{-5} sr.

The angular resolution of a four-detector network is about ten times better than that of a three-detector network, with an improved *sky coverage.* A network with four detectors can identify the correct source location except when two of the detector responses are simultaneously below the threshold of the noise in the receivers. Further details of the simulations for four-detector networks are given in our forthcoming paper.[2]

References

1. Y. Gürsel and M. Tinto, *Phys. Rev.* D **4o**, 3884 (1989).
2. M. Tinto and Y. Gürsel. In preparation
3. S. V. Dhurandhar and M. Tinto, *Mon. Not. R. astron. Soc.*, **234**, 663, (1988).
4. M. Tinto and S. V. Dhurandhar, *Mon. Not. R. astr. Soc.*, **236**, 621, (1989).

DATA ANALYSIS FOR RESONANT GRAVITATIONAL WAVE ANTENNAE: INTERNAL VETOES BY χ^2 LIKE TEST

S. VITALE, G. A. PRODI, J. P. ZENDRI

Department of Physics, University of Trento and I.N.F.N. Gruppo Coll. Trento Sezione di Padova, I-38050 Povo, Trento, Italy

A. ORTOLAN, L. TAFFARELLO, G. VEDOVATO

I.N.F.N. National Laboratories of Legnaro, via Romea 4, I-35020 Legnaro, Padova, Italy

M. CERDONIO and D. PASCOLI

Department of Physics, University of Padova and I.N.F.N. Sezione di Padova, Via Marzolo 8, I-35131 Padova, Italy

We discuss the practical possibility of performing a χ^2–like test of the goodness of the optimal data filtering in the data analysis of the subkelvin resonant gravitational wave antenna AURIGA. We show, with the help of a numerical simulation, how this kind of test can substitute the empirical consistency tests that are used to reject spuria events in single detectors. We also show how the test can help to discriminate against spurious coincidences in detector arrays.

1 Introduction

With gravitational wave (g.w.) detectors the signal extraction from noisy data is performed by some kind of optimal [1] or suboptimal [2] linear filtering. It is well known that optimal linear filtering and maximum likelihood filtering give the same results when applied to data buried in gaussian noise. The maximum likelihood criterion gives in addition a mean to test the hypothesises underlying the filtering algorithm. This last implication of optimal filtering has never been applied to test against spuria events in processing the data from one or more resonant detectors. This approach is often substituted, for a single detector, by empirical tests of consistency among amplitudes of different modes. Those tests will be not viable any longer once the postdetection bandwidth will become $\geq 20\ Hz$ and a single optimal filter must be implemented. It is worth to notice that spurious events with non gaussian distribution are ubiquitous in g.w. antennae.[3] There are many sources of these extra noise events: upconversion of the seismic noise, cosmic rays, electromagnetic disturbances in the electronic chain, etc. At least part of these events should lack the specific signature of a force pulse exciting the antenna as a g.w. burst would do. A quantitative rejection algorithm could efficiently suppress the spurious events rate. In this paper we work out the theoretical framework of a likelihood hypothesis testing for a single detector and for an array of detectors.

2 Maximum Likelihood Fitting

Let us assume that the data x_i from a g.w. detector, sampled at $t_i = i\Delta t$, with i an integer, are the superposition of a deterministic signal $f(t_i)$ and a stationary gaussian stochastic noise $\eta(t_i)$:

$$x_i = \eta(t_i) + Af(t_i - t_0, \vartheta) \ . \tag{1}$$

$\eta(t_i)$ has zero mean and correlation $\langle \eta(t_i)\eta(t_i)\rangle = \sigma_{ij}$ while $f(t_i - t_0, \vartheta)$ depends also on the arrival time t_0 and on the parameter set ϑ. In order to extract the value of the signal amplitude A, of t_0 and of the ϑ's, the maximum likelihood procedure searches, as a function of the parameter values, for the maximum of the probability to get a particular data set $\{x_1 \ldots x_N\}$. This joined probability is called the *likelihood.* A more practical way to look for this maximum is to search for the minimum of the negative of its logarithm $\Lambda(A, t_0, \vartheta)$, which is known as the log-likelihood function

$$\Lambda(A, t_0, \vartheta) = \frac{1}{2} \sum_{i,j=1}^{N} \mu_{ij}[x_i - Af(t_i - t_0, \vartheta)] \times [x_j - Af(t_j - t_0, \vartheta)] \ , \tag{2}$$

where μ_{ij} is the inverse of the correlation matrix σ_{ij}.

For a given choice of t_0 and ϑ the minimum of $\Lambda(A, t_0, \vartheta)$ as a function of A is found analytically at

$$A_{opt}(t_0, \vartheta) = \frac{\sum_{i,j=1}^{N} \mu_{ij} x_i f(t_j - t_0, \vartheta)}{\sum_{i,j=1}^{N} \mu_{ij} f(t_i - t_0, \vartheta) f(t_j - t_0, \vartheta)} \ , \tag{3}$$

and it amounts to:

$$\Lambda_{min}(t_0, \vartheta) = \frac{1}{2} \sum_{i,j=1}^{N} \mu_{ij} x_i x_j - \frac{A_{opt}^2(t_0, \vartheta)}{\sigma_A^2} \ . \tag{4}$$

The square of the uncertainty on $A_{opt}(t_0, \vartheta)$ turns out to be

$$\sigma_A^2 = \frac{1}{\sum_{i,j=1}^{N} \mu_{ij} f(t_i - t_0, \vartheta) f(t_j - t_0, \vartheta)} \ . \tag{5}$$

The linear combination of the data in eq. (3) is just the discrete Wiener optimal filter matched to the function $f(t)$. On the other hand, eq. (4) shows that, in order to minimize $\Lambda(A, t_0, \vartheta)$ as a function of t_0 and ϑ, one has to maximize the square of the signal to noise ratio. This means that once one has set up the Wiener filter for the data, the best estimate of the parameters is the one that maximize the filter output. As usually the dependence of σ_A on t_0 and ϑ is very weak, one has then just to maximize $A_{opt}(t_0)$. The minimum value of $2\Lambda(A, t_0, \vartheta)$ has to be distributed approximately as a χ^2 with $N - N_\vartheta - 2$ degrees of freedom, with N_ϑ the number of elements of the parameter set ϑ. To prove this last statement we notice that the signal to noise ratio for the signal amplitude A does not depend on any preliminary linear filter. As a resonant antenna is minimum-phase system we can apply to the data a well defined whitening filter [4] $y_i = \sum_{j=1}^{N} w_{ij} x_j$, where w_{ij} diagonalizes the covariance matrix $\sum_{jk=1}^{N} w_{ij}\sigma_{jk}w_{kl} = \sigma_w^2 \delta_{il}$. The new data set $\{y_i\}$, known as the innovation of the original process $\{x_i\}$, forms, in the absence of signals, a stochastic process with zero mean and correlation $\langle y_i y_j \rangle = \sigma_w^2 \delta_{ij}$. One can then write

$$\chi^2(t_0, \vartheta) \equiv 2\Lambda_{min}(t_0, \vartheta) = \sum_{i=1}^{N} \frac{y_i^2}{\sigma_w^2} - \frac{A_{opt}^2(t_0, \vartheta)}{\sigma_A^2} \tag{6}$$

which shows the χ^2 nature of the statistics.

Let now make the practical case where the only parameters to be evaluated are the amplitude A and the time of arrival t_0 of a g.w. burst. The procedure devised can then be made in the following steps. First the data are whitened with a whitening filter and the time series $\{y_i\}$ is generated. For resonant antennae, the whitening filter is just the causal part of the Wiener filter matched to a Dirac δ function at the antenna input. Second the anti-causal Wiener filter is set up. Both the filters can be implemented as simple ARMA filters.[1] As the exact arrival time of the δ-pulse is not known, the usual procedure for this step consists of building the time series $A_{opt}(t_0)$ and to take as candidate events those values where the two following conditions occur: i) $|A_{opt}(t_0)|$ exceeds some threshold value properly set to avoid too large a rate of false alarms and ii) $|A_{opt}(t_0)|$ is a local maximum in the time interval where the threshold is exceeded. Let call t_* a time where both conditions above are fulfilled. Consistency needs then that $\chi^2_{min}(t_*)$ is a sample of a χ^2 statistics with $N - 2$ degrees of freedom.

We have shown [5] that this procedure can be also extended to exert quantitative vetoes when we analyze the data coming from an array of M parallel antennae, located at different sites. Despite the arrival time of the g.w. signal in each antenna is not the same, we can set up a global χ^2 which tests the hypothesis of a plane g.w. burst with a unit wave vector $\vec{k}$ impinging the array.[5] We summarize here the main results of the minimization of the global likelihood function (defined as the product of the likelihood function of each detector) with respect to the amplitude A of the incoming g.w. signal:

i) for given values of the signal parameters t_0 and $\vec{k}$, the best estimate of the amplitude of the gravitational wave burst A_{opt} is the weighted mean of the best estimate of the amplitude of each detector $A_{opt} = [\sum_{\alpha=1}^{K} A_{opt}^{\alpha}/(\sigma_A^{\alpha})^2]/\sum_{\alpha=1}^{K}(\sigma_A^{\alpha})^{-2}$, with an error $\sigma_{A\ opt} = [\sum_{\alpha=1}^{K}(\sigma_A^{\alpha})^{-2}]^{-1/2}$, where A_{opt}^{α} is the best estimate of the signal amplitude in each detector and σ_A^{α} the corresponding error;

ii) the minimum of the global log-likelihood function evaluated at $A = A_{opt}$ is given by $\chi^2 = \sum_{\alpha=1}^{M} \chi_{\alpha}^2 + \sum_{\alpha=1}^{M} [A_{opt}^{\alpha} - A_{opt}]^2/\sigma_A^{\alpha\,2}$ where χ_{α}^2 is the χ^2 value for each detector. The second term of right hand side clearly shows that, in order to satisfy the global χ^2 test, the fluctuation of each estimated amplitude around the common weighted average value $A_{opt}^{\alpha} - A_{opt}$ should not exceed the standard deviation σ_A^{α}.

We have implemented the computation of χ^2 time series on the on-line data analysis of the AURIGA antenna and we have tested our procedure with the help of a numerical simulation. The antenna can be modelled as a linear electromechanical system [6] composed by two coupled mechanical oscillators, the bar and the transducer, a third "LC" resonator, and a linearized SQUID amplifier. The resulting equivalent lumped–element electrical circuit for the AURIGA detector has been solved numerically [6] and, within this linearized model, we have calculate the transfer functions for i) a δ–force pulse hitting the antenna (equivalent to a g.w. event), for ii) a δ–force pulse exciting the transducer as a mechanical creep event is expected to do, and for iii) an electromagnetic pulse entering in the electrical resonator respectively. We have then simulated $\approx 10^2$ events for each of these three kinds of signal; the resulting reduced χ^2, defined as $\chi_r^2 = \chi^2/(N-2)$, as a function of their signal-to-noise ratio $SNR \equiv A_{opt}/\sigma_A$ after the Wiener filter is plotted in fig. 1. The results we have obtained are quite general: as eq. (6) suggests the expectation value of χ_r^2 is fitted by

$$\langle\chi_r^2\rangle = 1 + \lambda\, SNR^2 \,, \tag{7}$$

where λ is a numerical constant which depends on the spurious and the antenna parameters. Obviously $\lambda = 0$ for the antenna events and $\lambda > 0$ for the extra noise events. The value of λ can be estimate directly from the experimental apparatus by generating spurious events in the antenna obtaining an estimate of the efficiency of the χ^2 test. In our numerical simulation we have found $\lambda = 9 \times 10^{-3}$ for the transducer events and $\lambda = 4.3 \times 10^{-2}$ for the electrical events that ensures, for instance, an efficient spurious rejection for $SNR \geq 8$.

The implementation of the test procedure for the actual AURIGA data

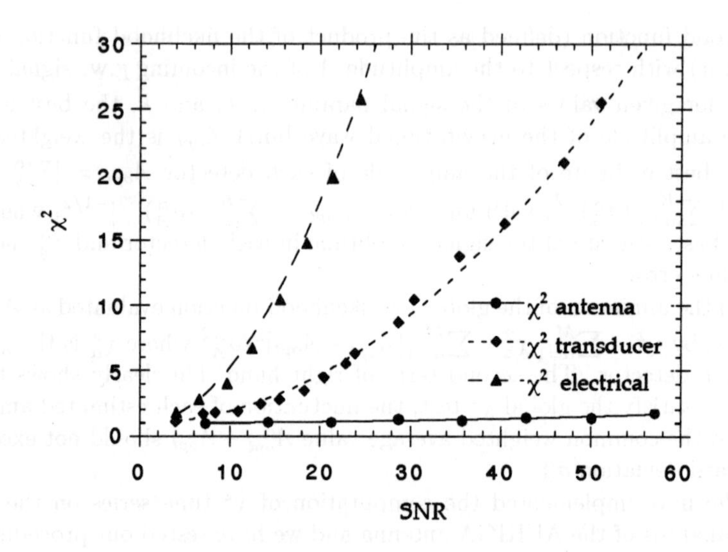

Figure 1: Expectation value of χ_r^2 as a function of SNR for the three kinds of event.

stream is currently under development.

References

1. S. Vitale *et al.* in *Proc. of the* 1th *E. Amaldi International Meeting on g.w. Experiments* (World Scientific, Singapore, 1995), p. 220.
2. P. Bonifazi *et al. Nuovo Cimento* **C1**, 465 (1978).
3. see e.g.: W. W. Johnson *et al.* as in ref. [1], p. 128; P. Astone *et al. Phys. Rev.* D **47**, 2 (1993).
4. A. Papoulis, *Probability, Random Variable, and Stochastic Processes*, (McGraw-Hill, London, 1984).
5. S. Vitale *et al. Nucl. Phys. B* (Proc. Suppl.) to be published.
6. M. Cerdonio, *et al.*, Phisica **B 194**, 3 (1994); M. Cerdonio, *et al.*,*Nucl. Phys.* B(Proc. Suppl.) **35**, 75 (1994).

STRATEGIES FOR ON-LINE DATA ANALYSIS IN THE VIRGO EXPERIMENT USING THE APE*mille* PARALLEL COMPUTER *

M. BECCARIA, G. CELLA, A. CIAMPA,
E. CUOCO, G. CURCI, A. VICERÉ

Dipartimento di Fisica dell'Università and INFN, sezione di Pisa
Piazza Torricelli 2, I-56126 Pisa, Italy

We discuss possible strategies for the analysis in real time of the data produced by the VIRGO experiment. We focus on signals from binary coalescences and from non axisymmetric neutron stars. The detection of coalescing binaries can be performed using Wiener filtering techniques; we estimate the required computing power and we show that a Single Instruction Multiple Data computer, like the APE*mille* system, can efficiently implement a filter bank capable to recover, *on-line*, 95% of the Signal to Noise Ratio. Signals from the pulsars of known position can be also monitored on-line, using a matched filter approach. Given the upper limits on the gravitational wave emission, the number of pulsars to be monitored is not very high; this situation is going to change with the on-going pulsar surveys, but we expect that this task will never require very high computational power, as is the case for the "blind" search.

1 The APE*mille* system

The APE (an acronym for Array Processor Experiment) project was started in 1986 by INFN, to build a computer for Lattice Quantum Chromodynamics.

The first stage of the project ended in 1989, and the APE 1 computer (peak power of 1 GFlop), was used for the computation of hadron masses; it was a *dedicated* computer, for instance its native data type was `complex`.

In the second stage of the project, started in 1990 and ended in 1995, it has been realized the APE100 system, a much more flexible machine [2]. It is based on a floating-point unit equipped with a local memory and a peak performance of 50 MFlops; up to 2048 FPUs can be embedded in a three-dimensional mesh, providing a peak 100 GFlops. The processors are arranged in cubes of 8 elements, linked by a data network optimized for nearest-neighbour communications. All the PE run the same program on different data (SIMD programming model), while the central CPU handles the global control. Local conditioning is possible, inhibiting memory modifications on those FPU where some statement is false.

*Talk given by A. Viceré

In 1995 the third stage started, to build in 4 years the APE*mille* system [3]. Each APE*mille* FPU will deliver 200 MFlops of peak performance and up to 4096 processors will be arranged in a three dimensional network with an improved data transfer mechanism, allowing efficient communications also among non-nearest neighbour nodes. The programming model will be still synchronous, however different processors will be able to address different local memory locations (SIMAMD scheme). Different *native* data types, `integer`, `real`, `double` will be available on the single Processing Element; each PE will be equipped with up to 10 MWords (40 MBytes) of local memory.

It is the purpose of this paper to discuss the application of this computing system to the data-analysis in the VIRGO experiment.

2 Coalescing binaries

The best understood and promising transient sources of GW are compact binaries, Neutron Stars (NS) and/or Black Holes (BH), spiralling one around the other because of the energy loss until coalescence [4,5,6,7].

The increasing accuracy of the theoretical predictions [8,9], makes a detection based on Wiener filtering (*optimal* for gaussian noise) very appealing [10]. The Wiener filtering is computationally expensive because it requires a large number of filters to map the parameter space; however it has long been recognized as particularly well suited for a parallel computer [11].

The strategy is well known [12,13,14]; the theoretical signal q_θ depends on a few continuum parameters, discretized in such a way to guarantee a SNR loss not exceeding a predefined amount. Each filter is realized taking the Fourier transform of the template and of a portion of the input data, $\tilde{s}(\nu)$: the statistic

$$C(\theta) = (\mathbf{q}(\theta) \mid \mathbf{s}) \equiv \int_{\nu_a}^{\nu_f} d\nu \, S_n^{-1}(\nu) \left[\tilde{q}_\theta(\nu)\, \tilde{s}^*(\nu) + \text{c.c.}\right] \tag{1}$$

is maximized over the θ lattice, and compared with a detection threshold. The finer the lattice, the smaller the SNR loss; on APE*mille* each PE would implement a subset of filters, evaluating C for different values of θ.

Considering a PN1 signal, it depends on two parameters, the masses of the coalescing objects, or equivalently the contributions, order by order in the post-newtonian expansion, to the time τ spent by the coalescing system between the instant t_a characterized by a GW frequency ν_a, and the merging:

$$\tau_0 \simeq 34 \left[\frac{\mathcal{M}}{M_\odot}\right]^{-5/3} \left[\frac{\nu_a}{40\text{Hz}}\right]^{-8/3}, \quad \tau_1 \simeq (0.75 + 0.94\frac{\mu}{M}) \left[\frac{\mu}{M_\odot}\right]^{-1} \left[\frac{\nu_a}{40\,\text{Hz}}\right]^{-2};$$

μ is the reduced mass, M the total mass and $\mathcal{M} = (m_1 m_2)^{3/5} / (m_1 + m_2)^{1/5}$ is the *chirp*-mass. In this approximation the templates are

$$\begin{aligned} \tilde{q}(\nu) &\propto \nu^{-7/6} \exp i \{2\pi\nu\chi(\nu) + \Phi\} \\ \chi(\nu) &= t_a + \frac{\nu_a}{\nu}\left[\frac{8}{5}\tau_0\left(1-\left(\frac{\nu}{\nu_a}\right)^{-5/3}\right) + 2\tau_1\left(1-\left(\frac{\nu}{\nu_a}\right)^{-1}\right) + \ldots\right], \end{aligned} \tag{2}$$

This scheme is most efficient when each filter fits in the local memory; as a coalescence provides a SNR in the frequency range $[\nu_a, \nu_f]$

$$\mathrm{SNR}^2 \propto \int_{\nu_a}^{\nu_f} d\nu \frac{|h(\nu)|^2}{S_n(\nu)} = \int_{\nu_a}^{\nu_f} d\nu \frac{\nu^{-7/3}}{S_n(\nu)}, \tag{3}$$

with $\nu_f \simeq$ 1kHz, about 95% of the total SNR is recovered, by VIRGO,[†] setting $\nu_a \simeq$ 40Hz. This cutoff, together with the smallest chirp mass (say, $0.25\, M_\odot$) considered, sets the higher limit on τ, of the order of 350 s. The needed frame length is $\approx 4\nu_c\tau$, where ν_c is a reduced sampling $\sim$ 2 kHz; one gets $\sim$ 3 MWords, well within the memory of the single APE*mille* PE.

To evaluate the number of filters, we consider $\mathcal{M} \in [0.25, 30]\ M_\odot$, corresponding to an area of $\sim 120\,\mathrm{s}^2$ in the τ_0, τ_1 plane. Each PN1 filter, detects the signals in a certain "recovery ellipse", whose area depends on the minimal % of SNR recovered, and for the VIRGO noise one has $\Delta A^{95\%} \simeq 0.0018\,\mathrm{s}^2$, $\Delta \mathrm{A}^{90\%} \simeq 0.0040\,\mathrm{s}^2$. These informations allow to evaluate the computing power

%SNR	$\mathcal{M} \in [0.25\, M_\odot, 30\, M_\odot]$	λ (MFlops)	N_f	Σ (GFlops)
0.95		.27	1.6×10^5	43.2
0.90		.27	7.0×10^4	18.9
	$\mathcal{M} \in [0.50\, M_\odot, 30\, M_\odot]$			
0.95		.25	2.3×10^4	5.8
0.90		.25	1.0×10^4	2.5

These estimates neglect the re-generation of the templates (which cannot be stored in memory), as well as the inefficiency; to play safe one may require around 80 GFlops *sustained*, which could be delivered for instance by a 1024 nodes APE*mille* system running at 50% efficiency.

[†]The model used for the VIRGO noise $S_n(\nu)$ is

$$S_n(\nu) = 10^{-47}\left[\alpha_1\left(\frac{\nu}{100\mathrm{Hz}}\right)^{-5} + \alpha_2\left(\frac{\nu}{100\mathrm{Hz}}\right)^{-1} + \alpha_3\left(\frac{\nu}{100\mathrm{Hz}}\right)^{2}\right]\mathrm{Hz}^{-1}, \tag{4}$$

where the three figures

$$\alpha_1 = 2.0,\ \alpha_2 = 91.8,\ \alpha_3 = 1.23 \tag{5}$$

characterize the relative weight of *pendola* thermal noise, *mirror* th. noise and shot noise.

3 Pulsars

An important source of continuous GW in the bandwidth of VIRGO and LIGO [15] detectors is a rapidly rotating NS. To date, 706 pulsars are known in a rotational frequency range $\sim 1-1000$ Hz. Their heliocentric distances are of some Kpc and only 5 are outside our galaxy. We will address the question of the computational power required by the on–line monitoring of these sources.

It is expected that pulsars emit grav. energy according to the equation

$$\frac{dE}{dt} = -\frac{G}{5c^5}\left(\frac{d^3Q_{\alpha\beta}}{dt^3}\right)^2 \tag{6}$$

at twice the rotational frequency, $\omega_p = 2\Omega_p$, but also at the rotational frequency Ω_p [16], depending on the relation between the rotation axis and the quadruple moment $Q_{\alpha\beta}$ [17]. The characteristic amplitude h_0 generated by a source with mom. of inertia I and ellipticity ϵ at a distance r has the form

$$h_0 = 1.07 \times 10^{-29}\left[\frac{\omega_p}{\text{Hz}}\right]^2\left[\frac{\text{Kpc}}{r}\right]\left[\frac{I}{10^{38}\text{Kg}\,\text{m}^2}\right]\left[\frac{\epsilon}{10^{-6}}\right], \tag{7}$$

where ϵ characterizes the degree on non-axisymmetry of the NS [17].

We write the interferometric signal adding a noise term $\xi(t)$

$$X(t_n) = \sum_i h^{(i)}(t_n) + \xi(t_n), \tag{8}$$

and the signal of the i-th source is a sum over two polarization states σ

$$h^{(i)}(t_n) = \sum_{\sigma=+,\times} F_\sigma^{(i)}(t_n)A_\sigma^{(i)} \cos\left(\omega_p^{(i)}t_n + \delta^{(i)}(t_n) + \phi_\sigma\right) . \tag{9}$$

The Doppler shift due to the relative detector–source motion is encoded in δ, while $F(t)$ models the sensitivity of the antenna in dependence on the rotation of the earth and the source position.

Due to F and δ, the pulsar's signal is not monochromatic and direct spectral methods do not permit a detection. This problem can be solved using a generalization of the Lomb–Scargle transform [18,19] to evaluate the demodulated power spectrum in a frequency band around ω_p. The interferometric data are reinterpreted as an unequally spaced time sequence, weighted with the instantaneous sensibility of the apparatus. A key point is the number of demodulations required for each pulsar; each demodulation can be seen as a filter matched for the known parameters in Eq. 9, hence we must take into account the indeterminations on position and frequency of the pulsars. To extract the signal we need a real filter close to the optimal matched filter, measuring the separation in terms of the reduction of signal extraction efficiency; for each pulsar we need a number of matched filters

$$N_{\mathrm{F}}^{(i)} = N_{\mathrm{ang.}}^{(i)} \times \frac{\Delta\omega_p^{(i)}}{2\pi} \times T \; ; \tag{10}$$

$N_{\mathrm{ang.}}^{(i)}$ is the number of filters covering the angular indetermination with the desired efficiency, while T is the observation time. For a required minimal efficiency of 0.95, 3 years of observation and a signal frequency > 4 Hz, the known sources can be divided in three groups: in the first there are 85 pulsars requiring a single filter. In the second group we find 180 pulsars with a larger parameter indetermination, requiring an average 10^5 filters. In the third we collect the sources requiring a huge number of demodulations, $10^{10} - 10^{15}$.

To discuss observability, we need to estimate the ellipticity: its distribution is very model dependent, so we limit ourselves to a *very* conservative upper bound, equating the kinetic energy loss (measured in terms of $\dot{\Omega}_p$) to the gravitational emission, i. e. assuming no other braking mechanisms. We obtain

$$\epsilon_{\mathrm{max}} = 6\sqrt{\left[\frac{10^{38}\mathrm{Kg\,m}^2}{I}\right]\left[\frac{P}{\mathrm{ms}}\right]^3 \dot{P}} \qquad \text{with} \qquad P = \frac{2\pi}{\Omega_p} \; . \tag{11}$$

For ~ 200 pulsars the value of $\dot{P}$ is too small to be measured. In the other cases we observe a strong correlation between ϵ_{max} and the frequency, well fitted by the law $\epsilon_{\mathrm{max}} \propto \omega_p^{-2}$. This relation cancels the ω_p^2 term in the equation 7, hence the estimate of h_0^{max} obtained setting $\varepsilon = \varepsilon_{\mathrm{max}}$ is independent on the frequency. This correlation does not imply that the expectation value of the signal amplitude is frequency independent, as the deviation from the upper bound depends on the balance between the different energy loss mechanisms, which depends on ω_p. Combining this result with the VIRGO noise spectrum $S_n(\nu)$ we get the SNR distribution for an integration time T:

$$\mathrm{SNR}^{\mathrm{max}} = h_0^{\mathrm{max}} \left(\frac{T}{S_n(\frac{\omega_p}{2\pi})}\right)^{1/2} \; . \tag{12}$$

The number of pulsars with a maximum expected value of SNR greater than one is 22, and the number of required matched filters is ~ 1000. Each of the proposed filters requires roughly 0.25 MFlop/sec, summing to a total computational power of ~ 250 Mflops/sec. With the future pulsar surveys this request will grow, but we do not think that our main conclusions will change: the real "grand challenge" is the "blind search" of continous sources over all the sky.

Acknowledgments

We thank Prof. A. Giazotto, Prof. F. Fidecaro and Dr. D. Passuello for many useful discussions. We also thank Prof. R. Tripiccione for details on APE*mille*.

References

1. C. Bradaschia *et al.*, *Nucl. Instrum. Methods* A **289**, 518 (1990).
2. A. Bartoloni *et al.*, *Int. J. Mod. Phys* C **4**, 955 (1993).
3. A. Bartoloni *et al* (APE collaboration), APE*mille*: *a parallel processor in the teraflop range*, internal INFN note (1995).
4. K. S. Thorne, in *Proceedings of the Snowmass 95 Summer Study on Particle and Nuclear Astrophysics and Cosmology*, edd. E.W. Kolb and R. Peccei (World Scientific, Singapore, 1996).
5. *Etude d'algorithmes rapides de recherche d'un signal d'onde gravitationnelle provenant de coalescences d'etoiles binaires*, D. Verkindt, Thèse, Université de Savoie, 93-CHAM-S003.
6. L. S. Finn, *Phys. Rev.* D **53**, 1996 (2878).
7. E. Poisson, preprint gr-qc/9508017, to appear in *Proceeding of the Sixth Canadian Conference on General Relativity and Relativistic Astrophysics.*
8. T. Damour, in *Three hundred years of gravitation*, edd. S.W. Hawking and W. Israel (Cambridge Univ. Press, 1987).
9. L. Blanchet *et al.*, *Phys. Rev.* D **51**, 1995 (5360).
10. B. F. Schutz in *The detection of gravitational waves*, ed. D. G. Blair (Cambridge University Press, Cambridge, 1991).
11. B. S. Sathyaprakash, S. V. Dhurandhar, *J. Comp. Phys.* **109**, 215 (1993).
12. C. Cutler *et al*, *Phys. Rev. Lett.* **70**, 2984 (1993).
13. L. S. Finn and D. Chernoff, *Phys. Rev.* D **47**, 2198 (1993).
14. B. S. Sathyaprakash and S. V. Dhurandhar, *Phys. Rev.* D **44**, 3819 (1991); *Phys. Rev.* D **49**, 1707 (1994).
    B. S. Sathyaprakash, *Phys. Rev.* D **50**, R7111 (1994).
15. A. Abramovici et al, *Science* **256**, 325 (1992).
16. M. Zimmerman and E. Szedenits, *Phys. Rev.* D **20**, 351 (1979).
17. S. Bonazzola and E. Gourghoulhon, preprint gr-qc/9604029, submitted to *Astron. and Astrophys.*
18. J. D. Scargle, *Astrophys. J.* **263**, 835 (1982).
    W. H. Press and G. B. Rybicki, *Astrophys. J.* **338**, 277 (1989).
19. M. Beccaria, G. Cella, A. Ciampa, G. Curci, E. Cuoco, A. Viceré, VIRGO note NTS-9624.

THE ON-LINE AND DATA ARCHIVING SYSTEM OF VIRGO

F. BARONE on behalf of

X.GRAVE, F.MARION, R.MORAND, B.MOURS
LAPP, Chemin de Bellevue, B.P. 110, F-74941 Annecy-Le-Vieux Cedex, France

F.BARONE, A.GARUFI, L.MILANO, G.RUSSO
INFN Napoli and Univ. Napoli Federico II, *Dip. Scienze Fisiche, Mostra d'Oltremare, Pad.19, I-80125 Napoli, Italy*

F.CAVALIER, M.DAVIER, F.LE DIBERDER, P.ROUDIER
Lab. de l'Accelélérateur Linéaire, IN2P3-CNRS and Université de Paris-Sud, F-91405 Orsay, France

In this paper we will describe the basic architecture of the on-line system and data archiving system of the VIRGO antenna. We will give also a global description of the data structure (frames) in connection with the architectures of the Frame Builder, On-line Processing and Data Archiving systems.

1 Introduction

The On-line and Data Archiving system acquires all the useful information for the VIRGO[1-3] on-line and off-line data analysis. In fact, all the locking and alignment signals, together with monitoring and environment ones, are acquired, structured, processed and stored on disks and on tapes together with the monitoring and environment ones.

2 Architecture

The configuration of the interferometer requires the implementation of a distributed readout system as shown in Fig. 1a. In fact, in each building a Local Readout System collects the available data and send them to the Frame Builder via a digital optical link (DOL). All these tasks are synchronized by a timing system which controls the acquisition (at a rate of 20 kHz) and the data transfer. A Slow Monitoring Network (Ethernet and/or FDDI) is used for data transfer of quantities which are sampled at low rate (a few Hz). The Frame Builder collects all these *raw* data, structures them into frames and send them to the Raw Data Archiving System and to the On-Line Processing System as shown in Fig. 1b (a rate of 2 MByte/sec is expected).

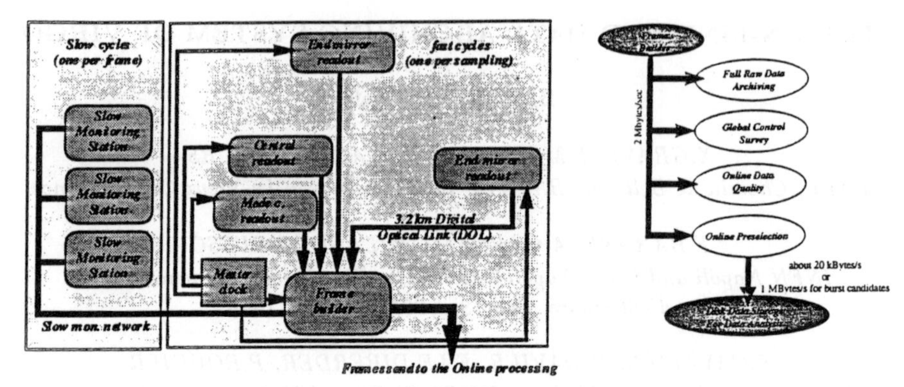

Figure 1: a) Distributed Data Acquisition System Architecture; b) On-Line Processing and Data Archiving Architecture.

The Raw Data Archiving System archives all the frames on DAT tapes, making it possible any retrieval and reprocessing of the original data. The On-Line Processing System converts the raw data to physical quantities, computes the corresponding h values and adds these information to the frame structure. It also runs on-line data analysis algorithms provided and validated by the VIRGO collaboration for the on-line selection of the frames in which a GW candidate may occur (e.g. bursts, binary coalescence). All the selected frames and subsets of the non-selected ones are then collected by the Data Distribution System which stores them on disks (on-line data distribution) and on DAT tapes (DST - Data Summary Tapes) for the VIRGO collaboration off-line data analysis.

The quality of the data is permanently surveyed by the On Line Data Quality Section, while the Global Control Survey reacts to faulty behaviours affecting the locking of the interferometer.

3 Data Structure

The frame is organized as a set of C structures described by an header holding pointers to additional structures and values of the parameters. The header is followed by an arbitrary number of additional structures, each holding the values of a rapidly varying parameter. The frame is divided into three main structures, as shown in Fig 2:
[1] Structures filled by the Frame Builder.
[2] Structures filled by the On-Line Processing or by the Off-Line Reprocessing.
[3] Structures filled by the simulation.

All this information is archived in the following way. All the sections [1] are stored by the Raw Data Archiving System on DAT tapes. The Data

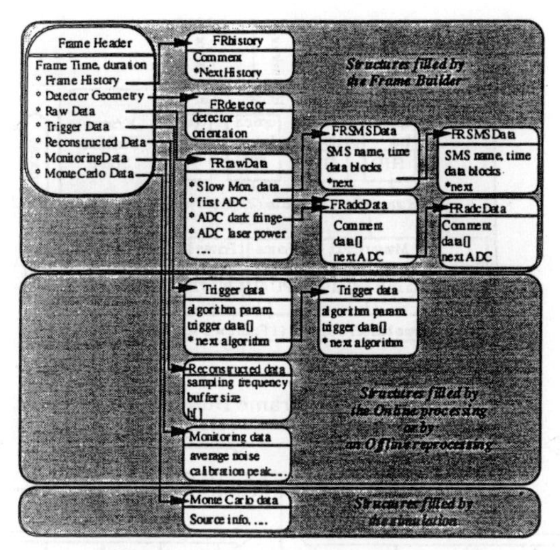

Figure 2: The frame structure.

Distribution System archives the sections [1][2][3] of the selected frames and the section [2] of the non-selected ones. If there is the need of reconstructing the whole frame, then the Data Distribution System reprocesses the sections [1] stored by the Raw Data Archiving System.

4 Implementation

The software-hardware structure is built within the framework of the client-server model. Communications among clients and servers is made via the multitask communication package, Cm [3].

4.1 Frame Builder

The Frame Builder architecture is shown in Fig 3. Synchronized by a clock, a processor collects the data from the FIFO memories for one frame, while the other one assembles the previous one and send it to the storage and on-line processing. The slow monitoring data are acquired by the master CPU and included in the frames.

4.2 Raw Data Archive

We have adopted a modular solution consisting in the parallel stage of the data on disks (up to 10 MByte/sec data flow) and then in the copy on DAT tapes. This solution makes it easy to reconfigure the system on the basis of the actual data flow. In Fig. 4 the architecture of the Raw Data Archive is shown [3].

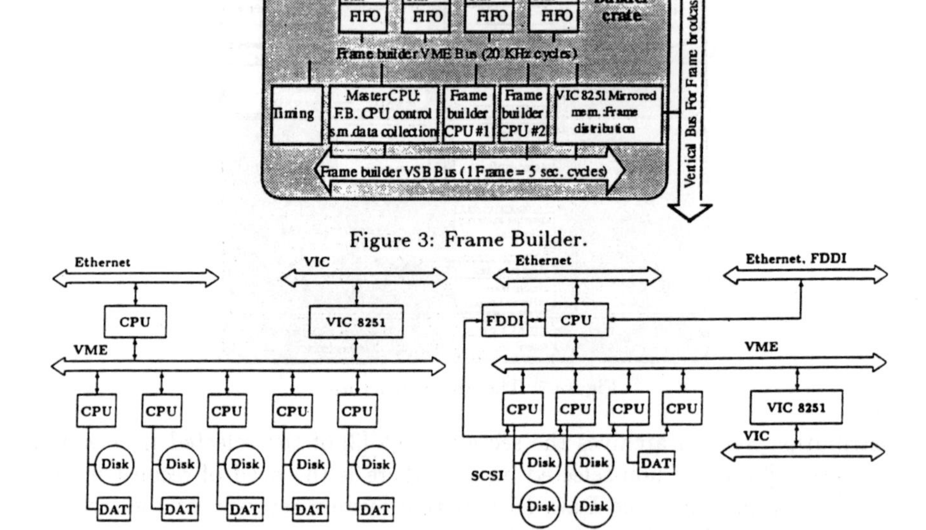

Figure 3: Frame Builder.

Figure 4: a) Raw Data Archiving System. b) Data Distribution System.

4.3 Data Distribution

The data acquisition and the data distribution follows two different channels. The data acquisition and storage is obtained using an open architecture multiprocessor system, with multiple buses for I/O and multiprotocol (VIC, VME, PCI) while the data distribution uses standard network access to the data (Ethernet, FDDI, VME, PCI). We foresee to have 500 GByte on line as a first step (80 disks 9 GByte each), but this structure can be easily expanded[3].

Acknowledgments

We acknoledge the Virgo collaboration for many helpful discussions and suggestions.

References

1. *VIRGO: Proposal for the construction of a Large Interferometric Detector of Gravitational Waves* (INFN, Italy and CNRS, France) (1989).
2. *VIRGO: Final Conceptual Design* (1992).
3. *VIRGO: Final Design*, (1995).

OPTICAL SIMULATIONS
OF THE VIRGO INTERFEROMETER

CAVALIER Fabien, on behalf of

CAVALIER, F., HELLO, P.
Laboratoire de l'Accélérateur Linéaire, IN2P3/CNRS and Université de PARIS-Sud 91405 ORSAY Cedex, FRANCE

VINET, J.Y.
Groupe VIRGO, Bat 208, 91405 ORSAY Cedex, FRANCE and Laboratoire d'Optique Appliquée (Ecole Polytechnique) 91120 Palaiseau, FRANCE

CARON*, B., FLAMINIO, R., MARION, F., MEHMEL*, C.MOURS, B.
LAPP Chemin de Bellevue, B.P 110, 74941 ANNECY-LE-VIEUX Cedex, FRANCE
* *at LAMII/CESALP, B.P 806 74016 ANNECY, FRANCE*

The optical simulations are a powerful mean to study and design an interferometer. On the one hand, we will present the different simulations developed for the VIRGO interferometer. On the other hand, we will show how those tools are used in one of their main application field : the locking studies.

In order to design and understand the VIRGO interferometer, significant efforts have been made by the VIRGO collaboration in the field of simulation : optics, mechanics, thermal noise ...

The various optical simulation implemented for VIRGO will be described and some examples of simulation will be presented.

1 The Various Optical Simulations

1.1 Definitions

First, let's define the general layout of optical simulations. Figure 1.a shows the main components :
- the mirrors, defined by their operators of reflection and transmission
- the length of the cavity and the associated propagation operator
- the input and cavity fields

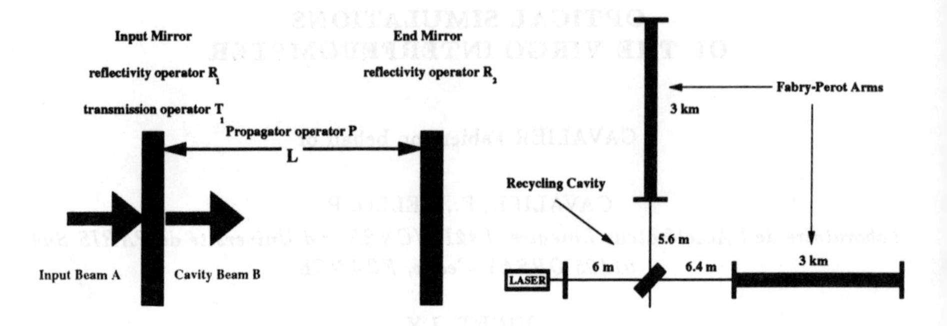

Figure 1: a. General Layout of a Cavity. b. VIRGO Layout

The cavity field obeys the following equation (see figure 1.a for the notations) :

$$B(t) = T_1 A(t) + R_1 P R_2 P B(t - \frac{2L}{c}) \tag{1}$$

This kind of formula can be generalised for the whole VIRGO interferometer (fig. 1.b). The simplest way to describe the fields is to use the plane wave approach (where the transversal structure of the beam is neglected) : fields and operators are complex numbers, and the propagation consists of a phase shift. So, the calculation is easy and fast.

If the transversal effects must be simulated, a modal[1] or a grid decomposition[2,3,4] must be used. In the modal decomposition, the field is a superposition of the Hermite-Gauss modes (TEM) of the cavity : $B = \sum a_{ij} TEM_{ij}$. In this case, the operators become matrices and the propagation is a phase shift, different for each mode. Unfortunately, the operators are computable only for small tilts of the mirrors.

If a precise description (surface map for example) of the mirrors must be handled, then the grid method is the right tool. In this case, the field is a function of the positions and its Fourier transform is used for the propagation : $B = B(x,y) \overset{FFT}{\longleftrightarrow} \tilde{B}(\nu_x, \nu_y)$. The reflection and transmission operators are phase shifts in the spatial domain and the propagation is a phase shift in the frequency domain. This method is the most accurate one but slow because of the Fourier transform.

1.2 Static and Dynamical Models

Once a beam description is chosen, the fields can be computed thanks to equation 1. But, iterations are needed each $\frac{2L}{c}$: $8 10^{-8}$ seconds for the VIRGO recycling cavity and $2 10^{-5}$s for the VIRGO Fabry-Perot arms (see figure 1.b for the VIRGO layout). Then, the simulation of a few seconds of operating time needs lot of iterations and takes hours on a ALPHA station.

In VIRGO, we have developed several algorithms in order to decrease the time consumption :
- propagation in the recycling cavity is neglected $\Longrightarrow$ fields computed each $\Delta t = 2 10^{-5}$s
- eq. 1 becomes a differential equation solved with $\Delta t \gg 2 10^{-5}$s
- eq. 1 translated in the frequency domain and computed with $\Delta t \gg 2 10^{-5} s$

The other way to save time is to suppose that the permanent regime is achieved : $B(t) = B(t - \frac{2L}{c})$. This equation can be solved analytically for the Plane Wave and the Modal approaches. So, it gives a very fast tool for the field computation, usable for locking studies. In the case of grid decomposition, iterations are needed until it converges, so it can not be used for locking studies but it is the right tool for the design and the specifications of the mirrors.

2 The Locking Studies

The majority of these algorithms are implemented in SIESTA [5], the VIRGO simulation program. SIESTA contains also simulations of mechanics and electronics, and its main use is the understanding of the locking problem : permanent locking, transient effects, start of locking, lost of locking, locking parameters (modulation frequency ν_{mod}, asymmetry between the two arms ...).

2.1 Permanent Locking in the Linear Regime

This kind of study can be performed with the Plane Wave approach and the static approximation. Thanks to their speed, one is able to simulate several minutes of the interferometer control. Figure 2.a shows the attenuation of the noise when a feedback is applied on the mirrors. One can easily change the parameters of the simulation (ν_{mod},asymmetry, mirrors where the forces are applied, frequency range of the feedback) in order to define the best strategy.

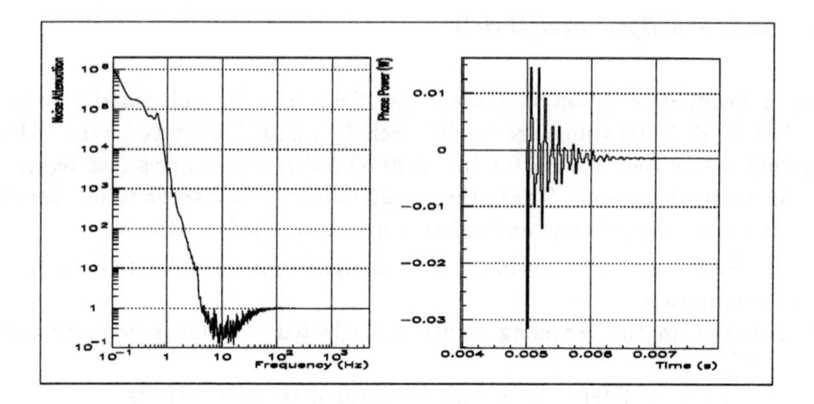

Figure 2: a. Longitudinal Locking. b. Transient Signal in a Cavity.

2.2 The Transient Effects

During the locking acquisition regime, the photo-diode signals can be more complicated than in the linear regime. Figure 2.b shows an example of a transient signal in the case of a simple cavity. The cavity is resonant at the beginning, and the end mirror is moved by 10^{-7} meters during 20 μseconds at t=.005 seconds. Once more, the complete control of the parameters allows to understand which physical information (speed, position) can be extracted from the photo-diode signals.

3 Conclusion

The optical simulations are extremely useful to design the VIRGO interferometer components, to understand its behaviour, and to define the possible locking strategies. They define a flexible framework where well defined problems can be studied in a controlled way.

References

1. Siegman, A. & Sziklas, E., Appl. Opt., **13**, 2775 (1974)
2. Sziklas, E. & Siegman, A., Appl. Opt., **14**, 1874 (1975)
3. Vinet, J.Y., Hello, P., Man, C.N. & Brillet, A., J. Phys. I (Paris), **2**, 1287 (1992)
4. Vinet, J.Y. & Hello, P., J. Appl. Opt., **40**, 1981 (1993)
5. Caron, B., Nucl. Inst. Meth. A, **375**, 360 (1995)

Pulsar Search & Doppler Effect

X.Grave, B.Mours
LAPP, BP 110
74941 Annecy-le-Vieux CEDEX, FRANCE

Periodic sources of gravitational waves (GW) like slightly asymmetric pulsars are expected to be rather weak, but the signal to noise ratio could be enhanced by integrating the data over long periods of time. We will detail the problems arising from this long integration focusing on those related to the Doppler effect coming from the motion of earth.

1 Introduction

One of the possible sources of GW for detectors like VIRGO are the pulsars. The expected amplitude is very weak but permanent and therefore the signal can be integrated for a very long time. This long integration time increase the sensitivity to small frequency variations of the signal. These variations can come from various effects like the very slow spin down of the neutron star due to the energy losses, the Doppler effect in the case of binary stars or the Doppler effect due to the earth motion. In this paper, we will focus on this last effect, especially in the case of blind search and we will discuss the constraint on a possible search algorithm.

2 Constraint on earth position

In the case of an known pulsar the Doppler effect due to the earth motion relative to the pulsar direction is only defined by the earth position. To properly integrate the GW signal we should know the signal arrival time within a fraction of period which is equivalent to know the wavefront or the earth position within a fraction of wavelength. With simple algebra one can show that to lose less than 5% of signal we need to know the earth position better than $15km(1kHz/\nu_{GW})$ where ν_{GW} is the signal frequency in kHz. Such a constraint requires the use of a detailed solar system model to compute the earth position. Fortunately a function of "Bureau des longitudes" in Paris [a] gives us this position with an accuracy of $2km$.

3 Constraint on pulsar position

In the case of an unknown pulsar (blind search) we have to assume a direction to be able to correct the data for the Doppler effect. Therefore the search algorithm splits the sky in small search area. The size of this search area and therefore the total number of search directions is determined by limiting the signal losses to a given number (5% in this study) in any direction within the

[a] *Astron. Astrophys. 128, 124-139 (1983)* G. Francou et al.

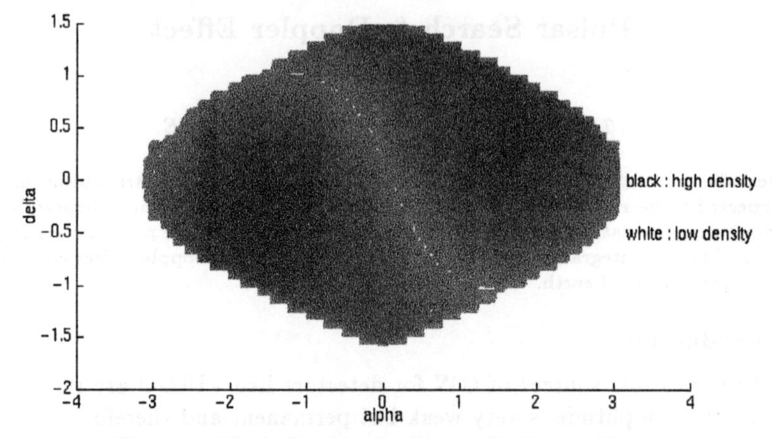

Figure 1: Search direction density in galactic coordinates

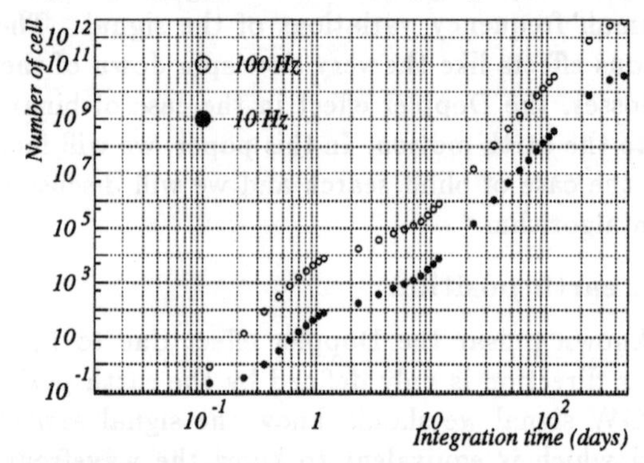

Figure 2: Total number of search direction versus the integration time

search area. Figure 1 shows the result of a numerical computation of the cell density in galactic coordinates for 90 days of integration. The low density area (light grey) correspond to the ecliptic plan and also include the galactic center. By integrating several of these maps one can get the fig. 2 which shows the total number of search directions versus the integration time for two GW frequencies.

4 Consequences on a search algorithm

The huge number of search directions shown by fig. 2 for several month of integration requires that other search techniques be investigated. One well known technique uses more simple trigger algorithms to search for candidates which are later on validated by a full analysis. In the case of periodic sources, instead of performing a single long FFT one can average the results of several shorter FFT.

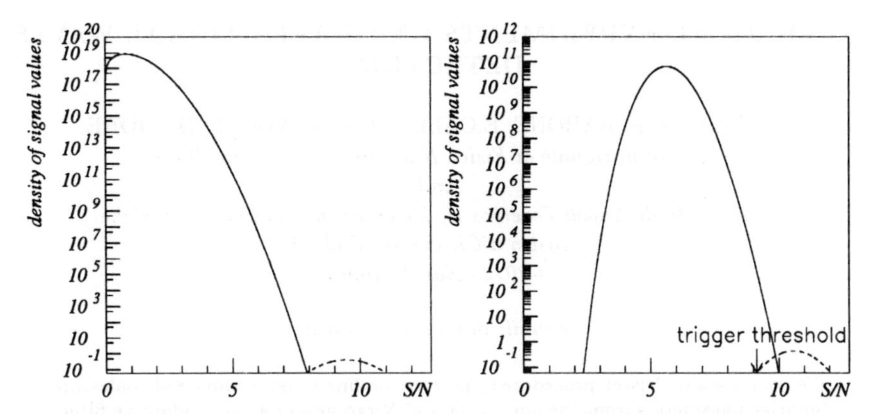

Figure 3: Signal values for 4 months of data for 1 FFT (left) or 32 FFT (rigth).

To illustrate this approach fig. 3_{left} shows the distribution of the power spectrum values of all the needed FFT to analize the full sky in the case of 4 months of data. This corresponds to 3.10^{10} search directions for signal up to 100 Hz with 10^9 entries per FFT and is computed using a white Gaussian noise. For a signal to be observed, its amplitude has to be above the tail of the noise distribution. This is 9.5 times the mean noise level in this case when we take into account the spreading due to the noise as presented by the dashed line on fig. 3_{left}. If we split the initial data sample in 32 subsamples the expected distribution of the noise power spectrum has an average value which increases by a factor $\sqrt{32}$ as shown by fig. 3_{right}. But the distribution becomes more symetric, the number of search directions and therefore the number of entries is largely reduced and the tail of this distribution shifts only slightly. The increase of the noise level move also the observed amplitude (dashed line) of our reference signal. Therefore if we do not want to miss any good signal we have to select all the candidates above the threshold indicate on fig. 3. This corresponds to only a few hundreds of frequency/direction couples. The frequency range and sky surface where the full analysis has to be performed is then reduced by several orders of magnitude.

This approach which dramatically reduces the required computing resources has now to be optimized and exercised with more realistic noises.

Acknowledgments

We would like to thank our Virgo colleagues for very useful discussions.

NUMERICAL EXPERIMENTS ON GRAVITATIONAL WAVES DETECTION

L. MILANO, F. BARONE, E.CALLONI, A.GRADO, L. DI FIORE
Istituto nazionale di Fisica Nucleare, sez. Napoli, Italia
and
Universitá di Napoli Federico II, Dipartimento di Scienze Fisiche
Mostra d'Oltremare, Pad.19
I-80125 Napoli, Italia

e-mail: milano@na.infn.it

We propose a multistep procedure to perform on line rough estimate of coalescing binaries parameters from the output data of Virgo antenna using adaptive filters (ALE and lock-in). The results of simulations are quite promising and the study must be completed with a full statistical analysis for the false alarm probability evaluation.

1 Introduction

It is well known the problem of data analysis that the class of GW sources, constituted by coalescing binaries, still poses in connection with the use of the matched filter technique: on line analysis of the signal is very difficult owing to the filters computational complexity coupled with the lack of robustness against possible signal missing because a lot of template must be implemented with mass parameters stepped by few percent each other and this means the implementation of thousands of filters. This was our starting point and we performed some numerical simulations of detection of gravitational wave signals corrupted by noise coming from coalescing binaries testing some very simple algorithms, that as we shall see, are rather interesting for a true real time analysis and an almost complete rough characterization of the signals we are interested in terms of mass parameters, coalescing time and distance.

We propose a multistep procedure to perform the signal extraction from the output data of Virgo antenna using suitable algorithms (adaptive filters) for on-line rough analysis (1° step) of interesting sequences of Virgo output data followed by a fine off-line data analysis (2° step) using more powerful but computationally more complex algorithms (matched filters) than on-line ones. The on-line algorithms must be efficient and robust: they must not loss signals that can be exctracted from off-line algorithms and they must be robust enough against false alarm detection like the off-line ones.

To test these two conditions of efficiency and robustness we simulated the output sensitivity response of Virgo antenna between 1 Hz up to 5 Khz taking account of the two main sources of noise in Virgo i.e. the low frequency thermal pendular noise and the high frequency shot noise according to the Virgo specifications (Virgo Coll., 1990,1992,1994). We performed the simulation working in the software environment of Simulink in Matlab using a sampling frequency of 10 KHz.

We assumed the following well known form of noise-free response of the detector to the signal :

$$s(t) = k \cdot H \cdot h(t) \cdot f(t)^2/3cos(\phi_0 - 16/5\pi \cdot f_0\tau \cdot ((1-t/\tau)^{5/8}) \tag{1}$$

$$h(t) = 1.19 \cdot 10^{-22} \cdot M/d \tag{2}$$

$$f(t) = f_0^{-8/3} - 1.53 \cdot 10^{-6}Mt))^{-3/8} \tag{3}$$

$$\tau = 6.53 \cdot 10^5 f_0)^{-8/3}/M sec \tag{4}$$

where M $= (m_1 \cdot m_2/(m_1 + m_2)) \cdot (m_1 + m_2)^{2/3}$ is the mass parameter of the coalescing binary and d is its distance in Megaparsecs (Mpc). The masses m1 and m2 of the binary components are expressed in solar mass units (Mo). ϕ_0 and f_0 are the phase and the frequency of the GW at t = 0, f(t) is the instantaneous frequency, h(t) the amplitude of the incoming GW, tau the duration of the coalescence, t is the time computed from the arrival time t0 of the GW signal, finally the parameter k ($0 < k < 1$) is a parameter depending on the source/detector orientation (in our simulation we assumed k = 1). In this form of the signal, tidal effects,post-Newtonian correction, Doppler shift and orbital eccentricity are neglected.

2 Algorithms of detection and results of the simulations

We performed two kind of tests at different distances :

a) detection of the well known standard signal constituted of two coalescing neutron stars both of 1.4 Mo and M = 1.4 with f_0 = 100 Hz and sampling frequency of 10 kHz at 10 Mpc.

b) detection of signals with different mass parameters, that is M = 3.7 , f_0 = 100 Hz and M = 36.7,f_0 = 40 Hz at 70 Mpc and 200 Mpc distances respectively.

All the simulations were performed with signals in Additive White Gaussian Noise of spectral linear density $\tilde{h} = 2 \cdot 10^{-23}$ Hz$^{-1/2}$ at 100 Hz. The class of algorithms we tested, for on-line detection of coalescing binaries, are the partially adaptive filters, used in such a way, to obtain a time-frequency output from the filtered signal. In other words we modified two classical filters, usually used for amplitude extraction or tracking of signals in AWGN, in such a way to obtain either the characterization of the frequency of the input signal or the amplitude determination (in progress) both in the time domain The adaptive filters, we tested, are digital lock-in's (Barone F et. al. 1994) in the classical quadrature configuration and IIR based Adaptive Line Enhancers (ALE); ALE is a constrained recursive center frequency adaptive filter for the enhancement of noisy bandpass signals with varying center frequency(R.V. Raja and R.N.Pal, 1990) .

We implemented a bank either of five digital lock-in's or of five ALE and tested their performances for the cases a) and b) we spoke above. In fig. 1 are shown the detected frequencies of the signals in the time domain (full line) and the theoretical trend of the frequencies (dashed lines) at the Newtonian order along the time of the coalescing systems emitting GW. A full statistical (significance) analysis to evaluate the false alarm probability connected with these two kind of filters is in progress, at this stage we performed some experimental tests on simulated data and we got on an Alpha Vax for both ALE and lock-in a coefficient of data rejection (WGN noise only as input signal) of nearly 95 % of the whole sample of 50 hours simulated data. The assumption we made for these tests was very simple, that is, we could have an alarm if the variance of the mean output frequency from the five lock-in bank or the five ALE bank is less than $4Hz^2$. It is a very simple, but strong treshold: anyway we must perform the full theoretical statistical analysis of the performances of these two kind of filters, especially concerning ALE filters, to evaluate the false alarm probability.

3 Conclusions

What we want to stress at the end of this short, and necessarily compressed communication , are the following results:

a) We tested two adaptive filters very simple and of practically negligible computational complexity: the implementation of a single ALE requires thirty multiplications; more or less the same number is required for a single digital lock-in, in other words we implemented a class of filters that can be easily used on-line and especially for ALE the performances are really very interesting and promising.

b) We recall the attention of the interested reader to the results that are shown in fig 1. In our opinion they represent a quite interesting and promising solution of the problem of data analysis for the class of coalescing binary signals: using ALE's we can completely characterize the signal because we have

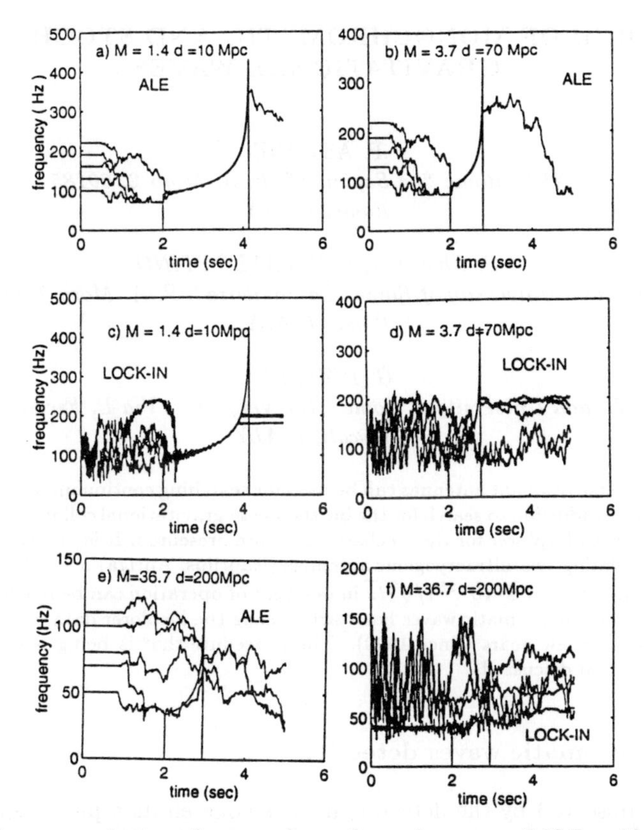

Figure 1: From top to bottom there are shown the performances of ALE and lock-in filters for different mass parameters of coalescing binaries at Newtonian order.

two sources of informations concerning either the trend of the frequency or of the amplitude evolution of the coalescence signal in the time domain with S/N ratios that can reach a limit of 0.2. The conclusion is that on-line rough and fast estimation of the parameters of the coalescence is possible and in any case will be useful to perform the initialization of off-line more powerful refined analysis.

4 References

1. Barone F.,Calloni E.,Di Fiore L., Grado A.,Milano L.,Russo G., *Rev. Sci. Instrum.* **66**, 3697 (1995).
2. Raja R.V. and Pal R.N., *IEEE Trans. Acoust. Speech and Sign. Processing.* **38**, 1710 (1990).
3. Virgo Collaboration, 1990,1992,1994 .

SEARCH OF MONOCHROMATIC AND STOCHASTIC GRAVITATIONAL WAVES

P. ASTONE
INFN, Rome "La Sapienza", P. A. Moro 2, 00185
Rome, ITALY

S. FRASCA, G.V. PALLOTTINO
INFN and University of Rome "La Sapienza", P. A. Moro 2, 00185
Rome, ITALY

G. PIZZELLA
INFN LNF and University of Rome "Tor Vergata" , Via E. Fermi 40,00044
Frascati, ITALY

The cryogenic resonant antennas can be used for searching continuous and stochastic waves, in addition to search for the bursts due to gravitational collapses. Results obtained with Explorer for the stochastic g.w. are presented. It is also shown that crosscorrelating two ultracryogenic antennas (Nautilus, Auriga) a spectral amplitude of the order of $8 \cdot 10^{-25} 1/\sqrt{Hz}$ in one year of operation can be reached. The search for monochromatic waves has started using the Explorer data, which cover a period of a few years (since 1990). The procedure that is being developed is presented and discussed.

1 Monochromatic waves detection

The signal observed by the detector, if the source emits a purely sinusoidal g. w., is frequency and amplitude modulated, due to the relative motion between the detector and the source [1,2]. If we know the location and the frequency of the source it is possible to demodulate the measured data and then to achieve a sensitivity that improves with the square root of the observation time [3] t_m: $h_0 = \sqrt{2S_h/t_m}$, where S_h is the (two-sided) spectral density of the dimensionless amplitude h that can be detected with $SNR = 1$. At the resonances the sensitivity becomes:

$$h_0 = 2.04 \cdot 10^{-25} \sqrt{\frac{T}{0.05\ K} \frac{2300\ kg}{M} \frac{10^7}{Q} \frac{900\ Hz}{\nu_0} \frac{1\ day}{t_m}} \tag{1}$$

Using eq.(1), t_m =1 year, and the parameters of Explorer [4] during the 1991 run ($T = 2\ K, Q = 10^6$) we obtain $h_0 = 2.3 \cdot 10^{-25}$, in a bandwidth of $\simeq 1\ Hz$ around the two resonances and about 10^{-24} in a bandwidth of 15 Hz between the resonances. With the 1994 parameters ($T = 2\ K, Q = 10^7$) $h_0 = 7.2 \cdot 10^{-26}$,

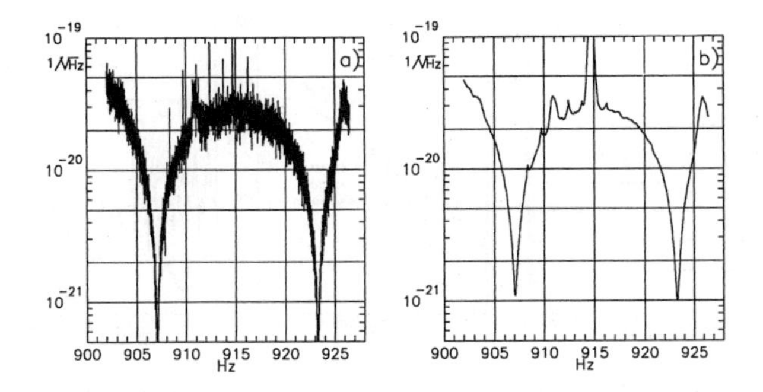

Figure 1: *a) One spectrum, $t_0 \simeq 0.66$ hours; b) Spectrum after the smoothing filter* (Data from the 1994 data base)

over a bandwidth of $\simeq 0.1\ Hz$ at the resonances. The planned sensitivity with the Nautilus detector is $h_0 = 1.1 \cdot 10^{-26}$. If we don't know the source location then we can do an "overall sky search". An approach to the analysis is to organize a spectral data base and then to analyze the spectra by combining their information. Each spectrum should be estimated in a period in which the Doppler effect is not important. Over $t_0 = 0.6617$ hours, with frequency step $\delta_\nu = 0.42\ mHz$, the maximum frequency variation is $\delta_{\nu_d} = 0.28\ mHz$, smaller than the step. We have 13239 spectra in one year of operation. Each spectrum of the data base has an **header** that contains the date and information for vetoing the experimental data. Fig.1a) shows the spectral amplitude of one spectrum of the 1994 data base. At the resonances the experimental sensitivity is $h_0 = 1.7 \cdot 10^{-23}$. By averaging different periodograms over a total time t_m we gain as $(t_m/t_0)^{-1/4}$, being $h_0 = 2\sqrt{S_h \cdot \delta_\nu / \sqrt{t_m/t_0}}$. A signal will appear in different channels of the spectra, with frequency and amplitude function of the source location. To obtain the amplitude it is necessary to eliminate the noise, subtracting from the spectrum its smoothed version (fig.1b)). The spectra can be averaged shifting the spectral channels to fit given locations in the sky. Actually we analyze the frequency and amplitude patterns of chosen [a] lines in order to identify, or to exclude, a certain locus of points as possible g.w. emitters. As an example let us consider, in fig.1a), the line at $\simeq 916\ Hz$. We study, within a chosen bandwidth, its frequency and amplitude for successive periodograms and then we fit them with the expected modulation for various sources. Using right ascension α and declination δ we divide the sky into equal

[a] We don't discuss here the selection criteria, that can be both statistical or based on the experimenter feeling.

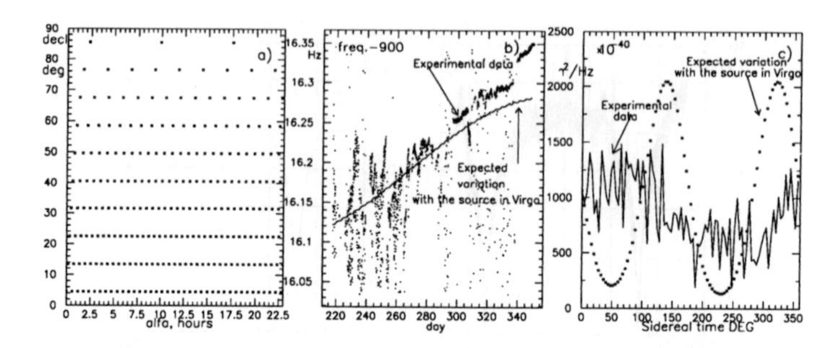

Figure 2: a) Points in the sky:$\delta = 0, 90$ *deg* (step 10 *deg*), $\alpha = 0, 23$ *hours* b) Frequency behaviour of the chosen line c) Squared amplitude

(Data from the 1991 data base)

solid angles, as shown in fig.2a). Fig.2b) shows the change in frequency of our selected line and also what we expect if the source is in the Virgo Cluster (which happens to be the best fit). The second step is to compare the behaviour of the amplitude with that expected for the same α and δ. Fig.2c) shows the averaged amplitudes, and the expected behaviour assuming the source be Virgo. The experimental data don't fit the expected behaviour. In the case of interesting results the further analysis may be to apply the "known source search" in that direction and the request to the astronomer of a deeper observation of that region of the sky.

2 Stochastic search

We started to study this problem in more detail after we learned of the model [5] that predicts relic g.w. in a frequency range accessible to our detectors. Using only one detector then its spectral amplitude $\tilde{h}(\omega) = \sqrt{S_h(\omega)}$ represents the **upper limit** [6]. From the Explorer data in fig.3a) we get $\tilde{h} \simeq 5.8 \cdot 10^{-22}\ Hz^{-1/2}$ at the resonances. The sensitivity at the resonances depends only on the bar temperature and on the merit factor [7]. We remark that if we increase the bandwidth of the detector the sensitivity doesn't degrade. Fig.3b) shows the expected strain sensitivity of the Nautilus detector ($T = 0.1\ K$, $Q = 8.5 \cdot 10^6$). The detection bandwidths are those we plan to reach by the year 2000, using an improved transducer. Using two identical, parallel and "near" detectors [b] both with this sensitivity, in operation for $t_m = 1$ year, with a bandwidth of 5

[b] The cross-correlation analysis is affected by a time delay between the detectors. The separation should be small if compared to the reduced wavelenght $\lambda/(2\pi)$ of the signal. In order to perform an analysis at 1 kHz the separation should be less than $c/(2\pi\nu) \simeq 50km$.

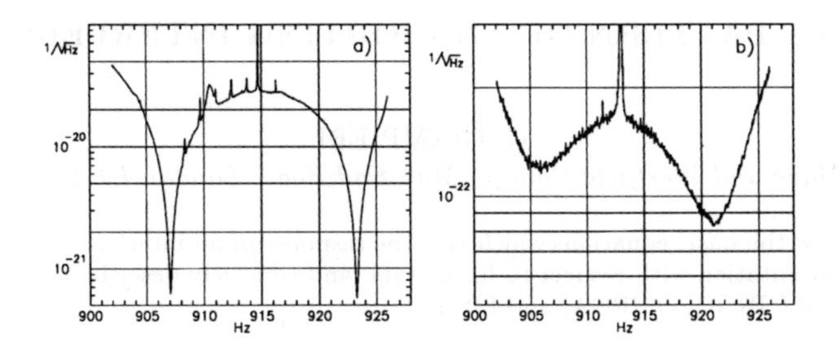

Figure 3: a) Explorer 1994; b) Nautilus 2000

Hz, the upper limit can be reduced to

$$\tilde{h}_{12}(\nu_+) \simeq 7.3 \cdot 10^{-25} \quad 1/\sqrt{Hz} \tag{2}$$

An interesting suggestion, due to the weak dependence of the sensitivity on the bandwidth is to cross-correlate systems made by bars and interferometers [8,9].

References

1. J. Livas in *Gravitational Wave Data Analysis*, ed. B. F. Schutz (Kluwer Academic Publishers, 1989).
2. B. F. Schutz, *The detection of gravitational waves*, (Max Plank Institut,AEI-003 1996).
3. G.V.Pallottino, G.Pizzella, *Nuovo Cimento* **7C**, 155 (1984).
4. G. V. Pallottino, in this volume.
5. R. Brunstein, M. Gasperini, M. Giovannini, G. Veneziano, *Phys. Lett.* B **361**, 44 (1995).
6. P. Astone, M. Bassan, P. Bonifazi, P. Carelli, E. Coccia, C. Cosmelli, V. Fafone, S. Frasca, S. Marini, G. Mazzitelli, P. Modestino, I. Modena, A. Moleti, G.V. Pallottino, M. A. Papa, G. Pizzella, P. Rapagnani, F. Ricci, F. Ronga, M. Visco, L. Votano, *Upper limit for a gravitational wave stochastic background measured with the Explorer and Nautilus gravitational wave resonant detectors* (submitted to *Physics letters* (1996)).
7. P. Astone, G.V. Pallottino, G. Pizzella, *Detection of impulsive, monochromatic and stochastic gravitational waves by means of resonant antennas* (LNF-96/001 IR 1996).
8. B. F. Schutz, in this volume.
9. P. Astone, J.A. Lobo, B.F. Schutz, *Class. Quantum Gravity* **11**, 2093 (1994).

ANTENNA PATTERN OF AN ORBITING INTERFEROMETER

G. GIAMPIERI

Queen Mary and Westfield College, Mile End Road, London E1 4NS, England

We review the basic equations which give the response of an interferometer changing its orientation with respect to fixed stars, and concisely describe the antenna pattern of VIRGO, OMEGA, and LISA.

1 Introduction

This contribution is based on some recent work aimed at describing the response of an interferometer changing its orientation with respect to a fixed reference frame. [1] We start by summarizing these results. The instantaneous detector's orientation is described by means of the Euler angles ξ, η, ζ in some fixed reference frame. The response R of the interferometer is a linear combination of the polarization amplitudes h_+ and $h_\times$, i.e.

$$R = F_+ h_+ + F_\times h_\times \tag{1}$$

The coefficients F_+ and $F_\times$ are called beam-pattern factors. They depend on the detector's angles ξ, η, ζ, and the source coordinates θ, ϕ, along with the polarization angle ψ. They are explicitely given by [1]

$$\begin{aligned} F_+ &= \sin(2\Omega)\,[A\cos(2\xi)\cos(2\psi) + B\cos(2\xi)\sin(2\psi) + \\ &\quad + C\sin(2\xi)\cos(2\psi) + D\sin(2\xi)\sin(2\psi)]\,, \qquad (2) \\ F_\times &= \sin(2\Omega)\,[B\cos(2\xi)\cos(2\psi) - A\cos(2\xi)\sin(2\psi) + \\ &\quad + D\sin(2\xi)\cos(2\psi) - C\sin(2\xi)\sin(2\psi)]\,. \qquad (3) \end{aligned}$$

In these expressions, 2Ω is the detector's aperture angle - 90° in the optimal case, naturally chosen for terrestrial interferometers, 60° in the proposed space-based detectors -, while the coefficients A, B, C, D, which can be found in Appendix A of Ref.[1], depend only on the angles ζ, θ, and $\delta \equiv \phi - \eta$.

The quantity describing the antenna directionality is the polarization-averaged power pattern

$$P \equiv \langle \left(\frac{R}{\sin(2\Omega) h} \right)^2 \rangle_\psi = \sum_{n=0}^{4} \{[\lambda_n + \mu_n \cos(4\xi)]\cos(n\delta) + \sigma_n \sin(4\xi)\sin(n\delta)\}\,. \tag{4}$$

The coefficients $\lambda_n, \mu_n, \sigma_n$, also given in Ref.[1], Appendix B, are constant in all cases of interest, so that Eq. 4 depends on time only through the sinusoidal functions of ξ and δ. For this reason, Eq. 4 is well suited for describing the detector's angular response as the antenna moves and changes its orientation with respect to the sources.

2 Ground-based Detectors

Eqs. 1-4 are valid in general, provided that the detector's motion is given in the form of the functions $\xi(t), \eta(t), \zeta(t)$. A first natural application is that of an interferometer located on the Earth surface. The diurnal motion of the Earth around its axis introduces an amplitude modulation in the response to a long-duration signal, as described by Eq. 1, with the beam-pattern factors given by Eqs. 2-3. Let us adopt the Equatorial frame as our fixed frame. Then the quantities ξ, η, ζ are simply

$$\zeta = \frac{\pi}{2} - \ell ; \qquad \eta = \eta_0 + \omega t ; \qquad \xi = \iota , \tag{5}$$

where ℓ is the terrestrial latitude, ω is the Earth angular velocity of rotation, and ι is the angle between the arms' bisector and the local parallel. Inserting Eq. 5 in Eq. 4 gives the instantaneous antenna power pattern. Averaging over the time of arrival of the signal, i.e. over ϕ, gives

$$\langle P \rangle_T = \lambda_0(\theta, \ell) + \mu_0(\theta, \ell) \cos(4\iota) . \tag{6}$$

For any particular value of θ, Eq. 6 gives the square of the r.m.s. power as a function of the antenna's position and orientation on the Earth's surface. [2] Fig. 1a shows the contour plot of $\langle P \rangle_T$ for $\theta = 102^o$, corresponding to the direction of the center of the Virgo cluster, and reproduces in a simple analytical way Fig. 8 of Ref.[2].

A curious result, readily obtainable from Eq. 6, is that if we build an interferometer at $\ell_{is} = \arcsin\left(\pm\frac{1}{\sqrt{3}}\right)$ and $\iota_{is} = \frac{1}{4}\arccos\left(-\frac{1}{5}\right)$, then the averaged antenna pattern $\langle P \rangle_T$ turns out to be perfectly isotropic, so that the chances of detecting a source with a given strength are the same, no matter where the source is located in the sky.

The averaged antenna pattern at VIRGO latitude and inclination ($\ell_{virgo} \simeq 43.5^o, \iota_{virgo} \simeq 26.30^o$) is shown in Fig. 1b. Since we can expect several terrestrial interferometers to be operative in the near future, the sky coverage of a single antenna is not an issue as critical as in the space-born case that we will now discuss.

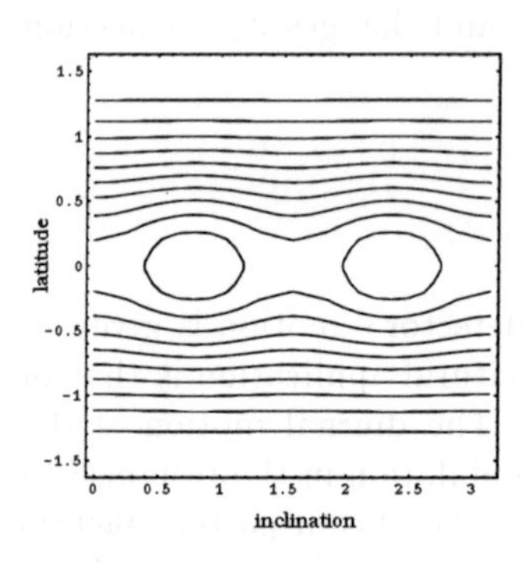

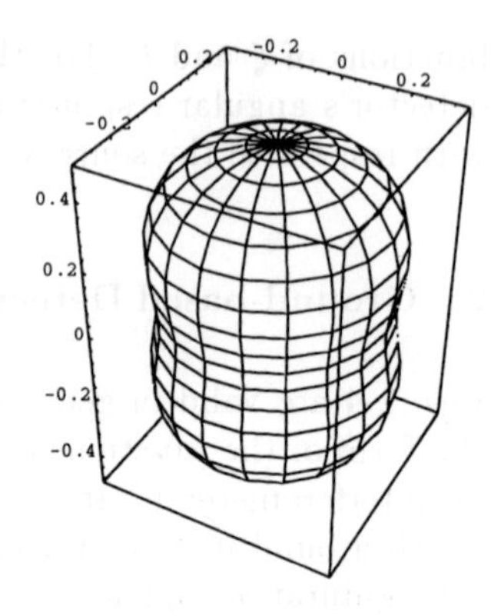

Figure 1: Ground-based interferometers. Left: r.m.s. response to a wave coming from the Virgo cluster, as a function of latitude and inclination. Right: Averaged antenna pattern of VIRGO.

3 Space-born Interferometers

The low-frequency regime ($f < 10^{-1}$ Hz) is only accessible to space-based detectors. In particular, Doppler tracking of interplanetary spacecraft has allowed for gravitational waves searches in the mHz frequency range, without reaching any detection due to the limited sensitivity of non-dedicated missions. Although better results can be expected from advanced Doppler missions, like CASSINI, the ultimate sensitivity can only be obtained through laser interferometry in space. In particular, two space-born interferometers have been recently proposed: LISA,[3] and OMEGA.[4] They both consist of six drag-free, laser-bearing spacecraft placed, in pair, at the vertex of a triangle. At each corner, the two spacecraft are phase locked through the exchange of a laser signal. Each of the two probes sends a laser beam to a probe located at each of the two other equilateral points, where the tracking signal is transmitted back by phase locked lasers, and the returning beams are eventually interfered.

From the point of view of this analysis, the two projects differ mainly in the details of the orbit. In fact, LISA will be put on a circular orbit around the Sun, whereas OMEGA will be orbiting around the Earth, beyond the Moon orbit, reducing in this way the mission's costs at a price in terms of performance. As we shall see, the different orbits imply a considerably different sky coverage for the two antennas, and therefore we will consider them separately.

3.1 OMEGA Antenna Pattern

OMEGA orbit is particularly easy to analyze due to the fact that the interferometer's plane is almost coincident with the Ecliptic one. In other words, in the Ecliptic reference frame, the motion is approximately given by

$$\zeta = \eta = 0\,; \qquad \xi = \xi_0 + \omega t\,, \tag{7}$$

where $\omega \simeq 1.4 \times 10^{-6}$ sec^{-1} is the angular frequency of rotation, corresponding to a period of ~ 53 days. As a consequence of Eq. 7, Eqs. 2-4 simplify considerably. For instance, only λ_0, μ_4 and σ_4 are different from zero, and the averaged antenna pattern becomes

$$\langle P \rangle_T = \frac{1}{8}\left[1 + \cos^4\theta + 6\cos^2\theta\right]\,, \tag{8}$$

and is shown in Fig. 2a. Note that the antenna directionality is very close to that of a fixed interferometer, except in the detector's plane, where the uniform rotation wipes out the well know four null directions of the fixed antenna pattern.

3.2 LISA Antenna Pattern

Somewhat more complicate is the analysis of LISA motion and its consequences on the overall sky coverage. In fact, in order to keep the triangular constellation as stable as possible, an elaborated solar orbit has been designed, with each spacecraft orbiting a circle of radius 3×10^6 km over a period of 1 yr. The plane containing the six probes, during LISA orbit, will remain always tangent to the surface of a cone of 60° degrees aperture, and the detector itself will be rotating in its plane with same periodicity - 1 year - but opposite direction. With respect to the Ecliptic frame this motion is described by

$$\zeta = \frac{\pi}{3}\,; \qquad \eta = \eta_0 - \omega t\,; \qquad \xi = \xi_0 + \omega t\,. \tag{9}$$

Note that η and ξ are counter-rotating, with same periodicity. In this case the 14 coefficients $\lambda_n, \sigma_n, \mu_n$ are all different from zero, and they can be found in Appendix C of Ref.[1]. The one-year average $\langle P \rangle_T$ gives the antenna's mean sky coverage during its orbit. The result is

$$\langle P \rangle_T = \lambda_0 + \frac{1}{2}\left(\mu_4 + \sigma_4\right)\cos[4(\phi - \eta_0 - \xi_0)]\,, \tag{10}$$

and is shown in Fig. 2b. As we can see, the orbital motion render the antenna's directionality almost nil.

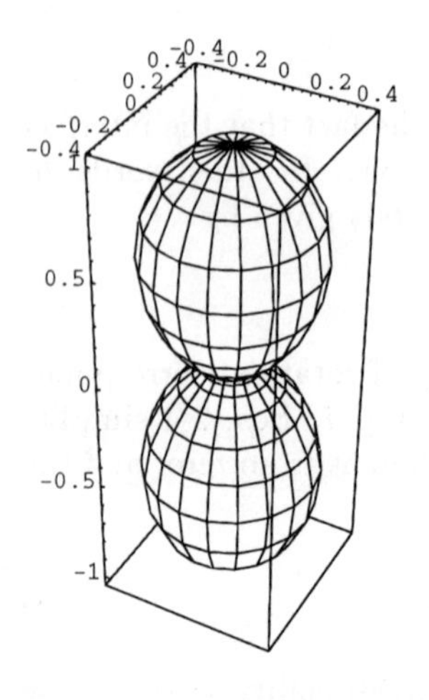

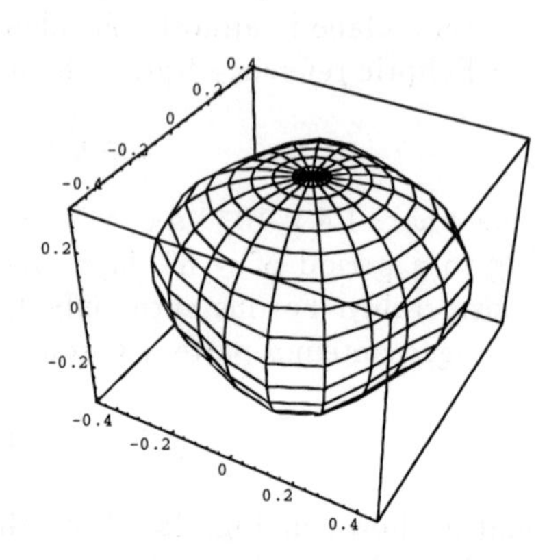

Figure 2: The antenna power pattern, averaged over time, of two proposed space-born interferometers, OMEGA (left) and LISA (right).

Acknowledgments

This research was partially carried out at the Jet Propulsion Laboratory, California Institute of Technology, while the author held an NRC/NASA fellowship.

References

1. G. Giampieri, "On the Antenna Pattern of an Orbiting Interferometer", submitted.
2. B.F. Schutz and M. Tinto, *Mon. Not. R. astr. Soc.* **224**, 131 (1987).
3. K. Danzmann et al., *LISA: Proposal for a Laser Interferometric Gravitational Wave Detector in Space*, Report No. MPQ 177, Max-Planck Institut für Quantenoptik (Garching bei München, Germany, 1993).
4. R.W. Hellings et al., *OMEGA: Orbiting Medium Explorer for Gravitational Astrophysics*, Midex Proposal (Jet Propulsion Laboratory, Pasadena, 1995).

EFFICIENT GW CHIRP ESTIMATION VIA WIGNER-VILLE REPRESENTATION AND GENERALIZED HOUGH TRANSFORM

M. FEO, V. PIERRO, I.M. PINTO, M. RICCIARDI
D. I.[3] E., Univ. of Salerno, via Ponte don Melillo,
I-84084 Fisciano (SA), Italy

The structure and performance of a working code implementing Wigner-Ville time-frequency detection and Hough-transform parameter estimation of gravitational wave chips from coalescing binaries is illustrated.

The possible relevance of gravitational wave chirp detection for Astrophysics and Cosmology has been repeatedly emphasized [1]. In the simplest (newtonian) model the expected signal can be written [a]:

$$h(t) = A\left(1-\frac{t}{T_c}\right)^{-1/4}\cos[\phi(t)+\phi_0], \quad \frac{d\phi}{dt} = 4\pi F_0\left(1-\frac{t}{T_c}\right)^{-3/8}. \qquad (1)$$

In (1) the factor A depends on the source distance r and angular position (θ,ϕ), which can be estimated by combining the data from several antennas[3], and ϕ_0 is the unknown initial phase. Here we focus on the estimation of the initial (orbital) frequency F_0 and time to coalescence T_c, using a single detector. Several strategies have been envisaged for GW chirp detection/estimation: *i)* matched filtering (requires a huge amount of computations[4]); *ii)* adaptive Wiener filtering (*no* template required, relies on different signal vs. noise coherence time; statistical analysis still to be done[5]); *iii)* time-frequency (henceforth TF) analysis [b] pioneered by the VIRGO-Annecy group [10,11].

Our approach to GW chirp parameter estimation is based on Wigner-Ville (henceforth WV) TF analysis and Hough-transform [12] (henceforth HT). It was first communicated at MG7 [13]. The key features of our algorithm (a FORTRAN code performing all steps is available) are summarized below. See [14] for details. We split the (very long!) GW antenna output time series into successive chunks, and for each compute the (discrete) Wigner-Ville transform

[a] There is no room to touch the post newtonian chirp template issue[2] here. It's only worth mentioning that our method can be extended to post-newtonian models using a *hierarchical* scheme, as e.g., proposed by Schutz[2].

[b] TF analysis including Fourier and Gabor[6], Wavelet[7], Wigner-Ville[8] and Zak's[9] representations, is the standard tool for *nonstationary* waveforms.

[c] by Pei-Yang algorithm [16] via FFTs, using a fast Hartley transform [17] engine [d]. The time-shift [e] and time-windowing properties of WV [f]:

$$W_{h(t-t_0)}(\Omega,t) = W_{h(t)}(\Omega, t-t_0), \quad W_{h(t)w(t)}(\Omega,t) \sim w(t)\Delta_\sigma(\Omega - \dot{\phi}(t)), \quad (2)$$

Δ_σ being a delta-like function with frequency-width σ of the order of the reciprocal time-width of $w(t)$, allow to estimate the instantaneous frequency by [g]

$$\dot{\phi}(t) \approx (value\ of\ \Omega\ closest\ to\ 0\ where\ |W_h(\Omega,t)|\ is\ a\ maximum) \quad (3)$$

and to track its evolution by splicing the processed chunks one after another.

A typical result is shown in *Fig. 2*, which refers to the (whitened) signal produced in LIGO-I by a NS-NS binary at 60 MPc. The corresponding (unwhitened) time series is shown in *Fig. 1.* For each couple of points in *Fig. 2*, using (1) we can determine a point in the (F_0, T_c) parameter space. In the presence of noise (outliers) and (frequency) quantization error [19], these points will cluster into a multi-modal distribution, the HT, whose (main) mode can be extracted [20] to provide an estimate of the unknown source parameters. To reduce the HT computational burden [h] we use a (noise-trained) 2D Kolmogorov-Smirnov test [14, 22], to identify which portions of *Fig. 2* are unlike to be produced by pure noise, to a specified level of confidence (see *Fig. 3*).

As a result, the unknown source parameters are located in a *tiny* box of the (F_0, T_c) source parameter plane (see *Fig. 4*). The residual *uncertainty* in

[c] The Wigner-Ville transform:

$$W_h(t,\Omega) = \int_{\Delta T} \tilde{h}\,(t+\tau/2)\, \tilde{h}^*\,(t-\tau/2) e^{-j\Omega\tau} d\tau$$

displays the instantaneous energy content of $h(t)$, $\tilde{h}\,(t) = A(t)e^{j\phi(t)}$ being the complex pre-envelope of $h(t)$ [15]. Note that the unknown initial phase ϕ_0 in (1) does not affect W_h.

[d] The chunk size depends on the available FFT-processor. In view of the correlation-like structure of WV, we use half-overlapping chunks, so that the number of frequency samples is the same for all time bins.

[e] This important property is *not* shared by all TF representations, but only by the bilinear ones belonging to Cohen's class [18].

[f] We proved the second equation in (2) asymptotically [14] in the *large* parameter $cT_0/\pi r_g$, T_0 and r_g being the initial period and gravitational radius of the source.

[g] This procedure, proved to be *far* more robust w.r. to noise than the barycentric property

$$\dot{\phi}(t) = \left(\int_{\Delta\Omega} W_h(\Omega,t)d\Omega\right)^{-1} \int_{\Delta\Omega} \Omega W_h(\Omega,t)d\Omega$$

used in previous work related to TF representations [10, 11].

[h] The HT is highly parallelizable, and its complexity can be made independent from the number of couples [21].

the parameters should be invoked to explain the modest statistical significance of the optimum non coherent detector output[i] for a twin NS binary at 60 MPc (see *Fig. 5*), as compared to the sharp evidence of the chirp presence provided by *Fig. 2*.

References

1. B.F. Schutz *GRG*, **19**, 1163, (1987).
2. B. J. Owen, to appear on *Phys Rev. D*, 1996.
3. A. Giazotto, Phys. Rep. **182**, 365, 1989.
4. S.V. Dhurandhar, B.F. Schutz, *Phys. Rev.* **D50**, 2390, (1994).
5. D. Nicholson, Lectures delivered at Salerno Univ., 1995.
6. D. Gabor, *IEE J.* **93**, 429, (1946).
7. I. Daubechies, *IEEE Trans.* **IT-36**, 1961, (1990).
8. T.A.C.M. Claasen, W.F.G. Macklenbräucker, *Philips J. Res.*, **35**, 217, (1980); *ibid.*, **35**, 276, (1980); *ibid.*, **35**, 372, (1980).
9. A.J.E.M. Jannsen, *Philips J. Res.*, **43**, 23, (1988).
10. J.M. Innocent, J.Y. Vinet, *VIRGO Note*, (1992).
11. D. Verkindt, *Thesis, Univ. Savoye*, (a.y. 1992-93).
12. J. Illingworth, J. Kittler, *Computer Vision*, **44**, 87, (1988).
13. M. Feo V. Pierro, I.M. Pinto, M. Ricciardi, *Proc. MG7*, Stanford USA, (1994).
14. M. Feo, V. Pierro, I.M. Pinto, M. Ricciardi, subm. to *Phys. Rev. D*, (1996).
15. B. Picinbono, W. Martin, *Ann. Télécomm.*, **38**, 179, (1983).
16. S.C. Pei, I.I. Yang, *IEEE Trans.* **SP-40**, 2346, (1992).
17. R. Bracewell, *Proc. IEEE*, **72**, 1010, (1984).
18. L. Cohen, *J. Math. Phys.*, **7**, 781, (1986).
19. T.M. Van Veen, F.C.A. Groen, *Pattern Recognition*, 137, (1981).
20. L. O'Gorman, A.C. Sanderson, *IEEE Trans.* **PAMI-6**, 280, (1984).
21. K. Hanahara, T. Maruyama, T. Uchiyama, *IEEE Trans.* **PAMI-10**, 121, (1988).
22. G. Fasano, A. Franceschini *MNRAS*, **225**, 155, (1987).
23. R. Estrada, P. Kanwal, *Proc. Roy. Soc. London*, **A428**, 399, (1990).
24. C.E. Heil, D.F. Walnut, *SIAM Rev.*, **31**, 628, (1989).

[i]We use the Estrada-Kanwal expansion[23], together with our asymptotic expansion of W_h in the large parameter $cT_0/\pi r_g$ to compute the matched filter (non-coherent correlator) as a line integral along the (estimated) instantaneous frequency curve in TF plane, via Moyal theorem[24].

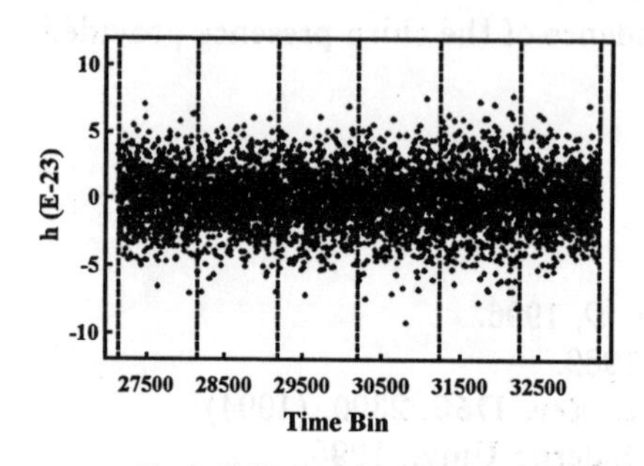

Fig. 1 - **Noise Corrupted GW Chirp**

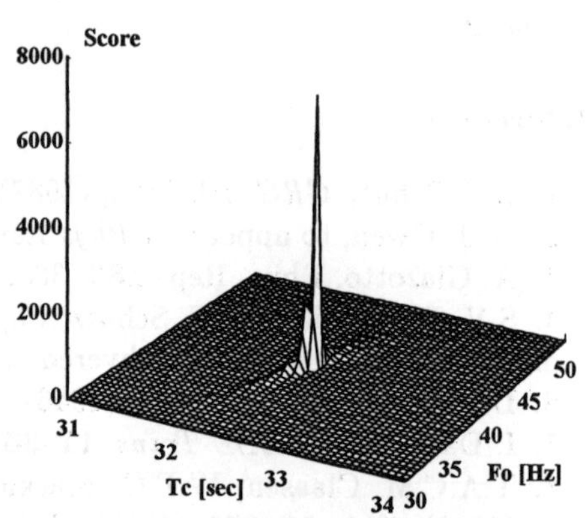

Fig. 4 - **Parameter Plane Histogram (Last Chunk with Signal)**

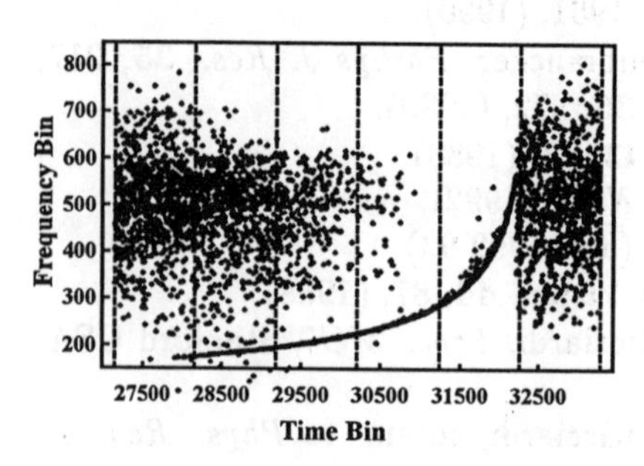

Fig. 2 - **Extracted Time-Frequency Line**

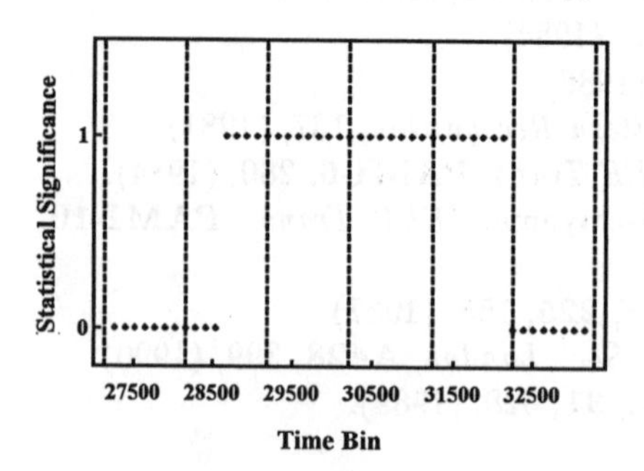

Fig. 3 - **2D K-S Sniffer [Threshold = 0.01]**

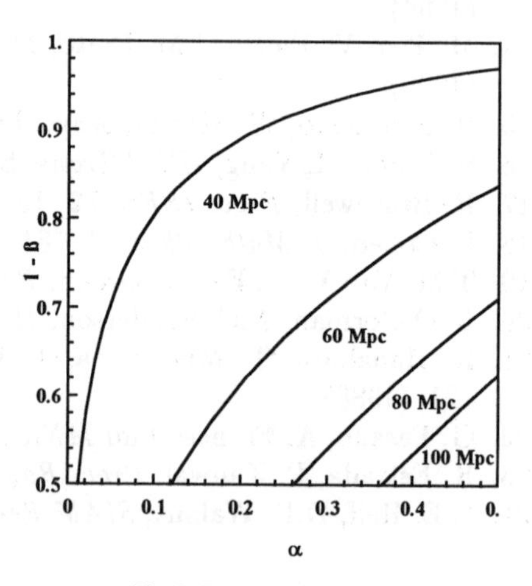

Fig. 5 - **Detection Characteristics (Twin NS Binary)**

Experimental Prototypes

VERY LOW TEMPERATURE MEASUREMENTS OF QUALITY FACTORS OF COPPER ALLOYS FOR RESONANT GRAVITATIONAL WAVE ANTENNAE

G. FROSSATI, H. POSTMA, A. DE WAARD AND J.P. ZENDRI
Kamerlingh Onnes Laboratory. Nieuwsteeg 18, 2311 SB Leiden,
The Netherlands

We present experimental results down to 20 mK on quality factors of different copper alloys aimed at finding a suitable material for the 3m diameter spherical detector GRAIL, being considered for construction in the Netherlands.

1 Introduction

The Dutch Science Foundation NWO has funded a R&D project aimed at defining the characteristics of a 3m diameter spherical resonant gravitational wave antenna with a quantum limited strain sensitivity $h/\sqrt{\mathrm{Hz}} \sim 4\times10^{-24}$ (project GRAIL[1]). One of the important problems involved in such a detector is the choice of the material of the sphere. For maximum sensitivity to gravitational waves the antenna must have a large mass M and a high sound velocity since the cross-section is proportional to Mv_s^2 . In order to be quantum limited in the amount of detectable gravitational energy one must also satisfy the following condition[2, 3, 4, 5]

$$T_{\mathrm{eff}} \cong T / \beta Q + 2T_N$$

where $k_B T_{\mathrm{eff}}$ is the minimum detectable energy, T is the thermodynamic temperature of the sphere, β is the ratio of gravitational energy deposited in the antenna to that converted in electromagnetic energy in the transducer, Q is the quality factor of the sphere and T_N is the noise temperature of the transducer. Since T_N is quantum limited to $h\nu/k_B$ where ν is the frequency of the gravity wave which we expect to be in the kHz range we have that $2T_N \cong 10^{-7}$K. From the above expression we see that $T/\beta Q < 10^{-7}$. β is a parameter which is smaller than one in passive transducers, and depend on the transducer type and design. Typical values are 0.001 with possibilities of improvement by one or two orders of magnitude[6]. Minimizing T/Q depends thus on cooling the antenna to as low as possible[5] and on choosing a material of low internal losses (high quality factor Q defined as $Q = \pi\tau\nu$ where ν is the frequency of a given resonance mode of the antenna and τ is the relaxation time of the excited mode). In the GRAIL project we contemplate cooling a sphere of 3m diameter to 10-20mK. If we assume $\beta \sim 0.01$ and T=20mK then $Q \sim 10^7$ which is indeed quite high for most metals[7]. If somewhat higher β and lower T are achieved then the condition on Q could be relaxed.

Having said that, there are other important factors in choosing the material for the sphere. First of all there should be an industry capable of casting the sphere. The material must thus be within the capability of the industry in question. The heat capacity of the material must be low enough at low temperatures that it can be cooled in a reasonable amount of time, thus alloys with magnetic impurities should be excluded. In order to achieve temperatures in the 10-20mK range the metal or alloy should not be a superconductor otherwise the poor thermal conductivity and the inevitable heat leaks will limit the minimum temperature. In fact, the Italian Nautilus and Auriga antennae, which are made of the well known 5056 aluminum alloy reach temperatures around 100 mK

Because of its high density and thermal conductivity we have chosen copper as a matrix and have done low temperature studies of different copper alloys, searching for compositions giving high Q values in the low mK temperature range.

2 Experimental techniques

We developed a technique for measuring Q which is reliable and imitates the final antenna configuration. Spherical samples 150mm in diameter were accurately machined and suspended from the center with a Cu rod. Several types of contacts, were used, either screwed, conical with large angle or conical with small angle, without significant difference in the results, except eventually for the lowest attainable temperature, where the higher contact pressure of the screw or small angle cone led to lower end temperatures. Piezoelectric crystals were used to excite and detect the different resonant modes. They were glued on small flats filed on the sphere surface using cyanoacrylate glue (instant glue). For more details see [8], where we describe measurements on bulk and explosion-welded Al5056. A low temperature hammer actuated by a superconducting coil was also used to excite the different modes of the sphere at temperatures down to 4K. The temperature was measured using commercial RuO2 resistors (Philips 2000 Ω SMD's).

Since the aluminum alloy 5056 was known to have Q's as high as 70-80million we started with that material and found values[9] as high as 120million below 100mK (fig.1), the highest ever reported in this alloy. Al5056 could be a good candidate for a sphere of 3m diameter (the diameter of 3m has been chosen so that the sphere will resonate around 900Hz like the existing resonant bar Weber antennae) nevertheless we would loose about a factor2.5 relative to Cu alloys as we will see below. The main problem was that it was not possible to find an industry capable (or willing) to cast a 3m 40ton sphere in the world. We examined the possibility of making a sphere by explosion-welding techniques, where layers a few cm thick would be explosively bond to each other[9]. The American company Explosive Fabricators, of Boulder made a sample for E.Coccia of the Roma group and we machined and tested it in our dilution refrigerator. The Q was significantly lower than that of the bulk sphere and the spreading of the five l=2 quadrupole modes was quite larger than that of the bulk sphere. After annealing it was clear that two of the five 30mm thick layers were not well bonded to each other. These results cast serious doubts on whether making a 3m sphere with more than $2000m^2$ of bonded area is at all possible, and are in contrast with the conclusions obtained by Duffy using a torsion oscillator.[10]

The Dutch company LIPS makes large propellers for ships and has the capability of melting 150tons of a particularly strong alloy, the so-called CuNiAl. They were quite interested in our project and made computer analysis of the melting and solidification processes involved in casting a sphere with their alloys. This was necessary because their experience is with nearly two-dimensional castings, which is the case of a propeller, while a sphere is of course a 3-D object. The main concern was with large holes left in the cast after solidification, because of its considerable contraction during solidification. A model of 50cm diameter was made by them, which showed that the defects can be confined, during solidification, into a region in the form of a neck on top of the sphere. This test showed that they could cast a 110ton sphere using their 150ton melting capability.

3 Experimental Results

3.1 CuAl alloys

The highest Q values of all the samples we measured are shown in fig.1. We see that CuNiAl reaches barely one million at 30mK (this alloy has in fact about 5% Fe, 5% Ni, 9% Al and 79% Cu). An alloy with 10% Al and 1% Fe showed a maximum Q of 2 million at 50mK, decreasing regularly with decreasing temperature. Pure CuAl had a reasonable Q of 2 million at 4K giving hopes that it would reach 4or5million at the lowest temperatures but below 4K an absorption appeared which lowered the Q and, despite a slight increase with decreasing temperature, never recovered the 4K value. This turned out to be a characteristic of all CuAlalloys we have measured. We have also seen that decreasing the Alconcentration to 6% gave a strong increase in the 4 K value of Q indicating that if the absorption at lower temperatures were not there, we might expect values in the ten million range. It is thus very important to understand the cause of the absorption . We haven't found any theory dealing with such processes at very low temperatures, and in particular, why is it that the absorption peak (Q^{-1}) does not disappear by lowering the temperature as in the well studied higher temperature absorption peaks[11]. Possibly this is due to dislocations, dislocation rings, or other processes with a broad energy distribution. The experimental course now is to find a non-magnetic third element capable of pinning such dislocations which will cause no decrease in Q. It must be non-magnetic and of low density so as to increase the sound velocity (about 4600m/s in the case of CuAl alloys and 5400m/s in Al5056).

3.2 *CuSn alloys*

We have also tested CuSn alloys, the so-called bell bronzes. These alloys have a rather low Q at room temperature but it increases regularly and reaches 9 million at 20mK. Five samples were prepared with concentrations around 20% varying by about 1% from 17% to 21%. Two were measured for the moment. Unfortunately the company forgot to identify which concentrations the samples had so we will have to analyze them later. The casts were completely full with holes about 1-2mm in diameter, showing that casting of a large sphere might be a problem. Despite the high Q there are two problems with these alloys. The sound velocity is low, around 3500 m/s, which means a cross section smaller than CuAl at equal mass. The tin alloy strongly pollutes the ovens so that LIPS is not willing in principle to cast a 3m. sphere. Cleaning their facilities would stop the company for one month. At any rate, this is a financial matter, not a technical one. Nevertheless there should be companies willing to cast the sphere, possibly in Russia. We decided to leave this alloy as a last possibility, particularly if it turns out that lower resonance frequencies are desirable from the point of view of sources.

3.3 *CuBe*

Ordinary CuBe with ~2% Be was measured by Duffy[7] to have a Q of eight million at 50mK. Two spheres, one with 5% Be and one with 10% Be were made and machined in Russia. These alloys have a very high Q, up to 30 million, and a very high sound velocity, above 5400 for the 10% alloy and above 4600 for the 5% alloy, which would make them ideal for the antenna if it wasn't for the fact that even in the former Soviet Union, the largest producer of Be in the world, the limitation in size is 2.5m. We believe that CuBe 5% is an ideal alloy for making the resonant masses of the transducers because of the high Q, high thermal conductivity and low heat capacity.

3.4 *Chrome-steel*

A steel ball with high Cr content has been measured and found to have a Q of 1.5 million. It was possible to cool it to 20 mK and it has the largest cross-section of all materials cited here, because of the high sound velocity (more than 5000 m/s) and high density. It is possible that even higher Q values can be found within the many high strength steels available. Steel seems an interesting possibility since it has the advantage that it is cheap and has a large cross-section, nevertheless the magnetism of these alloys might be a problem for the SQUID of the transducers, besides the expected high heat capacity. We will do more experiments in this line in the future.

4 Conclusions

In most Q measurements found in the literature the effect of the suspension is determinant in the result, and great care must be exercised in order to have meaningful results. The technique we have used here has the advantage of simplicity and reliability. In fact since there are very many resonances available for measuring the Q, it is a simple matter to excite all of them and see which ones have the highest Q. As seen with the Al5956, it is always possible to find some which are practically unaffected by the suspension itself, giving values so large that we can safely assume they are representative of the "intrinsic" values. The dip in the Q values of CuAl alloys is very annoying and the cause must be found so that the high values expected from the 4K results continue to increase with decreasing temperature as in the case of CuBe alloys, where a sharp absorption is also seen, at 11K but which disappears upon cooling.

Acknowledgments

This work was financially supported by the Dutch Science Foundation NWO, by the Dutch Foundation for Research on Matter FOM of the NWO, and by the University of Leiden. We are grateful to H.v.d.Graaf of the NIKHEF for providing several samples, and to E.ter Haar, D.Maas and J.Noordhoek for helping with some of the measurements. Stimulating discussions with E.Coccia, V.Fafone, and O.Deniz de Aguiar are gratefully acknowledged.

References

1. G. Frossati in *Proceedings of the 7th Marcel Grossman Meeting on G.R.*, Stanford 1994, (World Scientific, Singapore, 1996)
2. R.V. Wagoner and H.J. Paik in *Experimental Gravitation,* Proceeding of the Int. Symposium held in Pavia 1976 (Acc. Naz. Dei Lincei, Roma, 1977)
3. W.W. Johnson and S.M. Merkowitz in *Phys. Rev. Lett.* **70**, 2367 (1980)
4. E. Coccia in *Proceedings of the 14th Int. Conference on General Relativity and Gravitational Physics,* Florence (1994)
5. G. Frossati and E. Coccia in *Cryogenics ICEC supplement,* **34**, 9 (1994)
6. T. Stevenson in *Proceedings of the 7th Marcel Grossman Meeting on G.R.*, Stanford 1994, (World Scientific, 1996)
7. W. Duffy in *Cryogenics* **12**, 1121 (1992)
8. E. Coccia, V. Fafone, G. Frossati, E. ter Haar and M. W. Meisel preprint (1996)
9. E. Coccia, V. Fafone and G. Frossati in *Gravitational Wave Experiments,* proceedings of the First Edoardo Amaldi Conference, Frascati 1994 (World Scientific, Singapore, 1995)
10. W. Duffy and S. Dalal preprint (1996)
11. A. S. Nowick and B.S. Berry in *Anelastic Relaxation in Crystalline Solids*, (Academic Press, New York, 1972

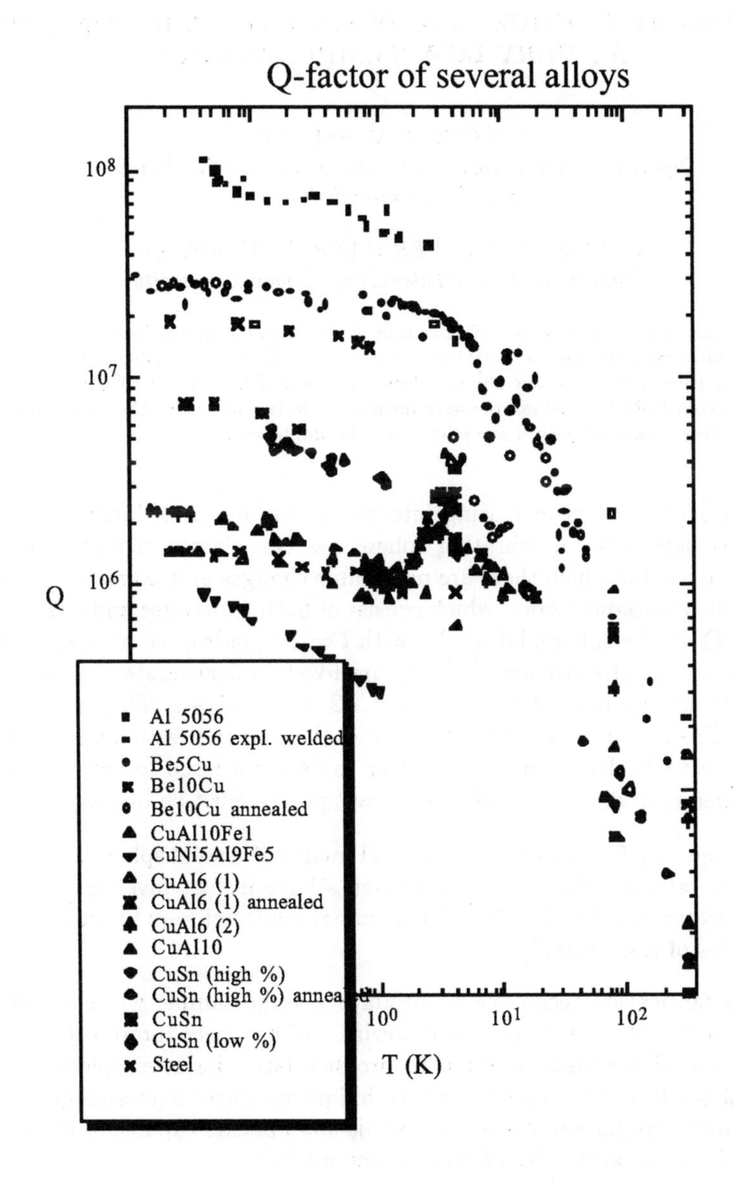

Figure 1: Q-factor of several alloys as a function of temperature.

TESTING ALUMINIUM ALLOY SPHERICAL RESONATORS AT VERY LOW TEMPERATURES

E. COCCIA, V. FAFONE
Dipartimento di Fisica, Universitá di Roma "Tor Vergata" and INFN sezione di Roma 2

G. FROSSATI, E. TER HAAR, M.W. MEISEL
Kamerlingh Onnes Laboratory, Leiden University

We report the quality factors of vibration of the lowest quadrupole modes of Al 5056 spherical resonators at very low temperatures. Q-values of spheroidal modes ranging from 3×10^6 to 2×10^7 in a bulk sample and from 1.7×10^6 to 8×10^6 in an explosively-bonded sample were measured. Independence of the quadrupole modes was measured within one part in 10^5 in amplitude.

At present there is increasing interest in the design of large spherically shaped gw detectors. A vibrating sphere has two classes of normal modes: toroidal modes, for which there are no volume changes and no radial displacements; and spheroidal modes, which consist of both transverse and radial components. Only the spheroidal modes with $l = 2$ (quadrupole modes) interact with a general relativistic gw [1,2]. They are five-fold degenerate and described by the spherical harmonics Y_{2m} with $m = +2, +1, 0, -1, -2$. They have a large and omnidirectional cross section and allows the source direction and the wave polarisation to be determined [2]. In order to design a large spherical detector (several meters in diameters) two important points must be investigated:

- The quality factor of the vibrational modes of a solid sphere at very low temperatures. Previous measurements have in fact investigated the Q of flexural modes of disks [3], longitudinal modes of bars [4,5] and torsional modes of resonators [6].
- The fabrication technique to produce a large sample while preserving the high inherent Q of the material. It has been proposed [7] to use an explosive welding method to produce large-diameter spheres out of explosively-bonded plates. This technique produces high-strength bonds between similar and dissimilar metals and appears capable of preserving the inherent attributes of each parent metal [8].

The Al5056 samples that we tested were obtained by a commercial extruded bar 300 mm diameter, without heat treatments, manufactured by Furumoto Kikoh Co. Two 153 mm diameter spherical samples were machined, both to a sphericity better than 20 μm. No post-machining surface cleaning or hetching was performed. The quality factor of the first sample was measured with two different suspensions. We will indicate these runs as referring to samples A and B. The second sphere (sample C) was machined from a block obtained by explosion welding 5 plates, approximately 30 mm thick, of the same parent alloy as for the first sample.

Sample A was suspended from the center using a 3.8 mm diameter, 160 mm long copper rod (ordinary non-annealed copper) screwed to the sphere. This suspension was chosen in an effort to provide good thermal contact since screwing the rod tightly provides a high pressure at the contact point and also breaks the Al oxide layer. A temperature of 44 mK was obtained in this way. Sample B and C were suspended by hanging them from the center by means of a 2 mm annealed copper wire 160 mm long put in a hole passing through the sphere. The center of the suspension hole in sample C was symmetric respect to the two nearest bond zones ($\simeq$ 15 mm distance). The bond zones were parallel to the vertical axis. With this suspension we gave up on good thermal contact and tried to decouple the spheres as much as possible from the suspension.

We used as electromechanical transducers piezoelectric ceramics (PZT's) glued on small flats filed on the surface of the spheres. The resonances were excited by applying a voltage at the correct frequency to one of the transducers by means of a synthesized signal generator. The signal amplitude was detected using a low noise preamplifier and a spectrum analyser. The maximum of the resonance peak was fed to a computer and the decay time τ was extracted from a fit to an exponential decay curve. The Q was calculated from $Q=\omega\tau/2$. The pressure of the exchange gas in the sample vacuum space was always lower than 1.5×10^{-6} mbar during the Q values measurements.

At low temperatures, the eigenfrequencies values depend very weakly on temperature, as reported for Al 5056 samples in previous works [3 6]. The reported values refer to temperatures $T \leq 4K$.

The independence of the modes in the multiplet was checked within one part in 10^5, by exciting with a voltage not exceeding 50-100 mVpp a transducer for a long enough time at one of the resonance frequencies and observing that none of the neighboring frequencies was excited.

The Q values of the lower quadrupole modes are shown in figs. 1 and 2. In fig. 1 we see that the Q values of sample A are somewhat lower than those of sample B. This may be due to the different amount and position of the PZT's coupled to the resonator and/or to the different suspension. The Q values of the spheroidal quadrupole modes of sample C are shown in fig 2. This sample gave slightly lower values than sample B although it had the same suspension.

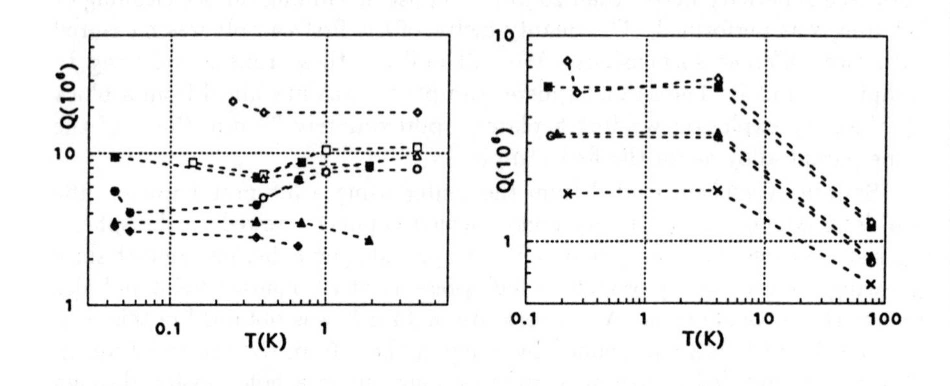

Fig.1 Q-values of the spheroidal quadrupole mode of sample A (■ = 18436.53 Hz, ▲ = 18437.89 Hz, ● = 18459.76 Hz, ♦ = 18460.28 Hz) and sample B (o =18289.61 Hz, □ = 18290.62 Hz, ◊ = 18383.99 Hz, Δ = 18500.36 Hz).

Fig. 2 Q-values of the spheroidal quadrupole mode of sample C, obtained by explosive welding (o = 18348.20 Hz, ■ = 18481.81 Hz, ▲ = 18538.57 Hz, ◊ =18590.45 Hz, x = 18683.10 Hz).

The result of our measurements is that Al 5056 spherical resonators have independent modes of vibration with high Q, as necessary for gw research. The explosively bonded sample showed Q-values higher than 10^6: the explosive welding method then appears as a suitable technique to fabricate large spherical detectors.

References

1. N. Ashby and J. Dreitlein, *Phys. Rev.* D 12, 336 (1975).
2. R.V. Wagoner and H.J. Paik in *Proc. of the Int. Symposium on Experimental Gravitation* (Accademia Nazionale dei Lincei, Rome, 1977).
3. T. Suzuki, T. Tsubono and H. Hirakawa, *Phys. Lett.* A67, 2 (1978).
4. P. Carelli et al, *Cryogenics*, 406 (1975).
5. E. Coccia and T.O. Niinikosky, *Lett. Nuovo Cimento*, 41, 242 (1984).
6. W. Duffy, *J. Appl. Phys* 68, 5601 (1990).
7. E. Coccia, V. Fafone and G. Frossati in *Gravitational Wave Experiments*, Proceedings of the First Edoardo Amaldi Conference, Frascati 1994, (World Scientific, Singapore, 1995), E. Coccia, G.Pizzella, F.Ronga (eds.).
8. J.G. Banker and E.G. Reineke ASM Handbook, Vol. 6 ASM Metal Park Ohio, 303 (1993).

POWER RECYCLING EXPERIMENT WITH SUSPENDED MIRRORS

S. MORIWAKI
Department of Applied Physics, University of Tokyo, Bunkyo, Tokyo 113, Japan

In order to develop design of a servo system with power recycling, we constructed a table-top simple Michelson interferometer with suspended mirrors. In such a multi-variable feedback system, interaction between the feedback loops causes nonlinear gain suppression or enhancement, so that the servo acquisition problem may arise. We designed a servo filter which kept appropriate phase margin under widely fluctuating gain and confirmed stable acquisition of the system. We also observed instantaneous power gain of 60 and averaged signal enhancement of about 50. A method to diagnose the imperfection of an interferometer is also discussed.

1 Control and Diagnostic Systems for Suspended Mirrors

A scheme of power recycling has been incorporated in several recent plans of ground-based interferometric gravitational wave detectors. The principle of power recycling has been demonstrated in prototype interferometers.[1-3] On the other hand, power recycling may bring some technical difficulties, for instance, locking acquisition, separation of control signals [4] and thermal distortion. [5] In order to study the property of the locking acquisition, we construct a table-top interferometer with suspended mirrors. This experiment is intended to develop servo systems and diagnostic methods.

2 Experimental Setup

The optical and control system is shown in Fig. 1. A stabilized He-Ne laser is used as a light source. The light is introduced to an interferometer through Faraday isolators and matching lenses. A Michelson interferometer consists of two arm mirrors and a beam splitter. Between the light source and the interferometer we put a recycling mirror whose intensity transmission and reflection were 0.72% and 98.9%. This interferometer has two degrees of freedom in displacement along the optical axes; one is differential displacement of two arm lengths and the other is common displacement which is equivalent to displacement of the recycling mirror. To obtain the error signals, we use 15kHz (OSC1) and 10kHz (OSC2) synchronous detections. All of the mirrors and the beam splitter are independently suspended as a double pendulum. Each final mass has four magnetic actuators; two of them are for servo control and other ones are used for modulation. Note that the beam splitter has no actuators.

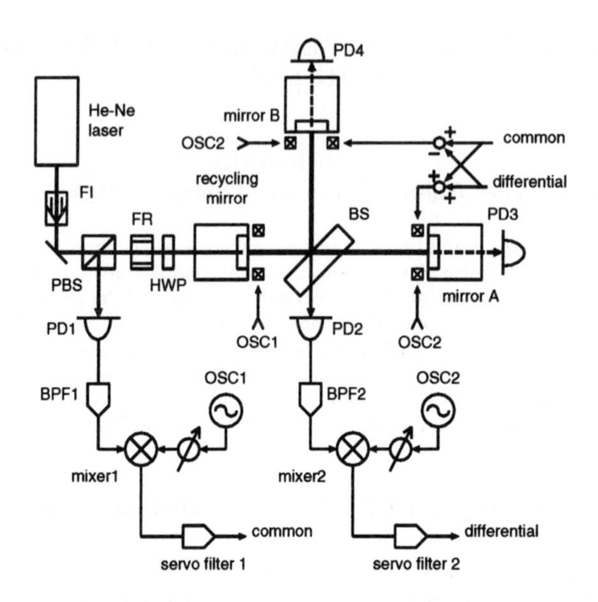

Figure 1: Experimental apparatus. All of arm mirrors, a recycling mirror and a beam splitter are independently suspended by a double pendulum.

3 Properties of Servo System

As mentioned above, the interferometer has two degrees of freedom in displacement of mirrors and they must be controlled by appropriate servo system since suspended mirrors have large seismic displacement around its resonant frequencies. We adopt a lead-lag filter for this purpose. Figure 2 shows a Bode plot of the open loop transfer function for differential displacement. Solid lines are calculated from the characteristic time constants of the servo filter. In the calculation we assume that displacement of the test mass has frequency dependence of f^{-2} to the current of the actuators. Measured frequency response is shown by dots in the figure. We can see good agreement between measured and calculated responses around the unity gain frequency of 700Hz. This transfer function has sufficient phase margin under gain variance of about 40dB. In such a configuration we confirmed that simultaneous locking of two servo loops occurred frequently. We observed averaged recycling gain of 50 by measuring the enhancement of open-loop gain for differential displacement. It is close to the value of instantaneous gain 60, which is obtained by monitoring the transmission intensity of two arm mirrors (PD3 and PD4 in Fig. 1).

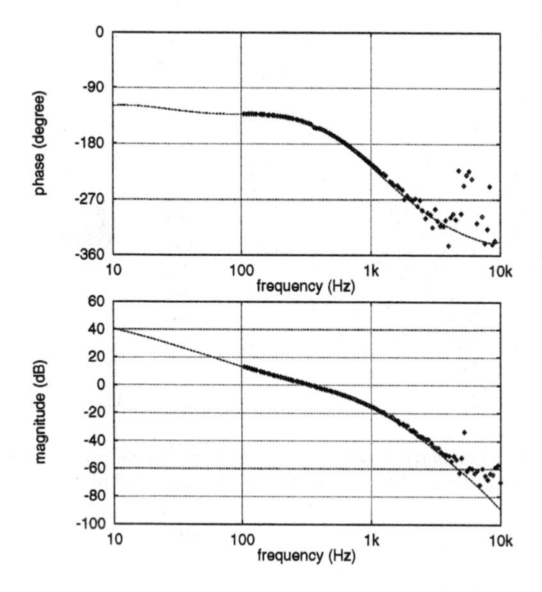

Figure 2: A Bode plot of open loop transfer function for differential displacement. The solid lines and the points show calculated and measured values.

4 Correlation of DC Photocurrent

In the recycled Michelson interferometer, it is not so easy to confirm that the servo system is working to realize a correct operating point, since the dc photocurrent at antisymmetric port (PD2 in Fig. 1) is not neither minimum nor maximum when the interferometer has maximum sensitivity. To monitoring the operating point, we found that we could utilize a correlation diagram of dc photocurrents at PD1 and PD2. Examples of such a diagram is shown in Fig. 3. Both axes mean intensity fractions at PD1 and PD2 relative to the incident light. According to displacement of mirrors, the point fills a region enclosed by a strait line and a fragment of an ellipse. A calculated boundary is shown by dotted line in Fig. 3. Dashed parallel lines represent the contours of loss in the interferometer. The boundary has two edges corresponding to the resonant and antiresonant states. We could see measured locus was localized near the resonant point when servo acquisition occurred. We could found empirically that the common phase change and the symmetric tilt of the arm mirrors caused increment of photocurrent at PD1, while the differential displacement and antisymmetric tilt increased that of PD2. Therefore this diagram was helpful when we trimmed the servo gain and aligned the mirrors.

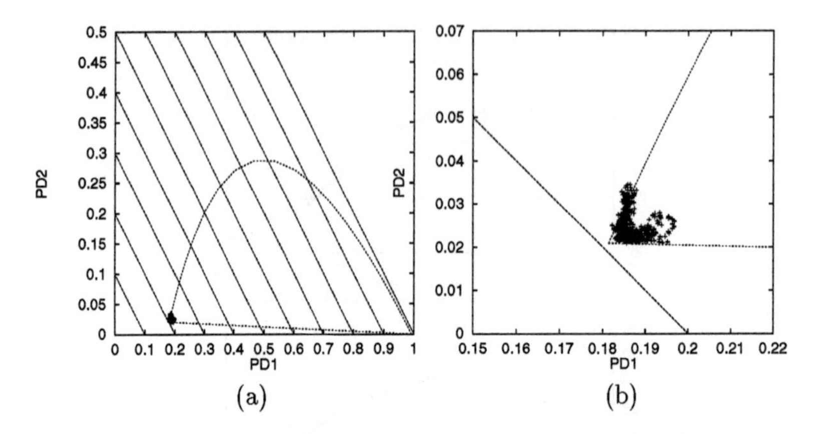

Figure 3: Examples of the correlation plot of photocurrent. (a) A correlation with two servo filters working. Measured point is shown by dots. (b) A magnified plot around the resonance point.

5 Conclusion

We have demonstrated the simultaneous locking acquisitions of differential and common displacements of suspended Michelson interferometer under the large gain variance caused by power recycling. A correlation diagram of photocurrent can be used to diagnose the servo performance and to support initial alignment of mirrors.

Acknowledgments

The author thanks N. Mio for helpful discussions and K. Tsubono for many useful suggestions especially in the design of the suspension system.

References

1. C. N. Man, D. Shoemaker, M. Pham Tu and D. Dewey, *Phys. Lett.* A **148**, 8 (1990)
2. K. A. Strain and B. J. Meers, *Phys. Rev. Lett.* **66**, 1391 (1991)
3. P. Fritschel, D. Shoemaker and R. Weiss, *Appl. Opt.* **31**, 1412 (1992)
4. M. W. Regehr, F. J. Raab and S. E. Whitcomb, *Opt. Lett.* **20**, 1507 (1995).
5. W. Winkler, K. Danzmann, A. Rüdiger and R. Schilling, *Phys. Rev.* A **44**, 7022 (1991)

LABORATORY EXPERIMENTS ON ADVANCED TECHNIQUES FOR GRAVITATIONAL WAVE DETECTORS

H. LÜCK and the GEO600-Team

Max-Planck-Institut für Quantenoptik, Aussenstelle Hannover, Appelstr. 2, 30167 Hannover, Germany

G. HEINZEL

Max-Planck-Institut für Quantenoptik, Hans-Kopfermann-Str. 1, 85748 Garching, Germany

D. MAASS, K.-O. MÜLLER

Inst. f. Atom- und Molekülphysik, Universität Hannover, Appelstr. 2, 30167 Hannover, Germany

Usually the sensitivity of interferometric gravitational-wave detectors *(IGWD)* is limited in a certain frequency regime by photon shot noise. In our labs at the University of Hanover and the "Max-Planck Institut für Quantenoptik" in Garching we investigate methods to lower that limit. This paper shortly presents the purposes and results of three diploma theses of K.-O. Müller, D. Maaß, and G. Heinzel

1 Introduction

The sensitivity of the large *IGWD* planned today (LIGO, VIRGO, GEO600, TAMA300) is usually shot noise limited in the upper frequency range of their projected bandwidth. According to Mizuno [1] the peak sensitivity of any shot noise limited interferometer is given by:

$$\tilde{h}_o \geq \sqrt{\frac{h\lambda}{\pi^2 c}\frac{\Delta f_{BW}}{\mathcal{E}}}. \tag{1}$$

whith λ is the laser wavelength, h the Planck constant, c the speed of light, Δf_{BW} the *detector bandwidth* and $\mathcal{E}$ the energy stored in the interferometer. The *detector bandwidth* is the approximate frequency range in which the spectral sensitivity is close to that at the peak. [1]

The sensitivity of shot noise limited interferometers can therefore be enhanced in two ways: Increasing the stored energy $\mathcal{E}$ by increasing the armlength or the circulated power gives enhanced sensitivity as well as decreasing the bandwidth.

1.1 Thermally Adaptive Optics

Increasing the power by means of Power Recycling *(PR)* is limited by thermal lensing that forms due to residual absorption inside optical elements transmitted by the laser beam. The wave front distortions in the beams returning from each of the interferometer arms are usually different. This reduces the achievable contrast at the output of the interferometer and therefore increases the losses in the *PR* cavity. Thermal lensing cannot be compensated ab initio by shaping the optical elements in an appropriate way. Such a precompensation would give a poor contrast without thermal lensing and therefore the power-buildup would be too low to form the thermal lenses the compensation was designed for. We therefore need a compensation which adapts itself to the wave front distortions as the thermal lens forms.

The high mechanical Q-values required to keep the thermal noise sufficiently low prohibit the use of standard adaptive optics which uses mechanical actuators such as piezo-electric elements. One way out is the control of the mirror surface through thermally induced expansion. Thermal expansion as a response of an optical element to energy absorbed from a laser beam leads to a phase shift in the light transmitted or reflected from that optical element. Scanning the laser beam across the surface while varying the power or the exposure time for a given position hence gives a method of adapting the optical element to the need of the interferometer without touching it.

The feasibility of that method was shown by Kai-Oliver Müller. He set up a table-top Michelson interferometer illuminated with an expanded He-Ne laser beam. The virtual interference pattern at the location of the end mirrors was recorded with a gated intensified CCD-camera. Dithering the axial position of one of the end mirrors and subtracting two images recorded at different well known mirror positions gave the two dimensional phase differences of the beams returning from the two interferometer arms. Thus the interferometer itself served as a wave front sensor. The beam of a 30W ND-YAG laser was scanned across the surface of one of the end mirrors with the help of an XY galvo-drive controlled by a computer. An additional X galvo-drive was used for blanking of the beam at positions of the mirror surface where no heating was required. In response to the dissipated power the substrate expanded and the mirror surface was locally translated. Hence the phase differences between the interferomter arms could be controlled and minimized.

The aluminium coated surface of the BK7-glass mirror absorbed about 10% of the ND-YAG beam. Thermal expansion of BK7-glass is much larger than of fused silica which would be a likely choice for *IGWD* test masses. These are welcome properties for a proof of principle as they give easily detectable

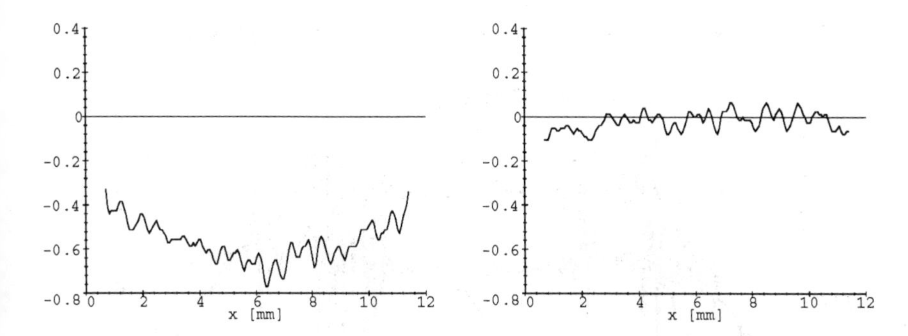

Figure 1: Profile of the wave front [rad] at the output port; before and after correction.

effects with moderate requirements for the control laser and the wave front sensor.

For good display of the effect a distorted mirror was chosen and intentionally shifted from the "dark fringe" position. A one dimensional cut through the wave front before and after correction is shown in fig. 1.

1.2 Signal Recycling by Schnupp Modulation

Besides increasing the light power circulating inside the interferometer the sensitivity can be enhanced by reducing the bandwidth by the use of e.g. Signal Recycling [2]. Obtaining a signal to control the position of the Signal Recycling mirror without modulating one of the mirrors or inserting additional optical elements into the interferometer is one of the key problems. If the interferometer arms have a slightly different lengths[a], phase modulation of the light injected into the interferometer (called Schnupp-, inline-, pre-, or frontal-modulation) will be seen at the output port of the interferometer which is usually kept close to a dark fringe. Using the phase shift of the modulation sidebands near some resonance of the Signal Recycling cavity with respect to a phase reference (a local oscillator or the residual carrier) can be used to generate an error signal for the Signal Recycling cavity. With an 18m delay-line (to give a reasonable free spectral range) table-top interferometer Dirk Maaßobtained good control signals and reached a sensitivity enhancement (see fig.2) which agreed well with theoretical considerations.

1.3 Resonant Sideband Extraction

Placing Fabry-Perot cavities *(FP)* in the interferometer arms enhances the carrier power, hence the energy stored in the detector and thus allows to lower the shot noise limit. Due to the high carrier power the signal sidebands also

[a] a few ten cm for modulation frequencies of about $10MHz$

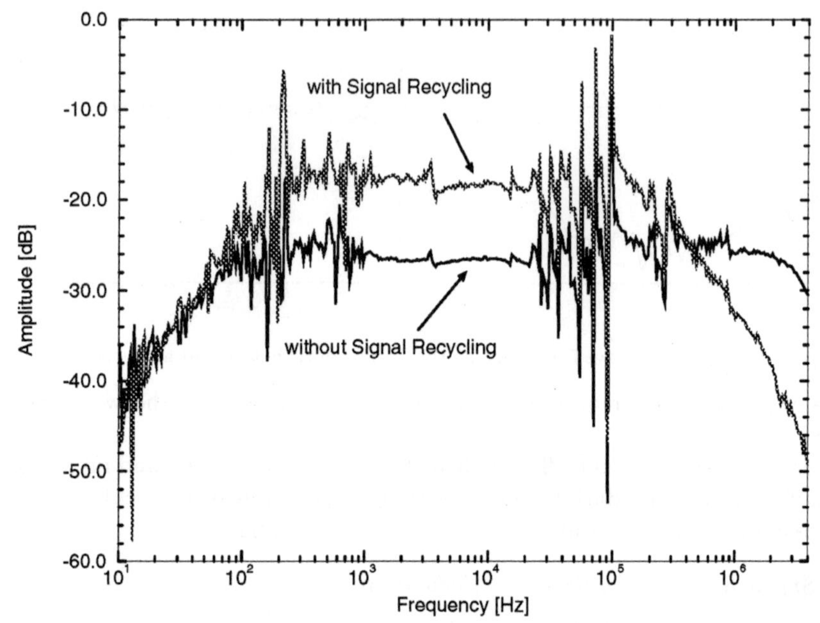

Figure 2: Transfer function of the interferometer with and without Signal Recycling.

gain, as a given part of the carrier power is converted into signal sidebands by the gravitational wave. Due to the small tranmission of the coupling mirror, necessary for the high buildup in carrier power, only an accordingly small part of these sidebands reaches the output port of the interferometer and can be detected. Only sidebands with a frequency spacing to the carrier less than half the linewidth of the *FP*s are also enhanced and therefore appear at the output port stronger than without *FP*s. To make full use of the high carrier power in the *FP*s a coupling mirror with a high reflectivity for the carrier frequency and a high transmission for the sideband frequency would be needed. An additional mirror placed in the output port of the interferometer can be used to extract the signals of the cavities, hence it's name: Signal Extraction Mirror.
Gerhard Heinzel used a table top experiment to show two different ways to get control signals for the Signal Extraction Mirror and obtained good agreement between experiment and theory (see fig. 3).

References

1. Jun Mizuno, *Comparison of optical configurations for laser–interferometric gravitational–wave detectors*, MPQ Report 203, (1995).
2. Brian Meers, *Recycling in laser-interferometric gravitational-wave detectors, Phys. Rev.* D **38**, 2317 (1988).

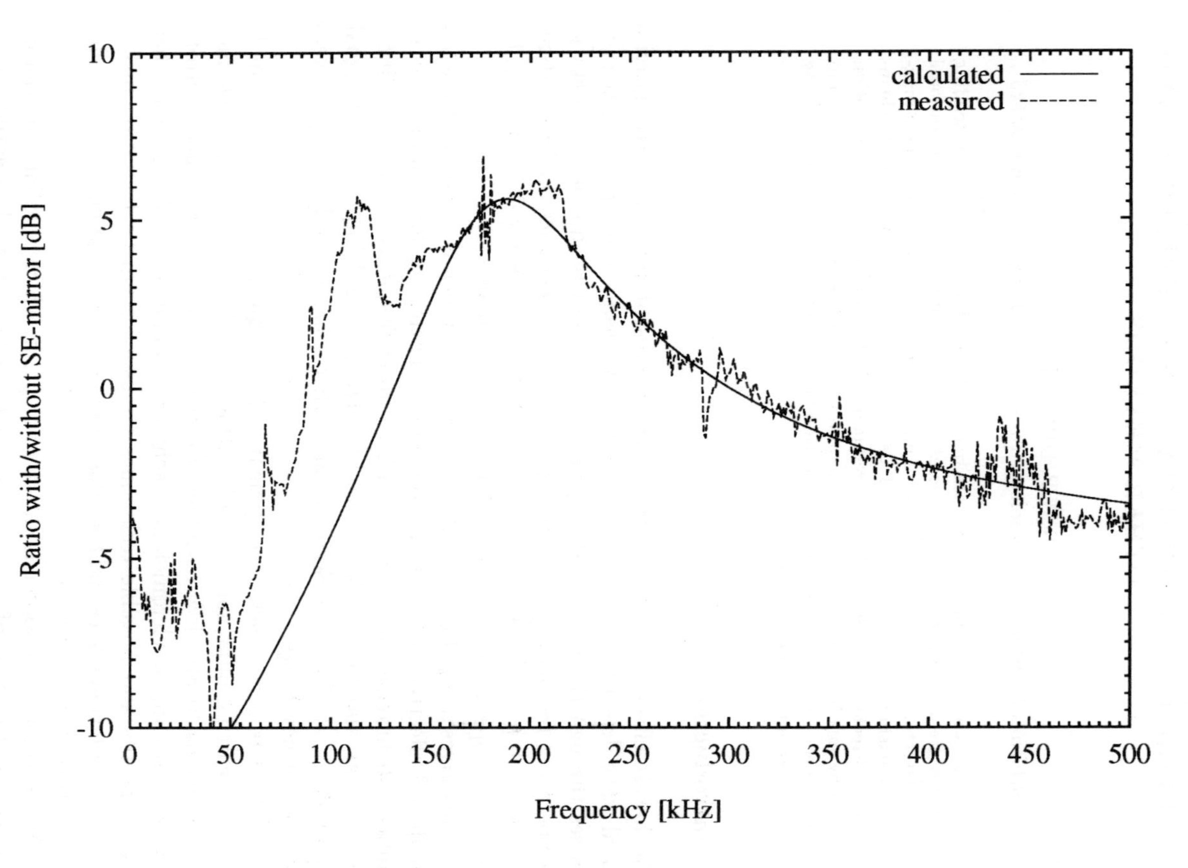

Figure 3: Signal enhancement by Resonant Sideband Extraction

TECHNIQUES FOR EXTENDING INTERFEROMETER PERFORMANCE USING MAGNETIC LEVITATION AND OTHER METHODS

R.W.P. DREVER

California Institute of Technology, 130-33, Pasadena, California 91125

Some new techniques and concepts are being developed with the aim of extending to lower frequencies the range of operation of laser interferometer gravitational wave detectors. These include the use of magnetic levitation with permanent magnets in seismic isolation systems, and also for suspending the test masses themselves. A technique is also outlined for reducing effects of differential ground motion by using auxiliary interferometers to monitor and control relative positions, and tilts if necessary, of elements in the isolation systems for the test masses defining each main interferometer arm.

1 Introduction

The lowest effective operating frequency of present interferometric gravitational wave detectors is usually determined by seismic noise reaching the test masses through the associated suspension wires and seismic isolation stacks. As operating frequency is reduced, ground motion increases, and the attenuation of each passive isolation stage is degraded by a factor depending on the ratio of the operating frequency to the resonance frequency of the stage.

This gives a rapid increase in instrument noise below a certain frequency, cutting off effective low-frequency operation. There are clearly important advantages in moving the seismic cut-off to lower frequencies. Detectability of neutron-star and black-hole coalescence signals would improve, and signals from pulsars, stochastic background and other sources could be sought over a wider frequency range.

There has been much development of low-frequency isolation systems by several research groups, some using isolation stages based on special mechanical configurations to give low resonance frequencies, and others using electronic servo-systems and accelerometers to achieve a similar effect. We introduce here some concepts and possible techniques for using magnetic levitation in low-frequency isolation stages and test mass suspension, and our current experience in exploring their feasibility.

We also outline another isolation concept, in which auxiliary interferometers couple the ends of each interferometer arm so that differential seismic motion can be reduced without requiring the same degree of absolute isolation.

2 Some Magnetic Configurations Being Investigated

2.1 General considerations

As an isolator, a magnetic suspension can be regarded as equivalent to a mechanical spring- mass system, at least to first order. However, there are at least two major potential benefits: it seems practicable to devise configurations that give low resonance frequencies in a fairly simple way; and the serious problems of resonance in unwanted modes of vibration in mechanical springs and suspension wires are avoided. There are naturally negative aspects also, and we will discuss some of these later.

We consider only room-temperature systems here, and in particular ones based on permanent magnets so that there is no significant power dissipation. Superconducting levitation systems have been used in several cryogenic bar gravitational-wave detectors, but at the present stage the simplicity of room-temperature operation is an important practical advantage for interferometers.

Static levitation arrangements of permanent magnets alone are in principle unstable. Stability can be achieved in various ways in room temperature systems. Diamagnetic materials can be used to stabilize a small passive system, and we have used this in tests with masses up to 10 gm, but the extreme weakness of diamagnetic phenomena causes practical difficulties with large masses. Stability is much more easily and robustly achieved with an electronic servo system which senses the position of the levitated object and adjusts the fields accordingly, and we have mostly used this approach.

2.2 Some examples of configurations

Some simple magnetic configurations we have been investigating are illustrated in concept in Figures 1 to 3. Only the permanent magnets are shown here, although in practice the levitated magnet would in most cases be carrying a load of non-magnetic material. (A possible exception might be a test mass made itself of magnetic material.)

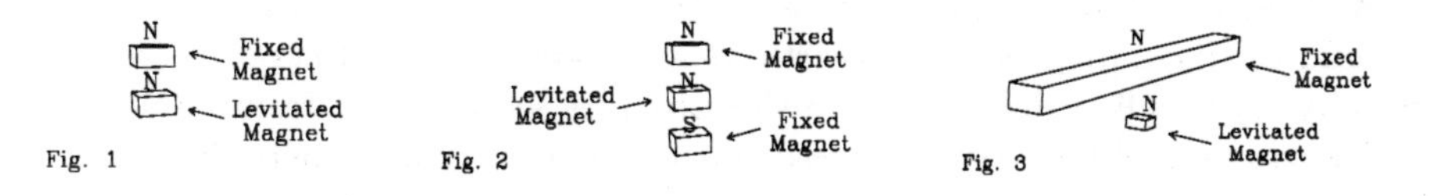

The simplest configuration is shown in Fig. 1. Here a fixed magnet provides the field gradient to support a levitated magnet below it, at a position where the magnetic attraction precisely cancels its weight. A servo system, not shown, may be used to maintain stability by sensing the vertical position of the levitated magnet and making small adjustments to the lifting field.

Fig. 2 shows an arrangement in which a second fixed magnet is located symmetrically below the levitated magnet, with its polarity reversed. At the location of the levitated magnet, there is a field gradient with small mean field. The region of near-uniform gradient can be larger than in the previous configuration, leading to longer natural periods of oscillation. In experiments and modeling of this configuration, we have found that there can be a rotational instability of the levitated magnet about a horizontal axis for finite horizontal displacements, even when the vertical position is stabilized. This instability can be overcome by connecting together three or more systems of this type.

Fig. 3 shows a configuration we have devised to achieve a long natural period of oscillation in one horizontal direction. Here the fixed magnet is made long compared with its height and width, and is magnetically polarized in a direction perpendicular to its length. Vertical polarization is shown as an example. The levitated magnet is polarized in the same direction, and is located below the fixed magnet, with a vertical sensing system to maintain stability. If the ratios of the length of the fixed magnet to its transverse dimensions are large, then the field experienced by the levitated magnet is to first order independent of its position in the longitudinal direction, when it is near the center of the system. Thus a long period is almost automatically obtained.

3 Applications

3.1 Magnetic levitation seismic isolation stage

A seismic isolation system must isolate against motions in all directions, and we have made some preliminary tests with a system consisting of three isolators of the type shown in Fig. 2, connected together. The arrangement is illustrated diagrammatically in Fig. 4, simplified to clarify the concept.

In this case we have used Hall-effect sensors to monitor the vertical position of each levitated magnet. These control small stabilizing currents in coils around the fixed magnets. Rare-earth magnets are convenient here. These are usually electrically conductive, and this provides some eddy-current damping, which can be useful for reducing resonant motions in a stage where the associated thermal noise from Johnson currents is insignificant. A small test system of this type, carrying a total load of 1 kg, had natural resonance frequencies for transverse motion of less than 0.5 Hz, and for vertical motion around 2 Hz. Significantly lower resonance frequencies are possible.

3.2 Magnetic levitation techniques for test mass suspension

Suspension of a test mass is a more difficult problem, since noise must be extremely small. Among noise sources to be avoided are thermal noise from the coupling of Johnson currents in conducting magnets with magnetic field gradients, and test mass motions induced by fluctuations in ambient magnetic fields. The configuration of Fig. 3 was designed to facilitate reduction of these noise sources. Any Johnson noise currents in the levitated magnet have relatively little effect, since as the variation of the field distribution from the fixed magnet is small in the direction of the laser beam, forces in that direction due to these currents are reduced. Johnson currents in the fixed magnet can still generate noise, but it may be sufficient for this magnet alone to be non-conducting.

To reduce noise from coupling to varying external magnetic fields we propose that two or more magnets are used on the test mass, arranged so that there is cancellation of the dipole, and possibly also of the quadrupole, moment of the suspended system. Then the residual coupling is only to higher order gradients of the external field, and can be significantly reduced. An example of this concept is illustrated in Fig. 5. In this case two magnets are used to support the test mass, with a pair of suitably polarized fixed magnets above them. In a preliminary test system of this type we have used a shadow sensor to monitor the vertical position of each levitated magnet. Photodiodes sense light passing above the levitated system, and each controls the current in a stabilizing winding around one of the fixed magnets. (Although two stabilizing systems were used here, in principle one is sufficient.)

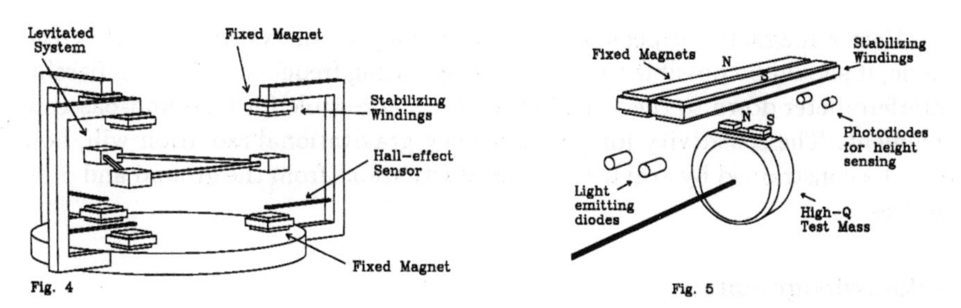

Fig. 4

Fig. 5

This configuration can be extended by using four or more suitably polarized magnets on the test mass, together with corresponding fixed magnets, to further reduce external coupling.

In preliminary tests with a system of this type it was found that the natural period of oscillation in the direction of the beam was determined more by nonuniformity in the fixed magnets used than by end effects. This can be partly compensated by small trimming magnets. Without any trimming, typical natural periods were around 10 seconds. Periods up to 20 seconds have been obtained with some simple trimming. Relaxation times of several hours have been observed in initial experiments in vacuum.

4 Tilt-Coupled Suspensions

We have earlier proposed that with pendulum suspensions the differential effects of residual seismic noise transmitted by seismic isolation stacks may be reduced by linking the suspension points in each arm by an auxiliary interferometer. This is used either as part of a servo system to cause the suspension points to track one another, or as a monitor to provide a seismic correction signal. With suspension systems which give periods longer than that of a simple pendulum, such as those described above or equivalent long-period mechanical systems, ground tilts can cause significant motions of the effective suspension points. We propose coupling the tilts as well as the horizontal position of appropriate stages in the isolation systems by a pair of servo-control or monitoring beams, as indicated in Fig. 6.

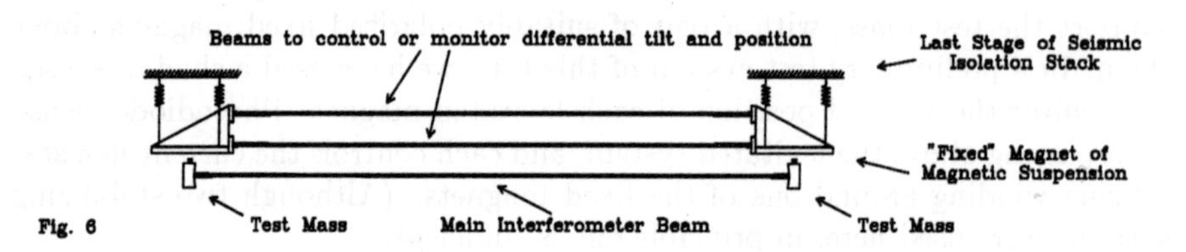

Fig. 6

Here a magnetic suspension is shown merely as an example. By this technique, it should be possible to extend the operating frequency of a long-baseline interferometer down to the lowest frequencies accessible for a ground-based instrument. The sensitivity for low frequency gravitational radiation will, however, be constrained by gravity gradient fluctuations from the ground and other sources.

Acknowledgments

I would like to acknowledge the valuable assistance of S. J. Augst, who built and tested several of the systems investigated. I would also like to thank E. W. Cowan for very helpful discussions and computer modeling; and J. L. Hall and C. W. Peck for much stimulation and encouragement. The experimental work was supported by the California Institute of Technology.

Detection in Space

SPACECRAFT DOPPLER EXPERIMENTS

L. Iess[a] and J.W. Armstrong
Jet Propulsion Laboratory, California Institute of Technology, 4800 Oak Grove Dr., Pasadena, Ca. 91109, USA

The exploration of the low frequency end of the gravitational spectrum has been so far performed using space detectors based upon precision Doppler tracking of interplanetary spacecraft. We review the fundamentals of this detection technique and the experimental achievements reached so far. We describe also the sensitivities of future, approved experiment to expected sources and data analysis strategies for their detection.

1 The detector

The idea of exploiting interplanetary Doppler tracking as a detector of gravitational waves dates back to the early seventies, when it was realized that transient perturbation of the local metric could affect in a measurable way microwave signals used for space navigation and telecommunications. A remarkable feature of this method of detection is indeed that it has been based mostly on existing instrumentation, both on the ground and onboard the spacecraft. A two-way communication link is part of the normal routine of every space mission, but what made possible its use as a detector of gravitational waves are especially the requirements of spacecraft navigation, with its need of accurate timekeeping and precise range rate measurement. Table I summarizes the experiments performed to date and those planned through the next years.

In the preferred radio link configuration (called two-way or coherent tracking) radiometric observables are obtained by using a highly stable oscillator sitting at a ground antenna. The oscillator output (typically at frequencies of a few MHz) is upconverted, amplified and transmitted to the spacecraft. The carrier frequencies allocated for deep space telecommunications are in S-band (2.1-2.3 GHz), X-band (7.2-8.4 Ghz) and Ka-band (32-34 GHz). The onboard electronics generates an amplified, phase-coherent replica of the incoming signal by means of a phase-locked-loop receiver and an amplifier. On the ground, the frequency ν_r of the spacecraft carrier is then measured by comparison with the frequency ν_0 of the master clock (in the case in which the receiving station is also the transmitting one) or with a clock of similar stability (when they are different, in which case one speaks of three-way tracking.) Space naviga-

[a]Permanent address: Dipartimento aerospaziale, Univ. di Roma La Sapienza, via Eudossiana 18, 00184 Roma, Italy

Table 1: Past and planned Doppler experiments.

1980	VOYAGER 1	Hellings et al.[1]	few passes S/X band
1981	PIONEER 10	Anderson et al.[2]	3 passes S band
1983	PIONEER 11	Armstrong et al.[3]	3 days S band
1988	PIONEER 10	Anderson et al.[4]	10 days S band
1992	ULYSSES	Bertotti et al.[5]	28 days S/X band
1993	MO/GLL/ULS	Armstrong et al.[6]	19 days X band (MO)
1994-5	GALILEO	Estabrook et al.[7]	40 days S band
1997	MGS		21 days X band
2002	CASSINI		120 days Ka band

tion makes use of the best existing frequency standards (H-masers), capable to ensure stabilities smaller than 1.10^{-15} over time scales of 1000 s.

The observable quantity in a range rate measurement is a frequency shift or its relative value

$$y = \frac{\nu_r - \nu_0}{\nu_0}. \tag{1}$$

Since the geometry and the spatial coordinates of both the spacecraft and the ground station change significantly during the time taken by photons to propagate along both legs, the quantity (1) is appropriately referred (and actually equal to) the time derivative of the difference T between the reception and the transmission times of the photons, a quantity usually called round-trip light-time (RTLT).

The performance of a Doppler measuring system depends of course on the stability of the reference frequency ν_0 and the noise introduced in ν_r during the propagation of the signal. What a good tracking system requires is indeed a good end-to-end coherence of the radio link. The most widely used figure of merit to characterize frequency standards and phase coherence is the Allan deviation, which may be seen as the structure function of time–averaged relative frequency shift:

$$\sigma_y(\tau) = \frac{1}{\sqrt{2}}\langle[\bar{y}(t) - \bar{y}(t+\tau)]^2\rangle^{0.5} \tag{2}$$

$$\bar{y}(t) = \frac{1}{\tau}\int_t^{t+\tau} y(t')dt' \tag{3}$$

where the brackets $\langle\rangle$ denote statistical average. The best end-to-end stabilities reached so far in spacecraft Doppler experiments are about $1.\ 10^{-14}$ over integration times of 1000 s, under favourable conditions of the propagation media. As it will be explained later, the clock and the electronics themselves would allow to get much better stabilities, but the limitation comes indeed from propagation noise.

The first proposal to use Doppler tracking systems as gravitational wave detectors was published by Braginsky and Gerthensthein[8] in 1967, but it was not until 1975 that the that Estabrook and Wahlquist[9] computed the precise relationship between the metric perturbation h induced by a linearly polarized gravitational wave and the physically observable quantity in radiometric measurements (eq. 1.) Their work was later extended to the case of arbitrary polarisation[10,11]. In the general case, the input excitation to the detector is provided by a linear combination of the two independent polarization states h_+ and $h_\times$ of the gravitational wave:

$$h(t) = (1 - \cos^2\theta)^{-1}\hat{\boldsymbol{n}} \cdot [h_+(t)\boldsymbol{e}_+ + h_\times(t)\boldsymbol{e}_\times] \cdot \hat{\boldsymbol{n}} \tag{4}$$

where θ is the angle between the unit vector $\boldsymbol{k}$ along the propagation direction and the unit vector $\hat{\boldsymbol{n}}$ along the Earth-spacecraft line of sight. $\boldsymbol{e}_+$ and $\boldsymbol{e}_\times$ are matrices which, in the frame of reference where the z-axis is along $\boldsymbol{k}$ may be reduced to

$$\boldsymbol{e}_+ = \begin{pmatrix} 1 & 0 & 0 \\ 0 & -1 & 0 \\ 0 & 0 & 0 \end{pmatrix}, \quad \boldsymbol{e}_\times = \begin{pmatrix} 0 & 1 & 0 \\ 1 & 0 & 0 \\ 0 & 0 & 0 \end{pmatrix}. \tag{5}$$

The observable relative frequency shift at the output of the detector may be seen as the convolution product

$$y(t) = h(t) * r(t) \tag{6}$$

between the quantity (4) and the impulse response $r(t)$ of the detector, which assumes a particularly simple form:

$$r(t) = -\frac{1-\cos\theta}{2}\delta(t) - \cos\theta\ \delta(t - 0.5T(1+\cos\theta)) + \frac{1+\cos\theta}{2}\delta(t-T). \quad (7)$$

Not surprisingly, the impulse response is determined only by two quantities, namely the angle θ and the round-trip light-time T. When $\theta = 0$ or $\theta = \pi$, the response vanishes, as it must be for transverse waves. A remarkable, although expected property of the response is that its integral over time is zero. This ensures that a static field or a wave of characteristic period τ much longer than T do not produce observable effects, as a consequence of the equivalence principle. In other words, the detector is a high-pass system and it can be proved easily that amplitude transfer function falls off linearly as the excitation frequency tends to zero. This means, for example, that bursts with memory[12] in which the local metric is permanently perturbed by a non-zero, static field after the transit of a gravitational wave, would be very difficult to detect. The critical frequency, after which the detector shows degraded performances, is approximately $1/T$.

The exact computation of the response function requires the integration of the equation of motion of the photons in the time-varying field of the wave, but a naive interpretation is possible if we think that the wave field changes both the master clock frequency and the geodesic separation between the earth and the spacecraft. At the Earth, the wave affects both the incoming and the outgoing photons, giving rise to the the first and last pulses, separated by T in the response function. When the wave "hits" the spacecraft, its buffeting give rise to the intermediate pulse. This analogy in terms of "clock speed-up" and buffeting of the two end masses can be pushed further to give exactly eq. 7.

In principe, the triple signature of the gravitational wave in the Doppler record looks attractive from the point of view of signal detection, as it is significantly different from what one might expect from noise. This is indeed for the case as long as the three pulses are clearly separated, but if the characteristic time of the wave τ becomes comparable to T or if the angle θ approaches 0 or π, the output may become quite unexpected, as it is shown in fig. 1.

A similar behaviour is found for sinusoidal signals[11,13]. In the case where T and θ are constant over the observing interval, the output of the detector is still a sinusoid with the same period of the excitation, but its amplitude may become extremely small in a discrete number of points of the frequency window, as shown in fig. 2. The total amplitude can be seen indeed as the sum of three phasors (one for each pulse)

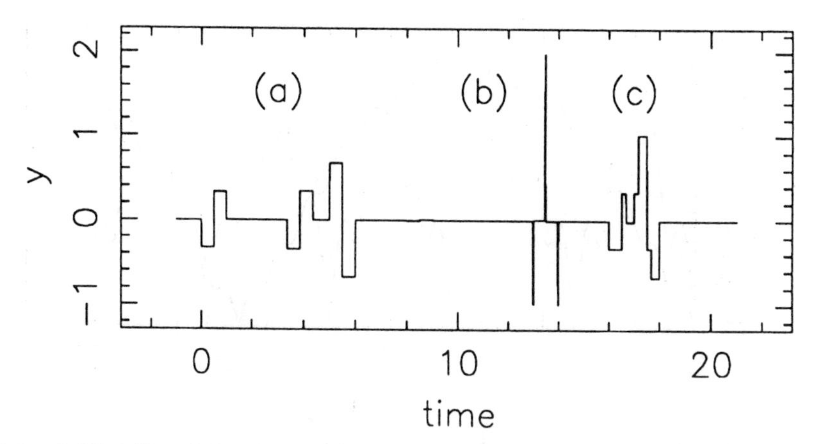

Figure 1: The effect of the response function on the observable. The same metric perturbation h of duration τ as seen from (a) detector of size $T = 5\tau$ and at an angle $\theta = 70°$; (b) $T = 5\tau$ and $\theta = 10°$; (c) $T = \tau$ and at an angle $\theta = 70°$.

$$Y = -\frac{1-\cos\theta}{2} - \cos\theta e^{-\pi i(1+\cos\theta)fT} + \frac{1+\cos\theta}{2} e^{-2\pi i fT}, \tag{8}$$

which may interfere constructively or distructively depending on θ and f. Eq. 8 allows also to evaluate the angular response. In general, the antenna pattern of the detector develops more and more sidelobes at increasing frequencies.

These characteristics have important consequences for the detection of signals from coalescing binaries, whose frequency increases with time. In this case one might expect strong variations of the amplitude at the detector output. As the excitation frequency drifts, the phasors in eq. 8 rotate at different pace and may combine constructively at a certain time and destructively at a later one (see fig. 3.)

Another effect which deserves attention is the change in size and orientation of the detector. In an observation period of duration $T_1 = 20$ days both T and θ may change enough to produce a non-negligible amplitude change to sinusoidal signal. Whether the system can still be considered as time-invariant (a feature that greatly simplifies the data analysis) depends of course on f, $\dot{T}$ and $\dot{\theta}$. In general, the effects become negligible at lower frequencies, when the change in phase, of order $2\pi f\dot{T}T_1$, is more likely to be much smaller than unity.

In order to assess how important the effect is, one can measure the amount of phase modulation induced by changes in T and θ. This can be quantified by introducing the bandwidth B, defined as an average over the spectral distribution $|S(f)|^2$ of the observed signal:

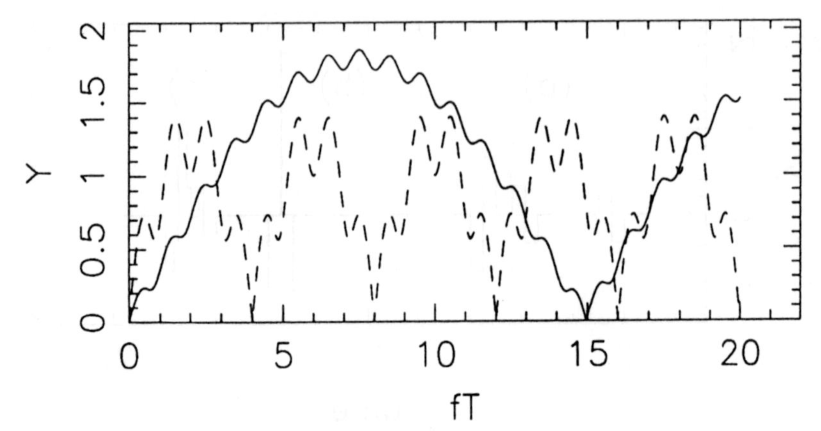

Figure 2: The effect of the response function sinusoidal waves. The response function (8) as a function of the normalized frequency fT for $\theta = 30°$ (solid line) and $\theta = 60°$ (dashed line.)

$$B^2 = \frac{\int (f - \langle f \rangle)^2 |S(f)|^2 \mathrm{d}f}{\int |S(f)|^2 \mathrm{d}f} \tag{9}$$

The integration is extended to the positive frequency domain. For a generic signal, the average frequency $\langle f \rangle$ may be also defined as an average over the spectral distribution. In the case of a sinusoidal signal $\langle f \rangle$ is very close to f_0. The quantity B is easily evaluated in the time domain, using the equivalence between the operators $2\pi i f$ and $\mathrm{d}/\mathrm{d}t$:

$$4\pi^2 \int f^2 |S(f)|^2 \mathrm{d}f = \int \left| \frac{\mathrm{d}}{\mathrm{d}t} y(t) \right|^2 \mathrm{d}t \tag{10}$$

For a generic signal $y(t) = A(t) \exp[i\phi(t)]$ the bandwidth is decomposed in a natural way into an amplitude (AM) and phase (PM) contribution:

$$4\pi^2 B^2 = \int [\dot{A}(t)]^2 \mathrm{d}t + \int (\dot{\phi}(t) - 2\pi \langle f \rangle)^2 A^2(t) \mathrm{d}t \tag{11}$$

In the case of a sinusoidal input, only the AM term survives and one can prove variations in T and θ give rise, respectively, to bandwidths

$$B_{\theta=const} = a_1 f_0 \dot{T}. \quad B_{T=const} = (b_1 + b_2 f_0 T \dot{\theta}) \tag{12}$$

where the numerical factors a_1,b_1 and b_2 are of order unity. When B is smaller than the spectral resolution $1/T_1$, the system can be considered as time invariant. For typical values of $\dot{\theta}$ and $\dot{T}$ due to interplanetary orbital motions, it

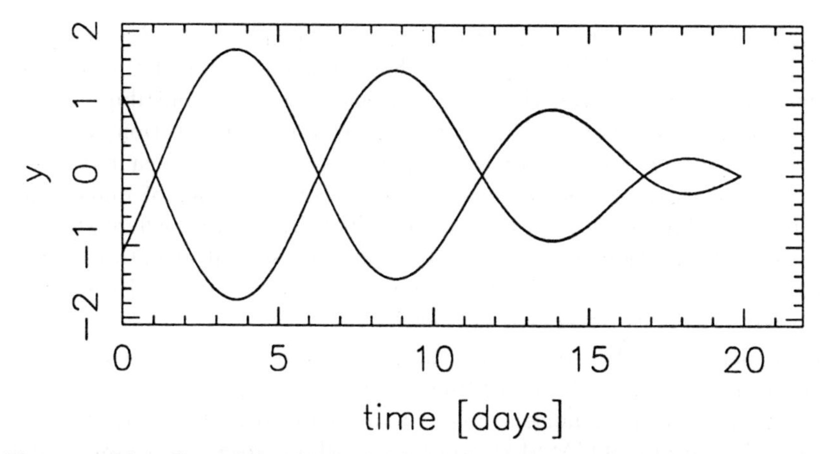

Figure 3: The effect of the change in size and orientation of the detector. Plot of the envelope of the observable y (see eq. 8) for a sinusoidal excitation of frequency 10^{-2} Hz and unit amplitude, for the orbital trajectory of Mars Observer during the 1993 experiment and a source in the galactic center ($\theta \approx 20°$.) Time is measured in days past 00:00 UTC, day 81, 1993. y is in arbitrary units.

turns out that $B_{\theta=const} > B_{T=const}$. Bandwidths larger than $1/T_1$ are usually found at frequencies larger than about $5\ 10^{-3}$ Hz, although the geometry is important in determining the exact frequency of transition between the two regimes.

In the case of a signal from a coalescing binary ("chirp"), the changes in size and orientations produce effects that are more difficult to evaluate analytically, but the qualitative picture is clear: chirps whose spectral content is concentrated mostly at low frequencies will be less affected. However, significant SNR degradations may be expected for some chirps and, therefore, an appropriate strategy for data analysis must be devised (see Sect. 3.)

2 The noise

Three different types of noise sources affect a Doppler detector: propagation media, electronics (including clock) and mechanical buffeting of ground and spacecraft antennas. This topic has been extensively treated in the literature[14,10,15] and we will present here a summary of the main results applicable to past and future experiments.

Electronic noise, being to a large extend controllable, is in general the most benign of the three. It is in general dominated by thermal noise of ground and onboard receivers, but significant contributions may come also from the transmitters. When expressed in terms of the power spectrum of

y, thermal, white phase noise has the usual f^2 dependence on frequency, so that it limits the sensitivity only in the high frequency end of the detector window. Instabilities of the H-maser frequency reference contribute to the Allan deviation at levels of $1.\,10^{-15}$ or lower at 1000 s integration time and therefore have never been important in past experiments, which have been limited by propagation noise at sensitivities one or two order of magnitudes larger. In future high sensitivity experiments (like CASSINI's), where accurate calibrations for propagation noise will be available, special attention will be devoted to control the thermal deformations of the cavity of the H-maser, in order to keep the clock noise at a safe level.

While mechanical deformations of the ground antennas due to wind, gravity and thermal expansion have not been significant in past experiments, they are potentially dangerous noise sources in the planned high sensitivity measurements. In the case of CASSINI experiment, whose target end-to-end Allan deviation is $\sigma_y(\tau) = 3.\,10^{-15}$ for $\tau = 1000$ s, all optical path variations at the level of $\Delta l \approx c\tau\sigma_y = 10^{-1}$ cm cause degraded performance. A large effort has indeed been devoted to design the ground antennas capable to satisfying such a challenging goal.

Spacecraft attitude motions have caused difficulties in past experiments, but they have never been a serious limitation. For spin stabilized spacecraft (like PIONEER and ULYSSES) the spacecraft rotation gives rise to lines in the power spectra of the observable, at the spin frequency and its harmonics. This lines may be aliased into the Nyquist band if the sampling rate is too low, but they can be easily identified. For three axis stabilized spacecraft, by far the preferred method of attitude control makes use of reaction wheels, rather than thrusters. However, reaction wheels have to be properly designed in order to ensure smooth rotations and not to generate spurious signatures in the data. MARS OBSERVER's reaction wheels, for example, experienced static friction when the direction of rotation was reversed. The effect was to move the spacecraft antenna phase center, producing Doppler shift that was typically ten times larger than the propagation noise. Fortunately, this effect could be modeled and removed using engineering telemetry data of the wheel rates.

Propagation noise due to fluctuations in the refractive index of the media crossed by the radio beam (troposphere, ionosphere and interplanetary plasma) are far more important. Past experiments performed at S-band had their limitation in the interplanetary plasma, which could be minimized by performing the observations when the spacecraft is close to solar oppositions[16]. What matters is indeed the transversal velocity $V_\perp$ of the density irregularities carried by the solar wind. The characteristic spatial scale of the turbulence for a measurement time τ_m is $\ell \approx V_\perp \tau_m$. Therefore, for a given τ_m the relevant spatial scales become smaller and smaller as $V_\perp \rightarrow 0$. Since the spectrum of interplanetary turbulence is a Kolmogorov-like, red one, the smaller spatial scales contribute with much less power than the larger scales. The scintillation noise is therefore smaller close to solar oppositions, where $V_\perp$ becomes small.

Being a dispersive medium, plasma can be effectively fought by using radio link at different frequencies or by increasing the carrier frequency. For the typical ionospheric and interplanetary environment, the refractive index of the plasma $n_r = [1 - (\nu_p/\nu_0)^2]^{1/2}$ is determined only by the ratio between the plasma frequency ν_p (proportional to the square root of the electron density) and the carrier frequency. Plasma noise can hence be arbitrarily reduced by using a higher frequency radio link. This is indeed the strategy which has been pursued for the MARS OBSERVER experiment, which has used a two-way X-band link. The experiment under preparation for the spacecraft CASSINI will be performed at Ka-band. At these frequencies the plasma noise is virtually eliminated at solar oppositions and the observations can be taken over a much wider range of solar elongation angles.

When higher frequency links are not available, one may use efficient plasma compensation schemes based on the simultaneous transmission of carriers in different bands. The difficulty in applying this method of calibration lies in the fact that, for past space missions, dual frequency transmission was available only in the downlink. Exact plasma compensation was therefore possible only for the return beam. This limitation may be partially circumvented by exploiting our knowledge of the interplanetary turbulence to devise a suitable Wiener filter for the estimation of the uplink plasma contribution from the downlink measurements[17]. In the case of the ULYSSES experiment performed in 1992 (see sect. 4), this method has led to a 30% reduction of the noise, with a final Allan deviation at 1000 s integration time of $7\ 10^{-14}$.

A much more challenging problem is given by tropospheric noise which, being non-dispersive, requires a completely different approach. Assuming to be able to compensate for the so called dry contribution by means of ground pressure measurements or GPS data analysis[18,19], the real difficulty is given by water vapour which, being poorly mixed with the other atmospheric components, requires *ad hoc* remote sensing instrumentation to be measured. Indeed a new generation of water vapour radiometers is being developed for the CASSINI experiment, where tropospheric noise is likely to be the main limitation to the sensitivity. In general, tropospheric noise is well below plasma noise for experiments performed at S-band, but become significant when X-band links are used[6]. Measurements taken at VLA[20] have provided estimates of Allan deviation at 1000 s ranging from 1.10^{-14} to no detection ($< 1.10^{-15}$), but one expects of course significant variations with the season and site. Sistematic water vapour measurements taken for a year at Deep Space Network site in California have confirmed these figures and shown that, with suitable ground instrumentation, the experimental goal of tropospheric compensation at the level of $1.\,10^{-15}$ (as in the case of CASSINI) can be achieved[21].

3 Waveforms and range of sight

Known sources of gravitational waves are out of reach of present Doppler experiments. The nearest compact binaries are known to emit in the low frequency band, but their signal, being at levels of $h \approx 10^{-21} - 10^{-20}$, will still be far from the sensitivities of the most accurate experiment planned so far, which could detect periodic sources at levels of $3 10^{-17}$ in a 40 days observation period of the spacecraft CASSINI. Doppler detectors may observe only the strongest sources, which must be nearby or involve large masses of galactic nature. These sources are inherently less abundant, so one has to expect a low event rate. The largest probability of detection is achieved for sources that are strong and long lived. Coalescing binaries of massive black holes deserve therefore special attention, especially after the recent observations of the Hubble Space Telescope and the indications of widespread galactic merging[22]. The range of sight of future detectors (like CASSINI) is indeed sufficiently large to include some astrophysically interesting regions, extending up to the Virgo cluster (see fig. 4.)

Doppler detectors are broadband instruments, which can be used to detect other types of signals, like broadband pulses generated by the formation of massive black holes and a stochastic background[26]. These three classes of sources generate signals with different localization properties in the time-frequency domain, which imply different data analysis approaches. In this section we confine ourselves to the characteristics of bursts and chirps, and discuss the the prospect of their detection with Doppler experiments.

3.1 Bursts

The asymmetrical collapse of a large mass leading to the formation of a massive black hole is the strongest source astrophysically conceiveble[23]. Most of the energy will be emitted in a burst of duration τ of order of the gravitational radius of the collapsing mass M. If a fraction ϵ of the total mass is converted into gravitational waves, the signal at a distance D has an amplitude

$$h_b \approx \sqrt{p\epsilon}\frac{M}{D} \tag{13}$$

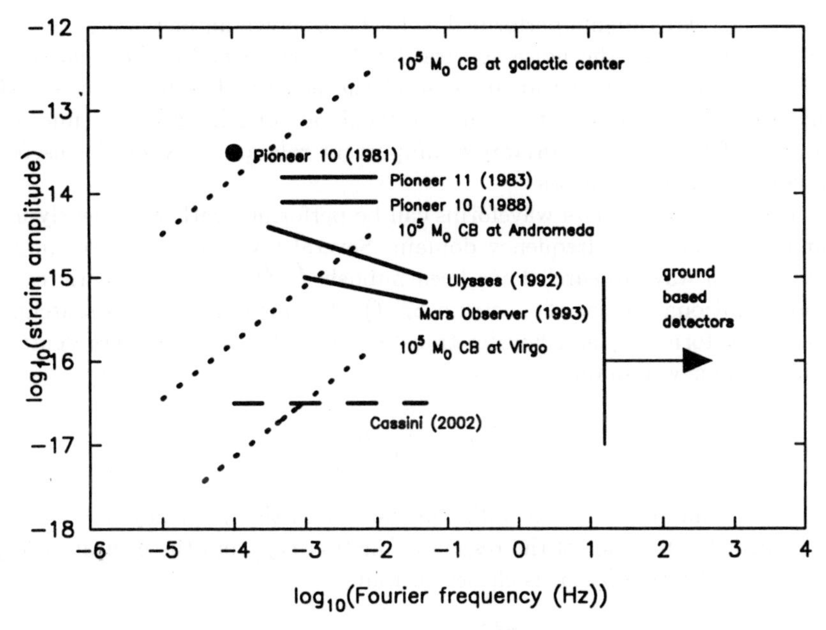

Figure 4: Sensitivities of Doppler detectors and evolutionary track of a coalescing binary system at different distances from the earth (dotted lines.) Only a small portion of the track would be accessible in a typical experiment lasting a few weeks.

The numerical factor p measures the characteristic time of the burst in units of M. Current estimates of p give values in the range 10-30 [24].

Eq. 13 provides the maximum range of sight for burst sources. For a generic direction of arrival and $\tau < T$, we may assume $\cos\theta = O(1)$, $y \approx h$ and determine D for a given sensitivity σ_y. In the case of the CASSINI experiment $\sigma_y \approx 3\,10^{-15}$ and the range of sight becomes

$$D_{max} = 10^3 \left(\frac{p}{20}\right)^{1/2} \left(\frac{\epsilon}{0.01}\right)^{1/2} \left(\frac{M}{10^8 M_\odot}\right) \left(\frac{\sigma_y}{3\,10^{-15}}\right)^{-1} \mathrm{Mpc} \qquad (14)$$

While this figure shows that Doppler detectors can reach distances of cosmological interest, one must note that a collapse of 10^8 solar masses generates a burst of duration $\tau \approx 10000$ s, for $p = 20$. This is essentially the edge of the low frequency cutoff of the CASSINI experiment. For burst durations longer than the round-trip light-time T, the response of the detector decreases linearly with $\tau = pM$ and the range of sight remains approximately constant, provided that the sensitivity σ_y does not change. This is because longer duration bursts involve larger masses ($h \propto \tau$.) The three-pulse cancellation at $\tau > T$ is therefore exactly balanced by a larger wave amplitude. In general, however, all experiments performed so far have shown an increase of the noise at low frequencies. the noise spectral levels increase at low frequencies performed so far have shown an increase of the noise at low frequencies. Also, control of systematic effects is more difficult at long integration time. The numerical factor p plays anyway an important role and it would be useful to refine the current estimates.

The search for bursts waveforms can be performed rather efficiently using matched filters in the frequency domain. Several reviews of this technique in gravitational wave research have been published[25,27]. For the generic case of colored, stationary noise of spectrum $S_n(f)$, the optimum filter for searching a known waveform $y(t)$ and Fourier transform $y(f)$ has a simple representation in the frequency domain

$$q(f) = N_f \frac{y(f)}{S_n(f)} \tag{15}$$

where N_f is an arbitrary normalization factor. In the frequency domain, $y(f)$ is of course the product of the response function $r(f)$ and the excitation $h(f)$. If the normalization factor is chosen so that

$$2 \int_0^\infty |q(f)|^2 S_n(f) df = 1 \tag{16}$$

the signal-to-noise ratio of the matched filter output at each lag t becomes

$$\rho(t) = 2 \int_0^\infty y(f) q^*(f) \exp(2\pi i t) df \tag{17}$$

which is a Gaussian process of zero mean and unit variance if the noise statistics of the time series is Gaussian[27]. The filter (15) effectively emphasizes the frequencies at which the detector is more sensitive and selects only the bandwidth covered by the signal.

As the exact waveform for the plunge is not known, the maximum SNR (17) will never be achieved and the degradation may in principle be large. The detection of bursts suffers indeed the twofold difficulty of searching unknown waveform with unknown parameters. A reasonable way to cope with this problem is to assume a rather arbitrary but not unrealistic set of waveforms, parametrized using a one or more physical quantities. The filter bank to be correlated with the data is then generated by convolving the waveforms with the instrumental response. Apart from an arbitrary amplitude factor (related to the distance and the polarization of the waves,) the natural parameter to generate the waveforms is the burst duration τ.

In the case of unknown direction of arrival, also the angle θ is required to parametrize the actual family of templates $y(t;\tau,\theta)$, which is obtained by convolving each $h(t;\tau)$ with the instrumental response $r(t;T,\theta)$. Doppler detectors are not very sensitive to the direction of arrival, so that a search over 10-20 different values of $\cos\theta$ is sufficient to avoid significant SNR degradation. However, if the range of sight is not large, the number of accessible sources may be so small that a targeted search becomes more attractive. By aiming at just two or three directions of known atsrophysical objects, the number of degrees of freedom of the output statistics is smaller and lower values of ρ are required to declare detection for a given confidence level.

Assuming the waveform as known, the matched filter algorithm allows to estimate a putative signal amplitude h_0 for each time of arrival t, direction in the sky θ and burst duration τ. By combining eq. 15, 16 and 17 one gets:

$$h_0(t) = \rho(t)\left[2\int_0^{\infty}\frac{|y(f)|^2}{S_n(f)}df\right]^{-1/2} = \rho(t)\left[\int_{-\infty}^{+\infty} y(t')q(t')dt'.\right]^{-1/2} \tag{18}$$

where $q(t')$ is the optimum template in the time domain. In the absence of signal, $h_0(t)$ is a Gaussian process if zero mean and variance determined, through eq. 18, by the spectral level of the noise and the energy of the template. The larger the signal energy in the denominator of eq. 18 (i.e. the longer the burst duration), the smaller is the value of h_0 that can be detected. However, due to the low frequency cutoff of the instrumental response and the typical noise spectra encountered in Doppler experiments, the optimum region for detection is limited to burst durations $\tau < T$ and to directions not too close to the nulls of $r(t)$.

Eq. 18 allows also to estimate in a quantitative way the range of sight as a function of τ and θ. In the simplified case (not too far from reality) of a white frequency noise of spectrum $S_n = \sigma_y^2(\tau)\tau$, the minimum detectable signal amplitude at a SNR level ρ becomes

$$h_{min}(\tau,\theta) = \rho\sigma_y(\tau)\sqrt{\frac{\tau}{E(\tau,\theta)}} \tag{19}$$

where $E(\tau,\theta)$ is the total energy of the template $y(t)$. $E(\tau,\theta)$ reaches the maximum value τ for $\theta = \pi/2$ and $\tau < T$, so that $h_{max} = \rho\sigma_y(\tau)$. In the long wavelength regime $\tau > T$, the signal energy decreases approximately as τ^{-1} and $h_{max} \approx \rho\sigma_y(\tau)(\tau/T)$.

The estimated dimensionless amplitude h_0 is a suitable quantity to select candidates or, more ambitiously, to base the decision on the presence of a signal. Several reviews have been published on this topic[28]. In addition to the fact that the waveform is not actually known, the main difficulty to a straightforward application of the matched filter concept lies in the non-stationarity of the noise. While the shape of the noise spectrum does not usually change much over the typical duration of an experiment, the spectral levels can show variations of factors 20 or more from pass to pass, so that the statistics of the matched filter outputs surely cannot be described only by the second moment of the probability density function. A way to cope with this difficulty is to identify a time scale T_s for systematic variations of the noise and divide the data set in batches of length T_s. Over such a time scale (which of course must be significantly larger than the size $T+\tau$ of the templates) one can then assume that the noise is locally stationary and apply the scheme described above. It is convenient to generate locally also the filter bank, using the instrumental response appropriate to the data batch under consideration. In doing so, the changes in size and orientation of the detector are straightforwardly accounted for, without any significant SNR degradation.

3.2 Chirps

Sinusoidal or chirping signals emitted by binary systems occupy a well defined portion of the time-frequency plane. For most of the source's lifetime the signal is localized in frequency and only in the final stages preceding the final plunge its bandwidth $B = \dot{f}/f$ becomes of order unity.

One of the reasons why these sources are attractive candidate for detection is due to the fact that the waveform is know from the quadrupolar approximation[29] and depends only on a limited number of parameters. This feature makes the search for the signal well suited for the matched filtering method. An important assumption is that the system is astrophysically "clean" and that the evolution is dominated by gravitational radiation. Only when the system is close to coalescence the waveform becomes more complicated (as described elsewhere in this volume) and additional parameters must be introduced for a full characterization of the signal.

If the binary system has undergone a sufficiently long evolution, the orbit is likely to be circularized (due to the larger emission of gravitational waves at perihelion), with a remarkable simplification of the data analysis. The frequency of emission (twice the orbital frequency) increases in time at a rate

$$\dot{f} = \frac{96}{5}\pi^{8/3}M_c^{5/3}f^{11/3} \tag{20}$$

which depend only on the so called chirp mass $M_c = \mu^{3/5}M^{2/5}$, a combination of the reduced mass μ and total mass M of the system. M_c and f determine a formal parameter (the Newtonian time to coalescence), defined as the time at which the frequency diverges:

$$t_n = t_0 + \frac{5}{256} M_c^{-5/3} (\pi f_0)^{-8/3} \tag{21}$$

where t_0 is the time at which the chirp reaches a frequency f_0. For a binary at distance D, the excitation (4) at the detector input reads:

$$h(t) = \frac{\sqrt{5}}{2} Q(\iota, \phi) \frac{M_c^{5/3}}{D} [\pi f(t)]^{2/3} \cos\left(\phi_0 + 2\pi \int_{t_0}^{t} dt' f(t')\right) \tag{22}$$

where ϕ_0 is the initial phase at the reference time t_0 and $Q(\iota, \phi)$ a geometrical amplitude factor

$$Q^2(\iota, \phi) = \frac{5}{16}[\cos^2(2\phi)(1 + \cos^2\iota)^2 + 4\cos^2\iota \sin^2(2\phi)] ; \tag{23}$$

which depends on the polarisation angle ϕ and on the angle ι between the earth-spacecraft direction and the normal to the orbital plane. The average of Q over ϕ and ι is unity.

The Newtonian approximation breaks down close to the final plunge. In the case of unequal masses $\mu \ll M$, this phase begins when the semimajor axis reaches the value $a = 6M$ (the so called innermost stable circular orbit) and the instantaneous frequency of emission is $f_{isco} = (\pi 6^{3/2} M)^{-1}$. A large effort is being devoted to the computation of the expected waveforms at higher PN orders, in view of the use of matched filtering techniques for signal detection and parameter determination for ground interferometers[30]. From the point of view of the data analysis is however remarkable that the waveform is degenerate in the mass ratio $q = M_1/M_2$ of the two components.

The plunge waveform is unknown, but its general characteristics may be easily understood. In the Newtonian phase the system is virialized and the energy E_{gw} radiated in gravitational waves is half the loss of potential energy $U = \mu M/a$ due to the shrinking of the semimajor axis a. If we assume that the ratio $E_{gw}/U \approx 0.5$ also throughout the plunge phase (it is not so, because it is a well known result that post-Newtonian corrections produce an anticipation of the plunge with respect to the Newtonian coalescence time, thus indicating that gravitational wave emission becomes more efficient), the system emits an energy $\mu/6$, in a time $\tau_p = p_1 M$ as the distance decreases from $a = 6M$ to, say, $a = 2M$. As in the case of the collapse of a large mass (see eq. 13) also here p_1 is a parameter which measures the actual duration of the plunge in units of M. The field perturbation at a distance D becomes therefore

$$h_p \approx \sqrt{0.3 p_1 \mu M} D^{-1} \tag{24}$$

larger than the amplitude in the last stable orbit by a factor of order $\sqrt{M/\mu}$. Therefore, in the limiting case of two equal masses the burst does not produce amplitudes significantly larger than final phases of the chirp, while in the opposite limit of very unequal masses the burst dominates. Notice however that the total energy emitted in the two phases is approximately equal.

The search for chirp looks more attractive not only because of the much better knowledge of the waveform, but also from the comparison of the signal energy (i.e. the quantity which determines the detectability of a signal):

$$E_{p,c} = \int_{\tau_{p,c}} h^2(t) dt \tag{25}$$

Here the integral is extended over the time domain where the chirp or the plunge occur. The relevant question is for how much time before the last stable circular orbit one needs to integrate $h^2(t)$ in the chirp phase in order to get a signal energy E_c which is, say, a fraction α of the signal energy E_p in the plunge. By integrating eq. 20 and using eq. 22 one gets:

$$E_c(t_1) \approx \frac{5\sqrt{5}}{64} \frac{\mu^{3/2} M}{D^2} t_1^{1/2} \tag{26}$$

where t_1 is the time of integration before the last stable circular orbit. Note that E_c does not diverges for a Newtonian template as $t \to t_n$, indicating that most of the signal energy comes from the precursor. Eq. 24 yields $E_p \approx h_p^2 \tau_p$, so that the required integration time becomes

$$t_1 \approx 3\alpha^2 p_1^4 \frac{M^2}{\mu} = 3\alpha^2 p_1^3 \tau_p \frac{M}{\mu} \tag{27}$$

Again, for binary systems of equal masses the chirp is by far easier to detect. The case of a smaller object falling into a large black holes may be much less favourable. For example, for $p_1 = 10$, $\alpha = 1$ and a binary system made up by a $10^3 M_\odot$ black hole orbiting around a galactic black hole of $10^6 M_\odot$, one would need to integrate for $t_2 = 1.5\ 10^8$ s to collect a signal energy equal to E_p.

The search for Newtonian chirps can be performed using the same tools described in the previous section, but some peculiar aspects make the data analysis significantly more complicated. Even in the Newtonian approximation, the excitation depends on a considerable number of parameters: the chirp mass M_c, the frequency f_0 and phase ϕ_0 at the reference time t_0 and the geometric factor Q. The response of the detector adds another unknown parameter to the actual waveform to be searched for, i.e. the angle θ of the source with the line of sight. Q is actually an amplitude factor which cannot be measured independently of D and therefore does not appear in the construction of the filter bank. The parameters f_0, ϕ_0 and t_0 are purely kinematical ones, while M_c is the crucial dynamical parameter which determines the evolution of the system. As in the case of the bursts, the output of the matched filter will provide a dimensionless amplitude from which, in turn, one could determine the ratio D/Q.

The strategy for the detection of a chirp waveforms is in principle similar to the one envisaged for ground interferometers[25]. To test for the presence of a chirp characterized by the set of parameters $\{M_c, f_0, \phi_0, t_0\}$, one has to generate the excitation $h(t; M_c, f_0, \phi_0, t_0)$, convolve it with the instrumental response $r(t; \theta, T)$, compute the optimum filter (15) and correlate it with the data. Of course one has to consider only that part of the source life in which the emitted frequencies fall within the detector bandwidth (or at least where one expects a good SNR.) One may consider, for example, all chirps with an initial frequency $f_0 = 1/T$. The filter bank is then parametrized using just M_c and ϕ_0. M_c determines the time to coalescence and therefore the length of the template. The evolutionary phase of the source at the time of the experiment is determined by the arrival frequency f_b at the start of the data acquisition, rather than through the parameter t_0. f_b is in turn determined, for a given M_c, by the correlation lag t between the template and the data. Each waveform could be truncated at or shortly before $f = f_{isco}$. This quantity depends on the mass ratio q and this would add in principle another parameter to be searched for. However, since little signal energy is contained close to the plunge, one could use a formal value of f_{isco} computed for $q = 1$ and expect a negligible SNR degradation.

To cope with the unknown phase ϕ_0, the best procedure is to generate two templates phase shifted by $\pi/2$ and correlate them simultaneously with the data. From the computational point of view, it is slightly more efficient to generate an analytic signal from the time series (which needs to be computed only once and then stored in memory) and correlate it with a single template of arbitrary phase (e.g. $\phi_0 = 0$.) For each lag t, one then has the projections $I_0(t)$ and $Q_0(t)$ of the time series on two orthogonal templates corresponding to the same M_c, f_0. The quantity upon which to base the decision on the presence of the signal becomes the squared SNR

$$\rho^2 = I_0^2 + Q_0^2 \tag{28}$$

With this scheme, the filter bank is effectively constructed using a single parameter, the chirp mass M_c. In the absence of signal and with the appropriate choice of the normalization factor N_f (eq. 16), ρ^2 is exponentially distributed with $\sigma = 1$. In the case of chirps, the presence of a new degree of freedom (the unknown parameter ϕ_0) changes completely the statistics of the matched filter output. The signal detection must occur now at a much larger number of σ with respect to the case of the Gaussian distribution that one would have obtained were ϕ_0 known.

Since the signal is not localized in time, the application of the scheme decribed above has to face a number of difficulties and limitations. First, for low frequency sources each template is usually longer or much longer than the data set itself. For example, for $f_0 = 10^{-4}$ Hz and $M_c = 10^6 M_\odot$, from eq.21 one gets a residual lifetime of about a month. The use of the fast Fourier transform to perform efficiently the correlations of the templates with the data requires a considerable amount of memory even if the templates are segmented into shorter batches.

Second, the presence of large data gaps, combined with the inherent variations of the template amplitude and bandwidth may introduce significant non-stationarities in the outputs.

Third, the varying instrumental response (due to the changes in T and θ) makes the system a time-variant one. Therefore it is not possible, in principle, to build a template (much longer than the data set) by convolving an excitation $h(t)$ with a constant response function $r(\theta, T)$. Since r is a function of time, for each M_c the template $y(t)$ depend now on the arrival frequency at the beginning of the experiment. From the computational point of view, this would imply that the correlations have to be performed in the time domain, at a huge cost of computer time.

As this difficulty makes any systematic search for chirps virtually impossible, the only alternative is to use a so called hierarchical search algorithm, in which the candidates are selected using a suboptimal filter built from a constant instrumental response. A more refined search using the exact templates and scalar products in the time domain could then be performed over a much smaller number of candidate waveforms. The threshold for the selection of candidates depends of course on the expected SNR degradation between the optimum and suboptimum templates. Numerical simulations have shown that for most waveforms the SNR degradation is not larger than 0.5, although for some extreme, high frequency chirps the use of suboptimum templates leads to a much more severe loss.

It must be pointed out that it is impractical and inefficient to use the matched filtering approach for the detection of weak (linear) chirps. Since the sources spend most of their lifetime in the sinusoidal or linear frequency drift regime, the length of the templates would be unmanageable. A much more efficient method for the detection of these sources is based on the linear dechirping algorithms[3,4,5]. Suboptimal search algorithms based upon the resampling-in-time method have been used also for the detection of strong chirps[4].

Two important, related questions have to be addressed next, namely 1) how large is the region of space from which a chirp signal can be detected and 2) what is the (discrete) parameter space used to generate the filter bank. These problems will be the topic of a paper in preparation[31] and here we will just outline the general approach and present some results. The expected SNR, provided again by eq. 17, can be evaluated using the Fourier tranform of the excitation (22)[32]

$$h(f) = \mathcal{A} f^{-7/6} e^{i\Psi(f)} \tag{29}$$

where the amplitude and phase factors are

$$\mathcal{A} = \frac{1}{\pi^{2/3}\sqrt{6}} Q(\iota, \phi) \frac{M_c^{5/6}}{D} \tag{30}$$

and

$$\Psi(f) = 2\pi f t_n - \phi_n - \frac{\pi}{4} + \frac{3}{4}(8\pi M_c f)^{-5/3} \tag{31}$$

ϕ_n is here the formal phase of the chirp at $t = t_n$.

For a coloured noise spectrum of the type $S_n(f) = \sigma_y^2(T) T (fT)^{-k}$ and a time invariant instrumental response, the squared SNR becomes

$$\rho_c^2 = \frac{5}{24\,\pi^{4/3}} \left(\frac{M_c}{T}\right)^{5/3} \frac{T^2}{\sigma_y^2(T)\,D^2} \int_{f_b}^{f_e} (T\,f)^{-7/3+k}\, |r(f;\theta,T)|^2\, df\,. \tag{32}$$

The geometric factor Q has been assumed equal to one and the integration is limited to the frequency range $[f_b, f_e]$ spanned by the chirp during the experiment. Indeed, the frequency components of the chirp outside this domain are small and do not contribute significantly to the integral. The end frequency f_e is determined by f_b (the arrival frequency) and the chirp mass M_c.

Eq. 32 combines the characteristics of the source (through the chirp mass and distance) with a kinematical parameter (the arrival frequency) and the characteristics of the detector (through σ_y, k, θ and T.) The same equation provides both the range of sight of a Doppler detector for a given SNR level and direction in the sky, and the region of the parameter space that can be successfully explored (see fig. 5.) Of course, if the distance of the source is small, the accessible domain of f_b, M_c becomes very wide. One has therefore

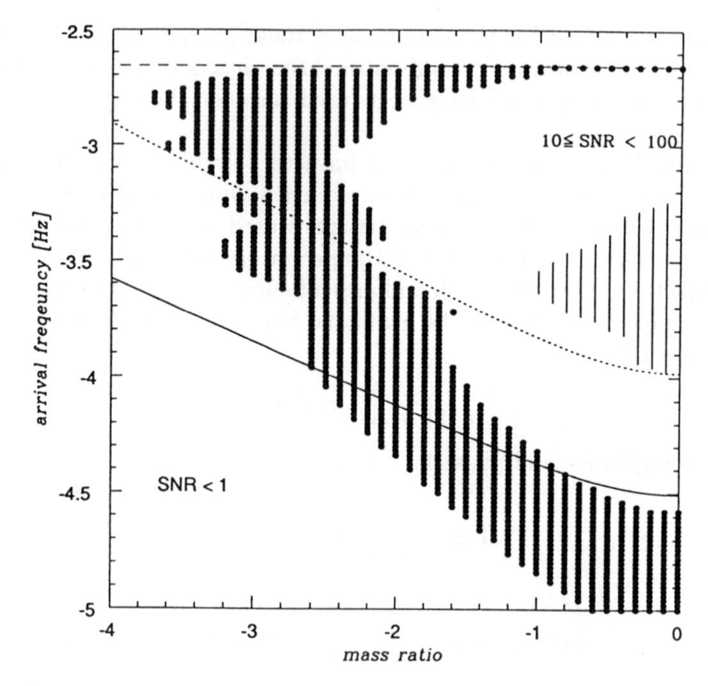

Figure 5: SNR levels to gravitational waves emitted by a binary system in the galactic center ($D = 10$ kpc,) as a function of the mass ratio and arrival frequency f_b, assuming a central object of $M_1 = 2 \times 10^6 M_\odot$, for the geometry and the noise levels of ULYSSES 1992 experiment ($\theta = 109^o$, $S_n = 2.5 \times 10^{-24}$ Hz $k = 1/2$, $T = 4430$ sec and $T_1 = 1.2 \times 10^6$ sec.) The dashed region corresponds to $SNR \geq 100$ and the bold dashed reagion to $1 \leq SNR < 10$. The long dashed line correspond to f_{isco}; the region of the plane (M_c, f_b) below the solide line contains sinusoidal signals; linear chirps lay between the solid and the short-dashed line, non-linear chirps above the short dashed line.

to decide a *minimum* distance for the sources, based upon our astrophysical expectations and guesses. For such a minimum distance one identifies then the widest domain of accessible chirps. The discrete spacing of the parameters filter bank is accomplished by requiring that the scalar product between two adjacent templates is sufficiently large. Of course, a very fine spacing of the parameters is very expensive in terms of computer time and cannot be afforded.

An important consequence of eq. 32 is the significant gain in SNR that is obtained by pushing the measurements at low frequencies. Indeed, for $k < 4/3$ the largest contribution to the integral in eq. 32 comes from the lower frequencies of the domain of integration, provided that $f_b > 1/T$. This shows that large detectors have not only access to a wider region of the parameter space, but also that at lower frequencies one expects to obtain the largest SNR (see fig. 5.) In the case $f_b = 1/T$, one may approximate $r(f)$ to unity and get, for a white frequency spectrum,

$$\rho_c \approx 5\,10^{-2} \frac{M_c^{5/3} T^{1/3}}{\sigma_y^2(T) D^2} \, . \tag{33}$$

Since $\sigma_y^2(\tau) \propto \tau^{-1}$ for the noise spectrum being considered, the previous equation shows that 1) the maximum SNR is achieved for sources with chirp mass $M_c = T$ and 2) that this optimum SNR is proportional to T^3. In the case of the planned CASSINI experiment, the maximum range of sight for SNR=1 amounts therefore to about 120 Mpc.

Acknowledgments

We thank B. Bertotti and F.B. Estabrook for their support. We are grateful to A. Vecchio for many discussions on the detection of chirps. This paper has been prepared while one of the authors (LI) held a NRC-NASA Resident Research Associateship at the Jet Propulsion Laboratory, California Institute of Technology, under a contract with NASA. JWA's contribution was supported by the UV/Visible/Gravitational Astrophysics Branch of NASA.

References

1. R.W. Hellings *et al.*, *Phys. Rev.* D **23**, 884 (1981).
2. J.D. Anderson *et al.*, *Nature* **308**, 158 (1984).
3. J.W. Armstrong, F.B. Estabrook and H.D. Wahlquist, *Astrophys. J.* **318**, 536 (1987).
4. J.D. Anderson, J.W. Armstrong and E.L. Lau , *Astrophys. J.* **408**, 287 (1993).
5. B. Bertotti *et al.*, *Astron. Astrophys.* **296**, 13 (1995).
6. J.W. Armstrong *et al.*, Paper 84.04, presented at the 182th AAS Meeting, Berkeley, USA, June 1993.
7. F.B. Estabrook *et al.*, to be submitted.
8. V.B. Braginskii and M.E. Gertsenshtein, *ZhETF Pis'ma* **5**, 348 (1967).
9. F.B. Estabrook and H.D. Wahlquist, *Gen. Rel. Grav.* **6**, 439 (1975).
10. R.W. Hellings, *Phys. Rev.* D **23**, 832 (1981).
11. H.D. Wahlquist, *Gen. Rel. Grav.* **19**, 1101 (1987).
12. V.B. Braginsky and L.P. Grishchuk, *Sov. Phys. - JETP* **62**, 427 (1985).
13. M. Tinto and J.W. Armstrong, *Astrophys. J.* **372**, 545 (1991).
14. H.D. Wahlquist *et al.* in *Atti Convegni Lincei*343351977.

15. L. Iess *et al.*, *Nuovo Cim.* **10C**, 235 (1987).
16. J.W. Armstrong, R. Woo and F.B. Estabrook, *Astrophys. J.* **230**, 570 (1979).
17. B. Bertotti, G. Comoretto and L. Iess, *Astron. Astrophys.* **269**, 608 (1993)
18. J. Askne and H.Nordius, *Radio Sci.* **22**, 379 (1987)
19. D.M. Tralli and S.M. Lichten, *Bull. Geod.* **64**, 127 (1990)
20. J.W. Armstrong and R.A. Sramek, Radio Sci. **17**, 1579 (1982).
21. S.J. Keihm, *TDA Progr. Rep., JPL Rep. 42-122*, Jet Propul. Lab., Pasadena, USA (1995)
22. Z. Tsvetanov *et al.*, *IAU-SYMP.*, 159, 289 (1994).
23. K.S. Thorne and V.B. Braginsky, *Astrophys. J.* **204**, L1 (1976).
24. T. Piran and R.F. Stark, in *Dynamical Spacetimes and Numerical Relativity*, ed. J.M. Centrella (Cambridge University Press, 1986).
25. B.F. Schutz, in *The Detection of Gravitational Radiation*, ed. D. Blair (Cambridge University Press, 1991).
26. K.S. Thorne, in *300 Years of Gravitation*, ed. S.W. Hawking and W. Israel (Cambridge University Press, 1987).
27. S.V. Dhurandar and B.S. Sathyaprakash B.S, *Phys. Rev.* D **49**, 1707 (1994).
28. M.H.A. Davies, in *Gravitational Wave Data Analysis*, ed. B.F. Schutz (Kluwer Acad. Publ., 1988).
29. P.C. Peters and J. Mattews, *Phys. Rev.* **131**, 435 (1963).
30. L. Blanchet *et al.*, *Class. Quant. Grav.* **13**, 575 (1996).
31. B. Bertotti, A. Vecchio and L.Iess, *Search for Gravitational Waves from Coalescing Binaries with Doppler Detectors*, to be submitted.
32. C. Cutler and É. E. Flanagan, *Phys. Rev.* D **49**, 2658 (1994).

Summary Talk

CONCLUDING REMARKS

Bruno Bertotti
Dipartimento di Fisica Nucleare e Teorica
Università di Pavia, via Bassi, 6
I-27100 Pavia (Italy)

1 LEARNING FROM THE PAST

After having been tensely engaged for a few days with the great challenge of the future – how to open a new and revolutionary band of astronomical research – it is sobering and instructing, I think, to first and briefly look back at the past history, with the false claims, the discarded choices and the pitfalls of this great adventure of ours. I hope the reader will excuse me if I give some preference to my own personal experience.

The story may begin with the ambitious title 'Evidence for discovery of gravitational radiation' [10] which J. Weber, the pioneer of our field, gave to his early paper, where, having found unlikely coincidences in two detectors at room temperature with a sensitivity of 10^{-16}, he claimed "This is good evidence that gravitational radiation has been discovered." This claim is surely unwarranted (and so are other later claims), but, as far as I know, no satisfactory and independent analysis of his instrumental noise and his data reduction has been done. I also recall that Anderson [1] has claimed significance for a coincidence between one of Weber's event at 0341 GMT on March 15, 1969 and Doppler data (with a fractional accuracy of a part in 10^{12}) obtained with the Mariner 6 spacecraft; but how difficult it would be now attribute to the same source two signals at frequencies five decades apart! Were Weber's claim true, it would have completely changed our views of relativistic astrophysics, requiring exceedingly frequent stellar collapses, with a very large efficiency of conversion of rest mass into gravitational waves. Many people, including myself, however, took it quite seriously; A. Cavaliere and myself [3], for example, pointed out that if the claimed bursts were generic – i.e., if they occurred more or less with the same rate at all times and in all galaxies – the corresponding energy density would by far dominate the cosmological dynamics and generate a large cosmological background. To explain the claimed detection Misner [6] proposed an enhancement mechanism called 'gravitational synchrotron radiation', according to which collapsed objects moving at relativistic speed near the galactic centre emit prevailingly in the forward direction; it had no sequel.

In the 70's the prevailing paradigm for the emission of gravitational waves bursts during the collapse of a mass M was very simple: neglecting numerical factors (which are, however, important for a quantitative understanding), the characteristic time scale is $\tau = M$ and, at a distance D, the dimensionless amplitude is

$$h = \sqrt{\epsilon}\frac{M}{D}; \tag{1}$$

here ϵ gives, in order of magnitude, the efficiency with which the rest energy M is transformed into gravitational waves; at the galactic centre this gives $h = \sqrt{\epsilon}\, 5\, 10^{-18}$, which shows how difficult it was to reconcile Weber's claim with the current and widely accepted (at that time) doctrine.

The same paradigm has been extensively used later also for the planning of low frequency experiments using Doppler tracking of interplanetary spacecraft [8]; in this case collapses of massive black holes in the galactic centres throughout the Universe were considered. The corresponding estimate looks much more favourable: for example, a collapse of $10^8\, M_\odot$ at a distance equal to the radius of the Universe, at $D = 3\ \ 10^{27}$ cm gives, according to the formula above, $h = \sqrt{\epsilon}\, 10^{-14}$, with a time scale of 1000 sec. These two numbers are in rough agreement with the accuracy of frequency standards (hydrogen masers) available at the ground stations and with the spectral sensitivity of Doppler measurements, which have a wide band around the reciprocal of the round-trip light-time T (about 5000 sec at 5 AU.)

This paradigm must now be regarded as inadequate and poor. To estimate the efficiency one should use the quadrupole formula, which says that a body of size a (with adequate asymmetry) emits gravitational radiation at a rate given by the square of the third time derivative of the quadupole moment $Q = Ma^2$, that is,

$$P = \frac{M^2 a^4}{\tau^6}, \tag{2}$$

giving an efficiency

$$\epsilon = \left(\frac{a}{\tau}\right)^4. \tag{3}$$

For a supernova collapsing to a neutron star the ratio a/τ should be of order of the thermal speed in the medium; the high exponent makes it rather difficult to get a large efficiency. Extensive calculations have indeed shown that supernovae are indeed very weak emitters and at present are not regarded as the main target of detection for ground systems in the high frequency band. The efficiency of a collapse to a massive black hole in galactic nuclei has never really been analyzed and it is hampered by the difficulty of getting a triaxial shape and the uncertainty of the rate of these events.

Theoreticians have also entertained serious doubts about the old quadrupole formula itself, a reflection of the difficulty encountered in solving to high accuracy Einstein's non linear field equations; Rosenblum [7], for example, claimed that 'the result differs by a factor 3.2 from the Einstein formula'. Thanks to the work by T. Damour, J. Ehlers and others, this misgiving is now forgotten and they can happily proceed to construct astrophysical machines to emit all the gravitational waves we need. Einstein's theory of general relativity has overcome all the tough experimental tests it has been subjected to and is now regarded as an accepted theory, like electromagnetism, ready for all kinds of applications.

Another example of devious theoretical path was the long protracted discussion about the energy of gravitational waves. People have for long looked for an an energy density ρ_{gr} whose total change in a volume V would give the energy loss through its boundary; it was really difficult to give up this tool, so familiar and useful for all other ordinary wave phenomena. But this, of course, is in principle impossible: in a generally covariant theory like general relativity only geometric quantities are lawful and it is just impossible to have a 'something' $T^{\mu\nu}$, proportional to the squares of the first derivatives of the metric tensor (in analogy with ordinary field theories), which behaves like a symmetric, covariant tensor. After all, one can always take a frame (an 'inertial frame') where all such derivatives vanish at a particular event. We all have painfully and slowly learned to reason in terms of terms of geometrical quantities only (coordinates are a private affair!): all we need to describe detection, for example, is to state how the vector joining two neighbouring test bodies changes in time under the action of the impinging curvature tensor $R_{\mu\nu\rho\sigma}$ ('equation of geodesic deviation'). This is a lawful geometrical quantity and is all we need to describe detection. Of course, one is free, at his own risk, to forget about general covariance and, working in a specific gauge (frame) as in a conventional field theory, freely deal with energy fluxes and cross section, thereby obtaining results which are generally valid; but for fine work more appropriate methods are needed.

The interaction between gravitational and electromagnetic waves has its pitfalls. P. G. Bergmann [2], assumed that the effect of gravitational waves on light is like a time-varying refractive index and therefore we should expect that a very distant optical source should scintillate – like a star seen through the atmosphere – if a random gravitational background is present in the Universe. Unfortunately the refractive index analogy is misleading and the effect does not exist [4]. Gravitational radiation is subtle and elusive!

2 LOOKING AT THE FUTURE

The early planning and operation of gravitational wave detectors was essentially aimed at increasing the sensitivity, without real concern for concrete astrophysical sources and with an eye to possible 'new', alternative theories and phenomena; now, as clearly shown also in this meeting, our programmes are driven by relativistic astrophysics and directed to specific sources. The slowing down of the binary pulsar PSR 1913 + 16 and its identification as a gravitational wave source has provided a milestone in which, at last, theory, observation and astrophysics are fully integrated.

As one can see from this conference, our discipline has now in great part divested itself from the past uncertainties, great progress has been made and serious business has begun. Both in the high frequency (kHz) and the low frequency (mHz) band a new kind of source has emerged, a binary system of collapsed objects in a circular orbit progressing toward final coalescence; in the two bands this is realized, respectively, by two neutron stars and by massive black holes in galactic centres. The signal consists in a well defined 'chirped' wave form, which increases in frequency, amplitude and band width. It is important to note that the information about the signal is spread over a wide frequency band, so that wide band detectors are the obvious and best choice. Its detection has raised important problems in two areas.

If the number of periods present in the record is large enough, in order to accurately follow the evolution of the phase, we need to know the relativistic corrections to the quadrupole formula; as the coalescence is approached, our ignorance about the signal and about the emitted power becomes more and more serious. More generally, there an urgent need to evaluate the emission of a pair of two (rotating) black holes, a very difficult problem which can be attacked only numerically. It involves following the evolution of three-dimensional geometry while it changes the topology (the two black holes merge into one); this requires highly innovative techniques in numerical analysis and riemannian geometry. The Grand Challenge (see the paper by Laguna) has been set up to do this and we look forward to its results.

The second area of ignorance is the way how to search for the chirped signal in a record. This involves taking into account the instrumental response and setting up appropriate and safe procedures to reject false alarms and false dismissals. The most common detection technique is based upon the scalar product (in the time domain or in the frequency domain) between the record and the expected signal (defined in an appropriate parameter space); detection is announced when the square of this scalar product is greater than an appropriate threshold. This is easier said than done: aside from our ignorance about the signal, the very large number of parameters involved, especially for long records, makes the computational load unbearable, unless special, hierarchical search techniques are devised (see the extensive work of the LIGO group and the paper by Iess.)

An important area of research is the study of known and stable sources, whose detection would be immediately connected to other bands of observations and produce 'standard gravitational candles' and enable the calibration of detectors. Unfortunately they are not easy to find; but the work by Gourgoulhon and Bonazzola on the emission by pulsar is quite interesting.

For a long time the search for a gravitational wave background of cosmological origin was carried out blindly, with only the guidance provided by the requirement that the expected energy density be smaller than the critical energy density required to close the Universe. Now, as discussed in this meeting by Veneziano and Grishchuk, we begin at last to make contact with the possible production of gravitational waves during the inflationary period. Both ground and space interferometers show great promises in this respect.

The instrumental progress in the detectors is fast, impressive and lasting. Low temperature resonant detectors have become reliable instruments with interesting sensitivity and are continuously improved. The cooling at 100 mK of NAUTILUS, an aluminum cyclinder weighing 2260 km and a resonant frequency of 915.8 Hz, has been a real technological feat. The prospects of much bigger resonant and multimode detectors of spherical shape is fascinating; Coccia has presented the concept of a a sphere of 3 m in diameters and a mass of 108 tons [5].

The construction and the design of ground interferometers is at last on its way. This is a major endeavour, at a scale of complexity and cost similar to elementary particle experiments. The challenge to decrease the low frequency cutoff and its solution for the VIRGO project is very valuable because, on general terms, it decreases the detectable energy flux, which is proportional to the square of the frequency. We have on record the astounding increase in the reflectivity of mirrors: from 1992 to 1995 the absorption coefficient has decreased from 20 ppm to 0.5 ppm and the scattering coefficient from 50 ppm to 0.6 ppm; this is crucial in eliminating the stray light in the vauum tube.

The space interferometer LISA is now a (far away!) cornerstone of the European Space Agency; its daring design and the wealth of expected, detectable sources continues to fascinate us and to attract the interest of people interested in new technologies. But the time being in the low frequency band we have only Doppler experiments – in particular, the experiments with the spacecraft CASSINI on its way to Saturn in 2002, 2003 and 2004, with a much smaller, and inadequate sensitivity – and their interesting variations [9] (see the paper by Tinto).

In the past I have won several bets with my colleagues with essentially the same wording: *I bet against ... that no gravitational waves will be discovered before .... The referee shall adjudicate the bet within one year of the deadline. The loser shall pay a dinner to the winner and the referee.* After this meeting I must say I find it more difficult meaningfully repeat the bet, with a reasonable chance to succeed. I have instead tried a 1/4-serious poll, worded as follows:

Which kind of instrument will first detect gravitational waves?

1. A space instrument (3).
2. A ground interferometer (18).
3. A mechanical resonator (10).

I polled 35 people at the conference, with the result indicated in brackets; 4 abstained.

I wish good luck to all those engaged in the endeavour; they deserve an early reward for their efforts. We can trust that in our future conference the ratio between results and projects will increase; and that reflection upon the past uncertainties will lead us safely to the goal.

References

1. A. J. Anderson, *Nature* **229**, 547 (1971).
2. P. G. Bergmann, *Phys. Rev. Lett.* **26**, 1398 (1971).
3. B. Bertotti and A. Cavaliere, *Nuovo Cim.* **2B**, 223 (1971).
4. B. Bertotti and D. Trevese, *Nuovo Cim.* **7B**, 240 (1972).
5. G. Frossati and E. Coccia, *Cryogenics* **34**, 9 (1994).
6. C. Misner, *Phys. Rev. Lett.* **28**, 994 (1972).
7. A. Rosenblum, *Phys. Rev. Lett* **41**, 1003 (1978).
8. K. S. Thorne and V. L. Braginsky, *Ap. J.* **204**, L1 (1976).
9. M. Tinto, *Phys. Rev.* **D 53**, 5354 (1996).
10. J. Weber, *Phys. Rev. Lett.* **22**, 1320 (1969).

Conference Programme

Tuesday, March 19th, 1996
Morning

Gravitational Waves: Theory Chairman: R. W. P. Drever

S. L. Shapiro Calculating Gravitational Waveforms - By Any Means Necessary! (60')
L. Grishchuk Gravitational Theories and Observations (60')
R. W. Hellings Gravitational Waves from White Dwarf Binaries (30')

Afternoon

Sources of Gravitational Waves Chairman: S. L. Shapiro

F. Pacini Neutron Stars and the Population of Collapsed Objects in the Galaxy (30')
C. Chiosi Supernova Rates (30')
V. Ferrari Gravitational Waves from Stars and Black Holes (60')
A. Di Fazio Primordial Supernova and Black Hole Formation (15')
K. Postnov Gravitational Radiation during Formation of Thorne-Zytkov Objects and Gravitational Wave Pulsars (15')
A. F. Zakharov Gravitational Radiation from Nonspherical Evolution of Pre-SN (15')
Z. Perjès The Evolution of Radiating Debris Trapped by a Black Hole (15')
W. Kluzniak Gravitational Waveforms from the Coalescence of a Neutron Star with a Black Hole (15')

Wednesday, March 20th, 1996
Morning

Numerical Relativity Chairman: S. Bonazzola

J.-A. Marck Gravitational Collapse and Radiation (60')
E. Gourgoulhon Gravitational Waves from Magnetized Neutron Stars (60')
P. Laguna Black Hole Collisions: a Computational Grand Challenge (60')

Afternoon

Interferometric Detectors Chairman: M. Cerdonio

A. Giazotto Status of VIRGO (30')
M. Coles Status of LIGO (30')
K. Danzmann Status of GEO 600 (30')
K. Kuroda Status of TAMA Project (30')
N. Kawashima Effort of Stable Operation and Noise Reduction of 100 m DL Laser Interferometer (TENKO-100) for Gravitational Wave Detection (15')
J.-M. Mackowski VIRGO Mirrors (15')
Cl. Boccara Mirror Metrology (15')
P. Rapagnani Mirror Suspensions (15')
J. Hough Suspension Development for GEO 600 (20')

Thursday, March 21st, 1996
Morning

Cosmology — Chairman: A. Di Giacomo

G. Veneziano String Cosmology and Relic Gravitational Radiation (60')
R. Brustein Spectrum of Cosmic Gravitational Waves Background from String Cosmology (15')
A. Degasperis Propagation of Gravitational Waves in Matter (60')

Resonant Bars — Chairman: G. Pizzella

G. V. Pallottino The Cryogenic Gravitational Wave Antennas EXPLORER and NAUTILUS (30')
G. A. Prodi The Ultracryogenic Gravitational Wave Detector AURIGA (30')

Afternoon

Detector Networks — Chairman: A. Giazotto

B. F. Schutz Interferometer-Bar Systems (30')
S. Frasca Arrays of Detectors (30')
D. G. Blair A New Source of Gravitational Waves Detectable by Cross Correlation of Gravitational Wave Detectors (15')

Experimental Perspectives — Chairman: A. Brillet

E. Coccia Sphere Detector for Coalescing Binaries (30')
D. G. Blair Sapphire Test Masses for Laser Interferometric Gravitational Wave Detectors - Achievable Parameters (15')
M. E. Tobar Improving the Sensitivity of the UWA Resonant-Mass Detector and the Detection of the Stochastic Background (15')
W. D. Walker Measure the Phase Speed of Gravity (15')
M. Bassan New Developments in Linear Transducer for Resonant GW Antennas (15')

Friday, March 22nd, 1996
Morning

Astrophysics — Chairman: B. Bertotti

S. Bonazzola Gravitational Waves from Rotating Neutron Stars (60')
K. Kokkotas Gravitational Waves and Pulsating Stars: What Can We Learn from Future Observations? (30')

Data Analysis — Chairman: B. F. Schutz

M. Tinto Data Analysis for Laser Interferometry (30')
S. Vitale Data Analysis for Resonant Gravitational Wave Antennas with Fully Numerical Processing: Optimal Filtering, Signal Timing, Rejection of Spuria and Correlation Analysis (30')
A. Viceré Strategies for Triggering and Data Analysis on the APE-1000 Parallel Computer (30')

Friday, March 22nd, 1996
Afternoon

Data Analysis Chairman: B. F. Schutz

F. Barone	On-Line System and Data Archiving of VIRGO (15')
F. Cavalier	Interferometer Simulation (15')
X. Grave	Pulsar Searches and Doppler Effect (15')
L. Milano	Numerical Experiments on Gravitational Wave Signal Detection (15')
P. Astone	Search of Monochromatic and Stochastic Gravitational Waves with the Cryogenic Antennas EXPLORER and NAUTILUS (15')
G. Giampieri	The Antenna Pattern of an Orbiting Interferometer (15')
I. M. Pinto	Efficient GW Chirp Estimation Via Wigner Representation and Generalized Hough Transform (15')

Experimental Prototypes Chairman: F. Fidecaro

G. Frossati	Low Temperature Measurements on Spherical CuAl and CuBe Samples (15')
V. Fafone	Eigenfrequencies and Quality Factors of Vibration of Aluminium Alloy Spherical Resonators (15')
R. Spero	Status Report on the 40 m LIGO Interferometer (15')
S. Moriwaki	Power Recycling Experiment with Suspended Mirrors (15')
H. Lück	Laboratory Experiments on Advanced Techniques for Gravitational Wave Detectors (15')
R. W. P. Drever	Techniques for Extending Interferometer Performance Using Magnetic Levitation and Other Methods (15')

Saturday, March 23rd, 1996
Morning

Theoretical Issues Chairman: F. de Felice

L. Blanchet	Inspiralling Compact Binaries (60')

Detection in Space Chairman: I. Ciufolini

L. Iess	Spacecraft Doppler Tracking (60')
P. Bender	Laser Interferometry in Space (60')

Summary Talk

B. Bertotti	Concluding Remarks (30')

List of Participants

ASTONE Pia
INFN/Univ. di Roma "La Sapienza"
P.le Aldo Moro, 2
I-00185 Roma
Italy
ASTONE@ROMA1.INFN.IT

BARONE Fabrizio
INFN/Università di Napoli
Mostra d'Oltremare pad. 19
I-80125 Napoli
Italy
FBARONE@NA.INFN.IT

BARONI Liana
Università di Bologna
Dip. di Fisica - Via Irnerio, 46
I-40125 Bologna
Italy
BARONI@PERUGIA.INFN.IT

BASSAN Massimo
INFN/Univ. di Roma "Tor Vergata"
Via della Ricerca Scientifica, 1
I-00133 Roma
Italy
BASSAN@ROMA2.INFN.IT

BECCARIA Matteo
Università di Pisa
Dip. di Fisica - P.za Torricelli, 2
I-56100 Pisa
Italy
BECCARIA@HPTH4.DIFI.UNIPI.IT

BENDER Peter L.
University of Colorado
JILA, CB 440
Boulder, CO 80309
USA
PBENDER@JILA.COLORADO.EDU

BERNARD Philippe
CERN
LHC Division
CH-1211 Geneva 23
Switzerland
PHBERNARD@CERNVM.CERN.CH

BERNARDINI Massimo
INFN - Sezione di Pisa
Via Livornese, 1291
I-56010 S. Piero a Grado (PI)
Italy
BERNARDINI@AXPIA.PI.INFN.IT

BERTOTTI Bruno
Università di Pavia
Dip. di Fisica - Via U. Bassi, 6
I-27100 Pavia
Italy
BERTOTTI@PAVIA.INFN.IT

BINI Donato
CNR - Ist. per Applic. della Matematica
Via P. Castellino, 111
I-80131 Napoli
Italy
BINID@VXRMG9.ICRA.IT

BLAIR David G.
University of Western Australia
Physics Department
Nedlands, WA 6009
Australia
DGB@EARWAX.PD.UWA.EDU.AU

BLANCHET Luc
Observatoire de Paris - Meudon
DARC - 5, Place Jules Janssen
F-92195 Meudon Cedex
France
BLANCHET@OBSPM.FR

BOCCARA Claude
ESPCI - Paris
Lab. Optique - 10, Rue Vauquelin
F-75005 Paris
France
BOCCARA@OPTIQUE.ESPCI.FR

BONAZZOLA Silvano
Observatoire de Paris - Meudon
DARC - 5, Place Jules Janssen
F-92195 Meudon Cedex
France
BONA@MESIOB.OBSPM.FR

BONIFAZI Paolo
CNR - IFSI
Via Galileo Galilei - CP 27
I-00044 Frascati (RM)
Italy
VAXROM::BONIFAZI

BRACCINI Stefano
INFN - Sezione di Pisa
Via Livornese, 1291
I-56010 S. Piero a Grado (PI)
Italy
BRACCINI@AXPIA.PI.INFN.IT

BRADASCHIA Carlo
INFN - Sezione di Pisa
Via Livornese, 1291
I-56010 S. Piero a Grado (PI)
Italy
BRADASCHIA@AXPIA.PI.INFN.IT

BRILLET Alain
LAL - Orsay
Univ. Paris Sud - Bât. 208
F-91405 Orsay
France
BRILLET@LALCLS.IN2P3.FR

BRUSTEIN Ramy
Ben-Gurion University
Dept of Physics
Beer-Sheva 84105
Israel
RAMYB@BGUMAIL.BGU.AC.IL

BUONANNO Alessandra
INFN - Sezione di Pisa
Dip. di Fisica - P.za Torricelli, 2
I-56100 Pisa
Italy
BUONANNO@SUN10.DIFI.UNIPI.IT

CALLONI Enrico
Università di Napoli
Dip. di Fisica, Mostra d'Oltremare pad. 19
I-80125 Napoli
Italy
CALLONI@NA.INFN.IT

CANTATORE Giovanni
Università di Trieste
Dip. di Fisica - Via A. Valerio, 2
I-34127 Trieste
Italy
CANTATORE@TRIESTE.INFN.IT

CAVALIER Fabien
LAL - Orsay
Univ. Paris Sud - Bât. 208
F-91405 Orsay Cedex
France
CAVALIER@FRCPN11.IN2P3.FR

CERDONIO Massimo
Università di Padova
Dip. di Fisica - Via F. Marzolo, 8
I-35131 Padova
Italy
CERDONIO@PADOVA.INFN.IT

CHIOSI Cesare
Università di Padova
Dip. di Astronomia - V. dell'Osservatorio
I-35122 Padova
Italy
CHIOSI@ASTRPD.PD.ASTRO.IT

CIAMPA Alberto
INFN - Sezione di Pisa
Dip. di Fisica - P.za Torricelli, 2
I-56100 Pisa
Italy
CIAMPA@FISICA.DIFI.UNIPI.IT

CIUFOLINI Ignazio
CNR - IFSI
Via Galileo Galilei - CP 27
I-00044 Frascati (RM)
Italy
CIUFOLI@NERO.ING.UNIROMA1.IT

COCCIA Eugenio
INFN - Sezione di Roma II
Via della Ricerca Scientifica, 1
I-00133 Roma
Italy
COCCIA@ROMA2.INFN.IT

COLES Mark
CALTECH
LIGO Project - MS 51-33
Pasadena, CA 91125
USA
COLES@LIGO.CALTECH.EDU

COLLINS Harry M.
University of Southampton
Dept of Sociology and Social Policy
Southampton, SO17 1BJ
UK
H.M.COLLINS@SOTON.AC.UK

CONTI Livia
INFN - LNL
Via Romea, 4
I-35020 Legnaro (PD)
Italy
CONTI@LNL.INFN.IT

COVARRUBIAS M. Guillermo
Univ. Metropolitana - Iztapalapa
Dep. de Fisica - Apdo postal 55-534
Mexico City
Mexico
GCOV@XANUM.UAM.MX

CRIVELLI VISCONTI Vasco
INFN - LNL
Via Romea, 4
I-35020 Legnaro (PD)
Italy
CRIVELLI@AXDLNL.LNL.INFN.IT

CUOCO Elena
INFN - Sezione di Pisa
Dip. di Fisica - P.za Torricelli, 2
I-56100 Pisa
Italy
CUOCO@HPTH1.DIFI.UNIPI.IT

CURCI Giuseppe
Università di Pisa
Dip. di Fisica - P.za Torricelli, 2
I-56100 Pisa
Italy
CURCI@FISICA.DIFI.UNIPI.IT

D'AMBROSIO Erika
Università di Pisa
Dip. di Fisica - P.za Torricelli, 2
I-56100 Pisa
Italy
ERIKA@HPTH1.DIFI.UNIPI.IT

DANZMANN Karsten
Universität Hannover - MPQ
Inst. Atom- Molekülphys., Appelstr. 2
D-30167 Hannover
Germany
KVD@MPQ.MPG.DE

DE FELICE Fernando
INFN - Sezione di Padova
Via F. Marzolo, 8
I-35131 Padova
Italy
VAXFPD::DEFELICE

DEGASPERIS Antonio
Università di Roma "La Sapienza"
Dip. di Fisica - P.le Aldo Moro, 2
I-00185 Roma
Italy
DEGASPERIS@ROMA1.INFN.IT

DESALVO Riccardo
INFN - Sezione di Pisa
Via Livornese, 1291
I-56010 S. Piero a Grado (PI)
Italy
DESALVO@AXPIA.PI.INFN.IT

DI FAZIO Alberto
Osservatorio Astronomico di Roma
Viale del Parco Mellini, 84
I-00136 Roma
Italy
DIFAZIO@OARHP1.RM.ASTRO.IT

DI FIORE Luciano
INFN - Sezione di Napoli
Mostra d'Oltremare pad. 20
I-80125 Napoli
Italy
DIFIORE@NA.INFN.IT

DI GIACOMO Adriano
Università di Pisa
Dip. di Fisica - P.za Torricelli, 2
I-56100 Pisa
Italy
DIGIACO@IPIFIDPT.DIFI.UNIPI.IT

DI VIRGILIO Angela
INFN - Sezione di Pisa
Via Livornese, 1291
I-56010 S. Piero a Grado (PI)
Italy
ANGELA@AXPIA.PI.INFN.IT

DIAMBRINI PALAZZI Giordano
Università Roma "La Sapienza"
Dip. di Fisica - P.le Aldo Moro, 2
I-00185 Roma
Italy
VAXROM::DIAMBRINI

DREVER Ronald W. P.
CALTECH
Physics 130-33
Pasadena, CA 91125
USA
RDREVER@CALTECH.EDU

ENARD Daniel
INFN - Sezione di Pisa
Via Livornese, 1291
I-56010 S. Piero a Grado (PI)
Italy
ENARD@GALILEO.PI.INFN.IT

FAFONE Viviana
INFN - Sezione di Roma II
Via della Ricerca Scientifica, 1
I-00133 Roma
Italy
FAFONE@ROMA2.INFN.IT

FERRANTE Isidoro
INFN - Sezione di Pisa
Via Livornese, 1291
I-56010 S. Piero a Grado (PI)
Italy
FERRANTE@AXPIA.PI.INFN.IT

FERRARI Valeria
Università di Roma "La Sapienza"
Dip. di Fisica - P.le Aldo Moro, 2
I-00185 Roma
Italy
VALERIA@VXRMG9.ICRA.IT

FIDECARO Francesco
Università di Pisa
Dip. di Fisica - Via Livornese, 1291
I-56010 S. Piero a Grado (PI)
Italy
FIDECARO@GALILEO.PI.INFN.IT

FORTINI Pierluigi
Università di Ferrara
Dip. di Fisica - Via Paradiso, 12
I-44100 Ferrara
Italy
FORTINI@FERRARA.INFN.IT

FRASCA Sergio
Università di Roma "La Sapienza"
Dip. di Fisica - P.le Aldo Moro, 2
I-00185 Roma
Italy
FRASCA@ROMA1.INFN.IT

FROSSATI Giorgio
Leiden University
Kamerlingh Onnes Lab., Nieuwsteeg 18
NL-2311 SB Leiden
The Netherlands
GIORGIO@QV3PLUTO.LEIDENUNIV.NL

GAMMAITONI Luca
INFN - Sezione di Perugia
Via A. Pascoli
I-06100 Perugia
Italy
GAMMAITONI@PERUGIA.INFN.IT

GARBEROGLIO Giovanni
Università di Pisa
Dip. di Fisica - P.za Torricelli, 2
I-56100 Pisa
Italy
GARBERO@SUN10.DIFI.UNIPI.IT

GEMELLI Gianluca
Università di Roma "La Sapienza"
Dip. di Matematica - P.le Aldo Moro, 2
I-00185 Roma
Italy
GEMELLI@MAT.UNIROMA1.IT

GIAMPIERI Giacomo
University of London
QMW College, Mile End Road
London E1 4NS
UK
G.GIAMPIERI@QMW.AC.UK

GIAZOTTO Adalberto
INFN - Sezione di Pisa
Via Livornese, 1291
I-56010 S. Piero a Grado (PI)
Italy
GIAZOTTO@PISA.INFN.IT

GOURGOULHON Eric
Observatoire de Paris - Meudon
DARC - 5, Place Jules Janssen
F-92195 Meudon Cedex
France
ERIC.GOURGOULHON@OBSPM.FR

GRADO Aniello
INFN - Sezione di Napoli
Mostra d'Oltremare pad. 20
I-80125 Napoli
Italy
GRADO@NA.INFN.IT

GRAVE Xavier
LAPP - Annecy
Chemin de Bellevue - B. P. 110
F-74941 Annecy-Le-Vieux Cedex
France
XAVIER@VIRGOA4.IN2P3.FR

GRISHCHUK Leonid
University of Wales - Cardiff
Dept of Physics and Astronomy
CF2 3YB Cardiff
UK
GRISHCHUK@ASTRO.CF.AC.UK

HELLINGS Ronald W.
JPL
Caltech, 4800 Oak Grove Drive
Pasadena, CA 91109
USA
RWH@GRAVITON.JPL.NASA.GOV

HOLLOWAY Lee
University of Illinois at Urbana-Champaign
Dept of Physics - 1110 West Green Str.
Urbana, IL 61801
Italy
LEH@UIUC.EDU

HOUGH James
University of Glasgow
Dept of Physics and Astronomy
Glasgow G12 8QQ
UK
J.HOUGH@PHYSICS.GLA.AC.UK

IESS Luciano
Università di Roma "La Sapienza"
Dip. Aerospaziale - Via Eudossiana, 18
I-00184 Roma
USA
IESS@NERO.ING.UNIROMA1.IT

KAWASHIMA Nobuki
ISAS
3-1-1 Yoshinodai
Sagaminara, Kanagawa 229
Japan
KNOBUKI@PLEIADES.SCI.ISAS.AC.JP

KLUZNIAK Wlodzimierz
University of Wisconsin - Madison
Physics Dept - 1150 University Ave.
Madison, WI 53706
USA
WLODEK@ASTROG.PHYSICS.WISC.EDU

KOKKOTAS Konstantinos D.
Aristotle University of Thessaloniki
Dept of Physics
GR-54006 Thessaloniki
Greece
KOKKOTAS@GRAVI.PHYSIK.UNI-JENA.DE

KOVALIK Joseph
INFN - Sezione di Perugia
Via A. Pascoli
I-06100 Perugia
Italy
KOVALIK@PERUGIA.INFN.IT

KURODA Kazuaki
Inst. for Cosmic Ray Res. - Tokyo
TAMA Pjt. - 3-2-1, Midoricho, Tanashi
Tokyo 188
Japan
KURODA@ICRR.U-TOKYO.AC.JP

LA PENNA Paolo
INFN - Sezione di Pisa
Via Livornese, 1291
I-56010 S. Piero a Grado (PI)
Italy
LAPENNA@AXPIA.PI.INFN.IT

LAGUNA Pablo
Pennsylvania State University
525 Davey Laboratory
University Park, PA 16802
USA
PABLO@ASTRO.PSU.EDU

LAURELLI Paolo
INFN - LNF
Via E. Fermi, 40
I-00044 Frascati (RM)
Italy
LAURELLI@LNF.INFN.IT

LORIETTE Vincent
ESPCI - Paris
Lab. Optique - 10, Rue Vauquelin
F-75005 Paris
France
LORIETTE@OPTIQUE.ESPCI.FR

LÜCK Harald
Universität Hannover - MPQ
Appelstraße, 2
D-30167 Hannover
Germany
HAL@MPQ.MPG.DE

LUITEN Andre Nicholas
University of Western Australia
Department of Physics
Nedlands, WA 6009
Australia
ANDRE@EARWAX.PD.UWA.EDU.AU

MACKOWSKI Jean-Marie
IPN Lyon
43, Bd. du 11 Novembre 1918
F-69622 Villeurbanne Cedex
France
MACKOWSKI@FRCPN11.IN2P3.FR

MAGGIORE Michele
INFN - Sezione di Pisa
Dip. di Fisica - P.za Torricelli, 2
I-56100 Pisa
Italy
MAGGIORE@IPIFIDPT.DIFI.UNIPI.IT

MAJORANA Ettore
INFN - Sezione di Roma I
P.le Aldo Moro, 2
I-00185 Roma
Italy
MAJORANA@ROMA1.INFN.IT

MARCK Jean-Alain
Observatoire de Paris - Meudon
DARC - 5, Place Jules Janssen
F-92195 Meudon Cedex
France
MARCK@OBSPM.FR

MAZZITELLI Giovanni
INFN - LNF
Via E. Fermi, 40 - CP 13
I-00044 Frascati (RM)
Italy
MAZZITELLI@LNF.INFN.IT

MILANO Leopoldo
INFN - Università di Napoli
Mostra d'Oltremare pad. 19
I-80125 Napoli
Italy
MILANO@NA.INFN.IT

MONTANARI Enrico
INFN - Università di Ferrara
Via Paradiso, 12
I-44100 Ferrara
Italy
MONTANARI@FERRARA.INFN.IT

MORIWAKI Shigenori
University of Tokyo
Dept of Applied Physics
7-3-1, Hongo, Bunkyo, Tokyo 113
Japan
MORIWAKI@T-MUNU.PHYS.S.U-TOKYO.AC.JP

MOURS Benoît
LAPP - Annecy
Chemin de Bellevue - B. P. 110
F-74941 Annecy-Le-Vieux Cedex
France
MOURS@FRCPN11.IN2P3.FR

NOVAK Jerôme
Observatoire de Paris - Meudon
DARC - 5, Place Jules Janssen
F-92195 Meudon Cedex
France
JEROME.NOVAK@OBSPM.FR

OGAWA Yujiro
KEK
1-1 Oho, Ibaraki-ken
Tsukuba 305
Japan
OGAWAYJ@KEKVAX.KEK.JP

ORTOLAN Antonello
INFN - LNL
Via Romea, 4
I-35020 Legnaro (PD)
Italy
ORTOLAN@LNL.INFN.IT

PACINI Franco
Osservatorio Astrofisico di Arcetri
Largo E. Fermi, 5
I-50125 Arcetri (FI)
Italy
FPACINI@ARCETRI.ASTRO.IT

PALLOTTINO Gian Vittorio
Università di Roma "La Sapienza"
Dip. di Fisica - P.le Aldo Moro, 2
I-00185 Roma
Italy
PALLOTTINO@ROMA1.INFN.IT

PAPA M. Alessandra
Università di Roma "Tor Vergata"
Dip. di Fisica - P.le Aldo Moro, 2
I-00185 Roma
Italy
PAPA@ROMA1.INFN.IT

PERJÉS Zoltán
Hungarian Academy of Sciences
KFKI - P. O. Box 49
H-1525 Budapest 114
Hungary
PERJES@RMK53ø.RMKJ.KFKI.HU

PINTO Innocenzo M.
Università di Salerno
Dip. Ing. Inform.- Via Ponte Don Melillo
I-84084 Fisciano (SA)
Italy
PINTO@SALERNO.INFN.IT

PIZZELLA Guido
INFN - Sezione di Roma II
Via della Ricerca Scientifica, 1
I-00133 Roma
Italy
VAXROM::PIZZELLA

POGGIANI Rosa
Università di Pisa
Dip. di Fisica - Via Livornese, 1291
I-56010 S. Piero a Grado (PI)
Italy
POGGIANI@AXPIA.PI.INFN.IT

POSTNOV Konstantin
Sternberg Astronomical Institute
13, Universitetskijpr.
119899 Moscow
Russia
PK@SAI.MSU.SU

PRODI Giovanni A.
Università di Trento
Dip. di Fisica
I-38050 Povo (TN)
Italy
PRODI@ALPHA.SCIENCE.UNITN.IT

RAPAGNANI Piero
Università di Roma "La Sapienza"
Dip. di Fisica - P.le Aldo Moro, 2
I-00185 Roma
Italy
RAPAGNANI@ROMA1.INFN.IT

RICCI Fulvio
Università di Roma "La Sapienza"
Dip. di Fisica - P.le Aldo Moro, 2
I-00185 Roma
Italy
RICCI@ROMA1.INFN.IT

ROWAN Sheila
University of Glasgow
Dept. of Physics and Astronomy
Glasgow G12 8QQ
UK
S.TWYFORD@PHYSICS.GLA.AC.UK

SCHUTZ Bernard F.
MPI für Gravitationphysik - Potsdam
A.-Einstein-Institut, Schlaatzweg 1
D-14473 Potsdam
Germany
SCHUTZ@HEL.AEI-POTSDAM.MPG.DE

SHAPIRO Stuart L.
Center for Astrophysics & Relativity
326 Siena Drive
Ithaca, NY 14850
USA
SHAPIRO@SPACENET.TN.CORNELL.EDU

SINIBALDI Alessandro
Università di Pisa
Dip. di Fisica - P.za Torricelli, 2
I-56100 Pisa
Italy
SINIBALD@IPIFIDPT.DIFI.UNIPI.IT

SPERO Robert
CALTECH
LIGO Project - MS 51-33
Pasadena, CA 91125
USA
ROBERT@LIGO.CALTECH.EDU

STRUMIA Franco
Università di Pisa
Dip. di Fisica - P.za Torricelli, 2
I-56100 Pisa
Italy
STRUMIA@IPIFIDPT.DIFI.UNIPI.IT

TAFFARELLO Luca
INFN - LNL
Via Romea, 4
I-35020 Legnaro (PD)
Italy
TAFFARELLO@LNL.INFN.IT

TINTO Massimo
JPL
Caltech, 4800 Oak Grove Drive
Pasadena, CA 91109
USA
MASSIMO@OBERON.JPL.NASA.GOV

TOBAR Michael Edmund
University of Western Australia
Physics Department
Nedlands, WA 6009
Australia
MIKE@PD.UWA.EDU.AU

TOSCHI Federico
Università di Pisa
Dip. di Fisica - P.za Torricelli, 2
I-56100 Pisa
Italy
TOSCHI@SUN10.DIFI.UNIPI.IT

TOURRENC Philippe André
Université P. et M. Curie - Paris
B. C. 142, Tour 22/12, 4e étage
F-75252 Paris Cedex 05
France
PHT@CCR.JUSSIEU.FR

TWYFORD Sharon
University of Glasgow
Dept of Physics and Astronomy
Glasgow G12 8QQ
UK
S.TWYFORD@PHYSICS.GLA.AC.UK

UNGARELLI Carlo
INFN - Sezione di Pisa
Dip. di Fisica - P.za Torricelli, 2
I-56100 Pisa
Italy
UNGAREL@IPIFIDPT.DIFI.UNIPI.IT

VENEZIANO Gabriele
CERN
TH Division
CH-1211 Geneva 23
Switzerland
VENEZIA@NXTH04.CERN.CH

VICERÉ Andrea
Università di Pisa
Dip. di Fisica - P.za Torricelli, 2
I-56100 Pisa
Italy
VICERE@SUN10.DIFI.UNIPI.IT

VINET Jean-Yves
LAL - Orsay
Univ. Paris Sud - Bât. 208
F-91405 Orsay
France
VINET@LALCLS.IN2P3.FR

VISCO Massimo
CNR - IFSI
Via Galileo Galilei - CP 27
I-00044 Frascati (RM)
Italy
VISCO@ROMA1.INFN.IT

VITALE Stefano
INFN/Università di Trento
Dipartimento di Fisica
I-38050 Povo (TN)
Italy
VITALE@ALPHA.SCIENCE.UNITN.IT

WALKER William D.
ETH Zurich
Institute of Mechanics, F373
CH-8092 Zürich
Switzerland
WALKER@IFM.MAVT.ETHZ.CH

WINTERFLOOD John
University of Western Australia
Department of Physics
Nedlands, WA 6009
Australia
JWINTER@EARWAX.PD.UWA.EDU.AU

ZAKHAROV Alexander F.
ITEP
B. Cheremushkinskaya, 25
117259 Moscow
Russia
ZAKHAROV@VITEP3.ITEP.RU

ZENDRI Jean-Pierre
Leiden University
Kamerlingh Onnes Lab., Nieuwsteeg 18
NL-2311 SB Leiden
The Netherlands
JEAN@QV3PLUTO.LEIDENUNIV.NL